Developments in Mathematics

Volume 52

More information about this series at http://www.springer.com/series/5834

Dragana S. Cvetković Ilić
Yimin Wei

Algebraic Properties
of Generalized Inverses

 Springer

Dragana S. Cvetković Ilić
Faculty of Science and Mathematics
University of Nis
Nis
Serbia

Yimin Wei
School of Mathematical Sciences
Fudan University
Shanghai
China

ISSN 1389-2177
Developments in Mathematics
ISBN 978-981-13-4862-4
DOI 10.1007/978-981-10-6349-7

ISSN 2197-795X   (electronic)

ISBN 978-981-10-6349-7   (eBook)

Mathematics Subject Classification (2010): 15A09, 65F20, 47L05, 47A05

Printed on acid-free paper

This Springer imprint is published by Springer Nature
The registered company is Springer Nature Singapore Pte Ltd.
The registered company address is: 152 Beach Road, #21-01/04 Gateway East, Singapore 189721, Singapore

Preface

This book gives a presentation of some current topics in theory of generalized inverses. As this theory has long been a subject of study of various authors, many of the problems and questions have been resolved. However, some have only been either partially solved or remain still open to this day. It was our goal to use this book to give a review of these efforts as well as to offer the reader possible directions for further study in this area of mathematics as well as hints at possible applications in different types of problems.

This book starts with definitions of various types of generalized inverses and listing the many sorts of applications of them to different branches of both mathematics, but also to some other scientific disciplines, which is aimed at providing motivation behind the study of this topic in general.

Chapter 2 gives an exposition of the so-called reverse order law problem, which is originally posed by Greville as early as in 1960, who first considered it in the case of the Moore–Penrose inverse of two matrices. This was followed by further research on this subject branching in several directions: products with more than two matrices were considered; different classes of generalized inverses were studied; different settings were considered (operator algebras, C^*-algebras, rings, etc.). We discuss the reverse order law for K-inverses, when $K \in \{\{1\}, \{1,2\}, \{1,3\}, \{1,2,3\}, \{1,3,4\}\}$ in different settings and present all recently published results on this subject as well as some examples and open problems.

In the next chapter, we first consider problems on completions of operator matrices and then proceed to present a particular approach to solving the problem on the reverse order law for $\{1\}$-generalized inverses of operators acting on separable Hilbert spaces which involves some of the previous research on completions of operator matrices to left and right invertibility. Although the reverse order law problem for $\{1\}$-generalized inverses of matrices was completely resolved by 1998, the corresponding problem for the operators on separable Hilbert spaces was only solved in 2015. So, we thus demonstrate usability of results on completions of operator matrices by showing how they can be applied to one of the topics in generalized inverses of operators that has seen a great interest over the years. Also, in Chap. 3, we consider the problem of existence of Drazin invertible completions of an upper triangular operator matrix and that of the invertibility of a linear combination of operators on Hilbert spaces.

In Chap. 4, we shift our attention from the problem of invertibility a linear combination of operators to that of some different types of generalized invertibility. Special emphasis is put on Drazin and generalized Drazin invertibility of linear combinations of idempotents, commutators, and anticommutators in Banach algebras. Also, some related results are presented on the Moore–Penrose inverse of a linear combination of commuting generalized and hypergeneralized projectors for which certain formulae are considered.

The problem of finding representations of the Drazin inverse of a 2×2 block matrix is of great significance primarily due to its applications in solving systems of linear differential equations, linear difference equations, and perturbation theory of the Drazin inverse. It was posed by S. Campbell in 1983, and it is still unsolved. In Chap. 5, we present all the partial results on this subject that have been obtained so far as well as the different methods and approaches used in obtaining them.

In the last chapter, we present some additive results for the Drazin inverse. Although it was already even in 1958 that Drazin pointed out that computing the Drazin inverse of a sum of two elements in a ring was not likely to be easy, this problem remains open to this day even for matrices. It is precisely on this problem when considered in rings of matrices and Banach algebras that we shall focus our attention here; i.e., under various conditions, we will compute $(P + Q)^D$ as a function of P, Q, P^D, and Q^D.

This book thus, as readers will surely see for themselves, only tackles some of the current problems of the theory of generalized inverses, but the topics that have been selected have also been thoroughly covered and a systematic presentation given of relevant results obtained so far as well as of possible directions in further research. We should mention that this book has come out as a result of a long and successful collaboration between the authors. Also we were inspired by the work of many colleagues as well as coauthors, some of the joint results with which appear in this book, to whom we are thankful for the experience of working with. Finally, we would like to thank professors Eric King-wah Chu from Monash University and Vladimir Pavlović from the Faculty of Science and Mathematics, University of Niš, who read this book carefully and provided feedback during the writing process.

This work was supported by the National Natural Science Foundation of China under grant 11771099 and International Cooperation Project of Shanghai Municipal Science and Technology Commission under grant 16510711200, School of Mathematical Sciences and Key Laboratory of Mathematics for Nonlinear Sciences, Fudan University, and by the bilateral project between China and Serbia, "Generalized inverses and its applications" no. 2–15.

Nis, Serbia   Dragana S. Cvetković Ilić
Shanghai, China   Yimin Wei

Contents

Chapter 1
Definitions and Motivations

1.1 Generalized Inverses: Definitions and Applications

The concept of generalized inverses seems to have been first mentioned in print in 1903 by Fredholm [1] who formulated a pseudoinverse for a linear integral operator which is not invertible in the ordinary sense. One year later, in 1904 Hilbert [2] discussed generalized inverses of differential operators. The class of all pseudoinverses was characterized in 1912 by Hurwitz [3] who used the finite dimensionality of the null-space of Fredholm operators to give a simple algebraic construction. Generalized inverses of differential and integral operators thus antedated the generalized inverses of matrices whose existence was first noted by Moore [4, 5], who defined a unique inverse, called by him the general reciprocal, for every finite matrix. Little notice was taken of Moore's discovery for 30 years after its first publication, during which time generalized inverses were defined for matrices by Siegel [6, 7] and for operators by Tseng [8–10], Murray and von Neumann [11], Atkinson [12, 13] and others. Revival of interest in the subject in the 1950s centered around the least squares properties of certain generalized inverses which were recognized by Bjerhammar [14, 15]. He rediscovered Moore's inverse and also observed the connection between generalized inverses and solving linear systems. In 1955 Penrose [16] extended Bjerhammar's results and showed that the Moore's inverse for a given matrix A is the unique matrix X satisfying the four equations:

(1) $AXA = A$ (2) $XAX = X$ (3) $(AX)^* = AX$ (4) $(XA)^* = XA$.

In honour of Moore and Penrose this unique inverse is now commonly called the Moore-Penrose inverse. Evidently, the Moore-Penrose inverse of a nonsingular matrix coincides with its ordinary inverse. Throughout the years the Moore-Penrose inverse was intensively studied, one of the primary reasons for that being its usefulness in applications to dealing with diverse problems such as, for example, that of solving systems of linear equations, which constitutes one of the basic but at the same time most important applications of this type of generalized inverse. Over the years,

© Springer Nature Singapore Pte Ltd. 2017

D. Cvetković Ilić and Y. Wei, *Algebraic Properties of Generalized Inverses*,
Developments in Mathematics 52, DOI 10.1007/978-981-10-6349-7_1

right alongside this one, considerable attention in research was given to generalized inverses that satisfy only some, but not all, of the four Penrose equations.

As we will be working with a number of different subsets of the above mentioned set of four equations, we need some convenient notation for the generalized inverses satisfying those certain specified equations: for any A, let $A\{i, j, ..., k\}$ denote the set of all matrices X which satisfy the equations (i), (j), ..., (k) from among the equations (1)–(4). In this case $X \in A\{i, j, ..., k\}$ is a $\{i, j, ..., k\}$-inverse of A and is denoted by $A^{(i,j,...,k)}$. With this convention one obviously has $A\{1, 2, 3, 4\} = \{A^\dagger\}$.

It is interesting to mention that about at the same time as Penrose, Rao [17], gave a method of computing what is called the pseudoinverse of a singular matrix, and applied it to solve normal equations with a singular matrix in the least squares theory and to express the variances of estimators. The pseudoinverse defined by Rao did not satisfy all the restrictions imposed by Moore and Penrose. It was therefore different from the Moore-Penrose inverse, but was useful in providing a general theory of least squares estimation without any restrictions on the rank of the observational equations.

For various historical references in literature concerning the generalized Green's function and Green's matrix for ordinary differential systems, in which the number of independent boundary conditions is equal to the order of the system, the reader is referred to paper of Rao [17], which was written before the author was aware of the E. H. Moore's general reciprocal. The construction of a generalized Green's matrix is considerably simplified with the use of the E. H. Moore's general reciprocal in designating the solution of certain algebraic equations expressing the boundary conditions, and this procedure was employed by Bradley on a class of quasi-differential equations [19], and on general compatible differential systems involving two-point boundary conditions [20].

Since 1955 a great number of papers on various aspects of generalized inverses and their applications have appeared. Generalized inverses pervade a wide range of mathematical areas: matrix theory, operator theory, differential equations, numerical analysis, Markov chains, C^*-algebras or rings. Numerous applications include areas such as: statistics, cryptography, control theory, coding theory, incomplete data recovery and robotics.

It is interesting to note that the applications that generalized inverses do have in many areas of mathematics and otherwise seem to have extensively been pointed out only in papers published in the 70s and 80s of the last century. Attesting to that are the proceedings of an Advanced Seminar on Generalized Inverses and Applications held at the University of Wisconsin-Madison on October 8–10, 1973, where we can find many applications of generalized inverses to analysis and operator equations, Numerical analysis and approximation methods, along with applications to statistics and econometrics, optimization, system theory, and operations research. In contemporary mathematics the theory of generalized inverses has been established as a well known and generally accepted tool of wide applicability, a fact that is no more anywhere explicitly being mentioned. The present text, which generally speaking is mostly of theoretical character, will try to put a strong emphasis on those problems that can in way or another extend the range of applications of generalized inverses. Such possibility of wide applications of this theory surely serving as a good motivation for the reader, we will now list some of them together with appropriate references.

As already mentioned, among some of the first applications of generalized inverses, that at the same time actually motivated their introduction, were those was related to problems concerning the Green's function and Green's matrix for ordinary differential systems, integro-differential equations and boundary problems, iterative methods for best approximate solutions of linear integral equations of Fredholm and Volterra [21, 22]. Also they were used in theoretical treatment of linear differential-algebraic equations with inconsistent initial conditions, inconsistent inhomogeneities and undetermined solution components and applied to iterated functions [23]. Further applications can be found as follows: in the constructions of algorithms for reproducing objects from their x-rays (see [24]), for the process of reconstruction of pictures and objects from their projections (see [25–28]); in computerized tamography [29]; in sampling theory, a theory being a topic with applications in several other fields such as signal and image processing, communication engineering, and information theory, among others. Generalized inverses are one of the basic tools in the quasi-consistent reconstructions, which are an extension of the consistent reconstructions (see [30]). Then there are applications in public key cryptography (see [31]); in statistics (see [32–36]); in electrical engineering [37, 38]; in control theory, specially in optimal control of linear discrete systems, in optimal control of autonomous linear processes with singular matrices in the quadratic cost functional, in control problem structure and the numerical solution of linear singular systems, in control in biological systems, in controllability of linear dynamical systems, [39–55]; in robotics [56–62]; in kinematic synthesis [63, 64]. The references we have listed are but a small portion of the literature pertaining to various applications of generalized inverses, but we hope that they will be sufficient to motivate the reader to further pursue the topics we have presented and perhaps also some related to them.

1.2 Drazin Inverse and Generalized Drazin Inverse: Definitions and Applications

In 1958, Drazin [65] introduced a different kind of a generalized inverse in associative rings and semigroups that does not have the reflexivity property but commutes with the element.

Definition 1.1 Let a, b be elements of a semigroup. An element b is a Drazin inverse of a, written $b = a^D$, if

$$bab = b, \quad ab = ba, \quad a^{k+1}b = a^k, \tag{1.1}$$

for some nonnegative integer k. The least nonnegative integer k for which these equations hold is the Drazin index $\mathrm{ind}(a)$ of a. When $\mathrm{ind}(a) = 1$, then the Drazin inverse a^D is called the group inverse and is denoted by a^g or $a^\#$.

The Drazin inverse is one the most important concepts in ring theory (see [66]), in particular in matrix theory and various of its applications [37, 67–70]. One of the reason for this are its very nice spectral properties. For example, the nonzero eigenvalues of the Drazin inverse are the reciprocals of the nonzero eigenvalues of the given matrix, and the corresponding generalized eigenvectors have the same grade [67].

Caradus [71], King [72] and Lay [73] investigated the Drazin inverse in the setting of bounded linear operators on complex Banach spaces. It was proved by Caradus [71] that a bounded linear operator A on a complex Banach space has a Drazin inverse if and only if 0 is a pole of the resolvent $(\lambda I - A)^{-1}$ of A; the order of the pole is equal to the Drazin index of A. Marek and Žitny [74] gave a detailed treatment of the Drazin inverse for operators as well as for elements of a Banach algebra. The Drazin inverse of closed linear operators was investigated by Nashed and Zhao [69] who then applied their results to singular evolution equations and partial differential operators. Drazin [75] studied some extremal definitions of generalized inverses that are more general than the original Drazin inverse.

The index of a matrix A, namely the least nonnegative integer k for which the nullspaces of A^k and A^{k+1} coincide, which is one of the key concepts in the theory of Drazin inverses of matrices [37, 67], turns out to coincide with its Drazin index.

In operator theory, the notion corresponding to the index of a finite matrix is that of the ascent (and descent) of a chain-finite bounded linear operator A (see [76, 77]). An operator A is chain-finite with the ascent (=descent) k if and only if 0 is a pole of the resolvent $(\lambda I - A)^{-1}$ of order k. If we want to translate results involving the index or the chain-finiteness condition to a Banach algebra $\mathscr{A}$ with the unit 1 we must interpret the index of $a \in \mathscr{A}$ to be 0 if a is invertible, and k if 0 is a pole of $(\lambda - a)^{-1}$ of order k. The set $\mathscr{A}^D$ consists of all $a \in \mathscr{A}$ such that a^D exists.

The theory of Drazin inverses has seen a substantial growth over the past few decades. It is a subject which is of great theoretical interest and finds applications in a great many of various areas, including Statistics, Numerical analysis, Differential equations, Population models, Cryptography, and Control theory, in solving singular, singularly perturbed differential equations and differential-algebriac equations, asymptotic convergence of operator semigroups, multibody system dynamics, etc.

There is a huge applicability of the Drazin inverse to a wide variety of problems in applied probability where a Markov chain is present either directly or indirectly through some form of embedding. Markov chains are discrete stochastic processes in which the probability of transition from one state to another is determined only by the current state. They are characterized by their transition matrix P, where P_{ij} represents the probability of transition from the state i to the state j. By characterizing the Drazin inverse of $I - P$, where P is the transition matrix of a finite irreducible discrete time Markov chain, we can obtain general procedures for finding stationary distributions, moments of the first passage time distributions, and asymptotic forms for the moments of the occupation-time random variables. It can be shown that all known explicit methods for examining these problems can be expressed in this generalized inverse framework. More generally, in the context of a Markov renewal process setting, the aforementioned problems are also examined using generalized inverses of $I - P$, where P is the transition matrix of the discrete time jump Markov chain (for more information see [78–80]. In [37] we can find many applications of the Drazin inverse involving Markov chains.

More generally, the group inverse has found applications involving expressions for differentiation of eigenvectors and eigenvalues [81–85] as well as in the study of M-matrices, graph theory, and general nonnegative matrices [38, 65, 86–95] and in the analysis of Google's PageRank system [96, 97]. Also, there has been a wealth of new results concerning the use of the group inverse to characterize the sensitivity of the stationary probabilities to perturbations in the underlying transition probabilities (see [88, 96, 98–106]). Many applications of the group inverse in the theory of Markov chains, Perron eigenvalue analysis and spectral graph theory can be found in [107]. This motivates our study in Chaps. 5 and 6.

For applications of the Drazin inverse in applied probability problems and in statistics see [78, 79, 108, 109].

The concept of the generalized Drazin inverse in a Banach algebra was introduced by Koliha [110]. The condition $a - a^2 x \in \mathscr{A}^{\mathrm{nil}}$ from the definition of the Drazin inverse, was replaced by $a - a^2 x \in \mathscr{A}^{\mathrm{qnil}}$. Hence, the generalized Drazin inverse of a is the (unique) element $x \in \mathscr{A}$ (written a^{d}) which satisfies

$$xax = x, \quad ax = xa, \quad a - a^2 x \in \mathscr{A}^{\mathrm{qnil}}. \tag{1.2}$$

It is interesting to mention that Harte also gave an alternative definition of a generalized Drazin inverse in a ring [111–113]. These two concepts are equivalent in the case when the ring is actually a Banach algebra. It is well known that a^{d} is unique whenever it exists [110]. For many basic properties of the generalized Drazin inverse see [110, 112–114].

The generalized Drazin and the Drazin inverse are used in solving equations with constant coefficients to give an explicit representation of the set of solutions, and also in solving singular systems of differential equations, singular, singularly perturbed differential equations and differential-algebraic equations [18, 68, 69, 90, 115–121]. Some applications can also be found to various control problems.

References

1. Fredholm, I.: Sur une classe d'equations fonctionnelles. Acta. Math. **27**, 365–390 (1903)
2. Hilbert, D.: Grundzüuge einer algemeinen Theorie der linearen Integralgleichungen, B. G. Teubner, Leipzig and Berlin, 1912, (Reprint of six articles which appeared originally in the Götingen Nachrichten (1904), 49–51; (1904), 213–259; (1905), 307–338; (1906), 157–227; (1906), 439–480; (1910), 355–417)

3. Hurwitz, W.A.: On the pseudo-resolvent to the kernel of an integral equation. Trans. Am. Math. Soc. **13**, 405–418 (1912)
4. Moore, E.H.: General Analysis. American Philosophical Society, Philadelphia (1935)
5. Moore, E.H.: On the reciprocal of the general algebraic matrix (abstract). Bull. Am. Math. Soc. **26**, 394–395 (1920)
6. Siegel, C.L.: Uber die analytische Theorie der quadratischen Formen III. Ann. Math. **38**, 212–291 (1937)
7. Siegel, C.L.: Equivalence of quadratic forms. Am. J. Math. **63**, 658–680 (1941)
8. Tseng, Y.Y.: Generalized inverses of unbounded operators between two unitary spaces. Dokl. Akad. Nauk. SSSR. **67**, 431–434 (1949)
9. Tseng, Y.Y.: Properties and classifications of generalized inverses of closed operators. Dokl. Akad. Nauk. SSSR **67**, 607–610 (1949)
10. Tseng, Y.Y.: Virtual solutions and general inversions. Uspehi. Mat. Nauk. **11**, 213–215 (1956)
11. Murray, F.J., von Neumann, J.: On rings of operators. Ann. Math. **37**, 116–229 (1936)
12. Atkinson, F.V.: The normal solvability of linear equations in normed spaces (russian), Mat. Sbornik N.S. **28(70)**, 3–14 (1951)
13. Atkinson, F.V.: On relatively regular operators. Acta Sci. Math. Szeged **15**, 38–56 (1953)
14. Bjerhammar, A.: Application of the calculus of matrices to the method of least squares with special reference to geodetic calculations, Kungl. Tekn. H11gsk. Hand. Stockholm. **49**, 1–86 (1951)
15. Bjerhammar, A.: Rectangular reciprocal matrices with special reference to geodetic calculations. Bull. Geodesique **52**, 188–220 (1951)
16. Penrose, R.: A generalized inverse for matrices. Proc. Cambridge Philos. Soc. **51**, 466–413 (1955)
17. Rao, C.R.: Analysis of dispersion for multiply classified data with unequal numbers in cells. Sankhyd. **15**, 253–280 (1955)
18. Reid, W.T.: Generalized Green's matrices for compatible systems of differential equations. Am. J. Math. **53**, 443–459 (1931)
19. Bradley, J.S.: Adjoint quasi-differential operators of Euler type. Pacific J. Math. **16**, 213–237 (1966)
20. Bradley, J.S.: Generalized Green's matrices for compatible differential systems. Michigan Math. J. **13**, 97–108 (1966)
21. Kammerer, W.J., Nashed, M.Z.: Iterative methods for best approximate solutions of linear integral equations of the first and second kinds. J. Math. Anal. Appl. **40**, 547–573 (1972)
22. Nashed, M.Z.: On moment-discretization and least-squares solutions of linear integral equations of the first kind. J. Math. Anal. Appl. **53**, 359–366 (1976)
23. Kunkel, P., Mehrmann, V.: Generalized inverses of differential-algebraic operators. SIAM J. Matrix Anal. Appl. **17**, 426–442 (1996)
24. Gaarder, N.T., Herman, G.T.: Algorithms for reproducing objects from their x-rays. Comput. Graphics Image Process. **1**, 97–106 (1972)
25. Gordon R., Herman G.T.: Reconstruction of pictures from their projections. Commun. ACM. **14**, 759–768 (1971)
26. Herman, G.T.: Two direct methods for reconstructing pictures from their projections. Comput. Graphics Image Process. **1**, 123–143 (1972)
27. Krishan, S., Prabhu, S.S., Krishnamurthy, E.V.: Probabilistie reinforcement algorithms lot reconsh-uction of pictures from their projections. Int. J. Syst. Sci. **4**, 661–670 (1973)
28. Krishnamurthy, E.V.: Mahadeva, R.T., Subramanian, K., Prabhu, S.S.: Reconstruction of objects from their projections using generalized inverses. Comput. Graphics Image Process. **3**, 336–345 (1974)
29. Rieder, A., Schuster, T.: The approximate inverse in action with an application to computerized tomography SIAM. J. Numer. Anal. **37**(6), 1909–1929 (2000)
30. Laura, A.M., Conde, C.: Generalized inverses and sampling problems. J. Math. Anal. Appl. **398**(2), 744–751 (2013)

31. Li, G., Guo, D.: One-way property proof in public key cryptography based on OHNN. Proc. Eng. **15**, 1812–1816 (2011)
32. Radhakrishin, C.R.: A note on a generalized inverse of a matrix with applications to problems in mathematical statistics. J. R. Stat. Soc. Ser. B. **24**, 152–158 (1962)
33. Radhakrishin, C.R.: Generalized inverse for matrices and its applications in mathematical statistics. Research papers in Statistics. Festschrift for J. Neyman. Wiley, New York (1966)
34. Radhakrishin, C.R.: Least squares theory using an estimated dispersion matrix and its application to measurement of signals. In: Proceedings of the Fifth Berkeley Symposium on Statistics and Probability, Berkeley and Los Angeles, University of California Press, vol. 1, pp. 355–372 (1967)
35. Chipman, J.S.: Specification problems in regression analysis, Theory and Application of Generalized Inverses and Matrices, Symposium Proceedings. Texas Technological College. Mathematics Series **4**, 114–176 (1968)
36. Bose, R. C.: Analysis of Variance, unpublished lecture notes, University of North Carolina (1959)
37. Campbell, S.L., Meyer, C.D.: Generalized Inverse of Linear Transformations, Pitman, London (1979). Dover, New York (1991)
38. Anderson Jr., W.N., Duffin, R.J., Trapp, G.E.: Matrix operations induced by network connections. SIAM J. Control **13**, 446–461 (1975)
39. Akatsuka, Y., Matsuo, T.: Optimal control of linear discrete systems using the generalized inverse of a matrix, Techn Rept., vol. 13. Institute of Automatic Control, Nagoya Univ., Nagoya, Japan (1965)
40. Albert, A., Sittler, R.W.: A method for computing least squares estimators that keep up with the data. SIAM J. Control **3**, 384–417 (1965)
41. Balakrishnan, A.V.: An operator theoretic formulation of a class of control problems and a steepest descent method of solution. J. Soc. Ind. Appl. Math. Ser. A. Control **1**, 109–127 (1963)
42. Barnett, S.: Matrices in Control Theory. Van Nostrand Reinhold, London (1971)
43. Beutler, F.J., Root, W.L.: The operator pseudoinverse in control and systems identification. In: Nashed, M.Z. (ed.) Generalized Inverses and Applications. Academic Press, New York (1976)
44. Campbell, S.L.: Optimal control of autonomous linear processes with singular matrices in the quadratic cost functional. SIAM J. Control Optim. **14**(6), 1092–1106 (1976)
45. Campbell, S.L: Optimal control of discrete linear processes with quadratic cost. Int. J. Syst. Sci. **9**(8), 841–847 (1978)
46. Campbell, S.L.: Control problem structure and the numerical solution of linear singular systems. Math. Control Signals Syst. **1**(1), 73–87 (1988)
47. Dean, P., Porrill, J.: Pseudo-in J. Math. Anal. Appl.verse control in biological systems: a learning mechanism for fixation stability. Neural Netw. **11**, 1205–1218 (1998)
48. Kalman, R.E.: Contributions to the theory of optimal control. Bol. Soc. Mat. Mexicana **5**(2), 102–119 (1960)
49. Kalman, R.E., Ho, Y.C., Narendra, K.S.: Controllability of linear dynamical systems. Contributions to Differential Equations, vol. I, pp. 189–213. Interscience, New York (1963)
50. Doh-Hyun, K., Jun-Ho, O.: The Moore-Penrose inverse for the classificatory models. Control Eng. Pract. **7**(3), 369–373 (1999)
51. Liu, X., Liu, D.: Recursive computation of generalized inverses with application to optimal state estimation. Control Theory Adv. Tech. **10**, 1485–1497 (1995)
52. Lovass-Nagy, V., Powers, D.L.: Matrix generalized inverses in the handling of control problems containing input derivatives. Int. J. Syst. Sci. **6**, 693–696 (1975)
53. Minamide, N., Nakamura, K.: Minimum error control problem in banach space, Research Report of Automatic Control Lab 16. Nagoya University, Nagoya, Japan (1969)
54. Wahba, G., Nashed, M.Z.: The approximate solution of a class of constrained control problems. In: Proceedings of the Sixth Hawaii International Conference on Systems Sciences, Hawaii (1973)

55. Wang, Y.W., Wang, R.J.: Pseudoinverse and two-objective optimal control in Banach spaces. Funct. Approx. Comment. Math. **21**, 149–160 (1992)
56. Doty, K.L., Melchiorri, C., Bonivento, C.: A Theory of Generalized Inverses Applied to Robotics. Int. J. Robotics Res. **12**(1), 1–19 (1993)
57. Schwartz, E.M., Doty, K.L.: Application of the Weighted Generalized-Inverse in Control Optimization and Robotics. Florida Atlantic University, Boca Raton FL (June, Fifth Conf. on Recent Advances in Robotics (1992)
58. Schwartz, E.M., Doty, K.L.: The Weighted Generalized-Inverse Applied to Mechanism Controllability. University of Florida, Gainesville, FL (April, Sixth Conf. On Recent Advances in Robotics (1993)
59. Liu, W., Xu, Y., Yao, J., Zhao, Y., Zhao, B.Y., Liu, W.: The weighted Moore-Penrose generalized inverse and the force analysis of overconstrained parallel mechanisms, Multibody System Dynamics, **39**, 363–383 (2017)
60. Lasky, T.A., Ravani, B.: Sensor-based path planning and motion control for a robotic system for roadway crack sealing. IEEE Trans. Control Syst. Technol. **8**, 609–622 (2000)
61. Tucker, M., Perreira, N.D.: Generalized inverses for robotics manipulators. Mech. Mach. Theory **22**(6), 507–514 (1987)
62. Lenarčić, J., Husty, M.: Latest Advances in Robot Kinematic. Springer (2012)
63. Angeles, J.: The application of dual algebra to kinematic analysis. In: Angeles, J., Zakhariev, E. (eds.) Computational Methods in Mechanical Systems, pp. 3–31. Springer, Heidelberg (1998)
64. Angeles, J.: The Dual Generalized Inverses and Their Applications in Kinematic Synthesis. In: Lenarcic, J., Husty, M. (eds.) Latest Advances in Robot Kinematics, pp. 1–10. Springer, Netherlands (2012)
65. Drazin, M.P.: Pseudoinverse in associative rings and semigroups. Am. Math. Monthly **65**, 506–514 (1958)
66. Hartwig, R.E.: Schur's theorem and the Drazin inverse. Pacific J. Math. **78**, 133–138 (1978)
67. Ben-Isreal, A., Greville. T.N.E.: Generalized Inverse: Theory and Applications, 2nd edn. Springer, New York (2003)
68. Campbell, S.L., Meyer, C.D., Rose, N.J.: Applications of the Drazin inverse to linear systems of differential equations with singular constant coefficients. SIAM J. Appl. Math. **31**, 411–425 (1976)
69. Nashed, M.Z., Zhao, Y.: The Drazin inverse for singular evolution equations and partial differential equations. World Sci. Ser. Appl. Anal. **1**, 441–456 (1992)
70. Simeon, B., Fuhrer, C., Rentrop, P.: The Drazin inverse in multibody system dynamics. Numer. Math. **64**, 521–539 (1993)
71. Caradus, S.R.: Generalized inverses and operator theory, Queen's Papers in Pure and Appl. Math. **50**, Queen's University, Kingston, Ontario (1978)
72. King, C.F.: A note on Drazin inverses. Pacific J. Math. **70**, 383–390 (1977)
73. Lay, D.C.: Spectral properties of generalized inverses of linear operators. SIAM J. Appl. Math. **29**, 103–109 (1975)
74. Marek, I., Zitny, K.: Matrix Analysis for Applied Sciences, 2, Teubner-Texte zur Mathematik, vol. 84. Teubner, Leipzig (1986)
75. Drazin, M.P.: Extremal definitions of generalized inverses. Linear Algebra Appl. **165**, 185–196 (1992)
76. Heuser, H.G.: Functional Analysis. Wiley, New York (1982)
77. Taylor, A.E., Lay, D.C.: Introduction to Functional Analysis, 2nd edn. Wiley, New York (1980)
78. Hunter, J.: Generalized Inverses and their application to applied probability problems. Linear Algebra Appl. **45**, 157–198 (1982)
79. Hunter, J.: Generalized inverses of Markovian kernels in terms of properties of the Markov chain. Linear Algebra Appl. **447**, 38–55 (2014)
80. Meyer, C.D.: The role of the group generalized inverse in the theory of finite Markov chains. SIAM Rev. **17**(3), 443–464 (1975)

81. Ben-Israel, A., Charnes, A.: Generalized inverses and the Bott-Duffin network analysis. J. Math. Anal. Appl. **7**, 428–435 (1963)
82. Ben-Israel, A.: Linear equations and inequalities on finite dimensional, real or complex vector spaces: a unified theory. J. Math. Anal. Appl. **27**, 367–389 (1969)
83. Ben-Israel, A.: A note on partitioned matrices and equations. SIAM Rev. **11**, 247–250 (1969)
84. Berman, A., Plemmons, R.J.: Cones and iterative methods for best least-squares solutions of linear systems. SIAM J. Numer. Anal. **11**, 145–154 (1974)
85. Cline, R.E.: Representations for the generalized inverse of a partitioned matrix. SIAM J. Appl. Math. **12**, 588–600 (1964)
86. Anderson, W.N. Jr., Trapp, G.E.: Shorted operators II. SIAM J. Appl. Math. **28**, 60–71 (1975)
87. Bart, H., Kaashoek, M.A., Lay, D.C.: Relative inverses of meromorphic operator functions and associated holomorphic projection functions. Math. Ann. **218**(3), 199–210 (1975)
88. Ben-Israel, A.: A note on an iterative method for generalized inversion of matrices. Math. Comput. **20**, 439–440 (1966)
89. Campbell, S.L.: The Drazin inverse of an infinite matrix. SIAM J. Appl. Math. **31**, 492–503 (1976)
90. Campbell, S.L.: Linear systems of differential equations with singular coefficients. SIAM J. Math. Anal. **8**, 1057–1066 (1977)
91. Campbell, S.L.: On the limit of a product of matrix exponentials. Linear Multilinear Algebra **6**, 55–59 (1978)
92. Campbell, S.L.: Singular perturbation of autonomous linear systems II. J. Differ. Eqn. **29**, 362–373 (1978)
93. Campbell, S.L.: Limit behavior of solutions of singular difference equations. Linear Algebra Appl. **23**, 167–178 (1979)
94. Faddeev, D.K., Faddeeva, V.N.: Computational Methods of Linear Algebra, (translated by Robert C. Williams). W.H. Freeman and Co., San Francisco (1963)
95. Gantmacher, F.R.: The Theory of Matrices, vol. II. Chelsea Publishing Company, New York (1960)
96. Cederbaum, I.: On equivalence of resistive n-port networks. IEEE Trans. Circuit Theory **CT-12**, 338–344 (1965)
97. Cederbaum, I., Lempel, A.: Parallel connection of n-port networks, IEEE Trans. Circuit Theory **CT-14**, 274–279 (1967)
98. Ben-Israel, A.: An iterative method for computing the generalized inverse of an arbitrary matrix. Math. Comput. **19**, 452–455 (1965)
99. Ben-Israel, A.: On error bounds for generalized inverses. SIAM J. Numer. Anal. **3**, 585–592 (1966)
100. Campbell, S.L.: Differentiation of the Drazin inverse. SIAM J. Appl. Math. **30**, 703–707 (1976)
101. Churchill, R.V.: Operational Mathematics. McGraw-Hill, New York (1958)
102. Golub, G., Kahan, W.: Calculating the singular values and pseudoinverse of a matrix. SIAM J. Numer. Anal. Series B **2**, 205–224 (1965)
103. Golub, G.H., Pereyra, V.: The differentiation of pseudo-inverses and non-linear least squares problems whose variables separate. SIAM J. Numer. Anal. **10**, 413–432 (1973)
104. Golub, G.H., Wilkinson, J.H.: Ill conditioned eigensystems and the computation of the Jordan canonical form. SIAM Rev. **18**(4), 578–619 (1976)
105. Golub, G.H., Reinsch, C.: Singular value decomposition and least squares solutions. Numer. Math. **14**, 403–420 (1970)
106. Greville, T.N.E.: Spectral generalized inverses of square matrices, MRC Tech. Sum. Rep. 823, Mathematics Research Center, University of Wisconsin, Madison (1967)
107. Kirkland, S.J., Neumann, M.: Group Inverses of M-Matrices and their Applications. CRC Press (2012)
108. Rao, C.R.: A note on a generalized inverse of a matrix with applications to problems in mathematical statistics, J. R. Stat. Soc. Ser. B Stat. Methodol. **24**(1), 152–158 (1962)

109. Riaza, R.: Differential-Algebriac Systems. Analytical Aspects and Circuit Applications, World Scientifics, River Edge, NJ (2008)
110. Koliha, J.J.: A generalized Drazin inverse. Glasgow Math. J. **38**, 367–381 (1996)
111. Harte, R.E.: Spectral projections. Irish Math. Soc. Newslett. **11**, 10–15 (1984)
112. Harte, R.E.: Invertibility and Singularity for Bounded Linear Operators. Marcel Dekker, New York (1988)
113. Harte, R.E.: On quasinilpotents in rings. PanAm. Math. J. **1**, 10–16 (1991)
114. Han, J.K., Lee, H.Y., Lee, W.Y.: Invertible completions of 2×2 upper triangular operator matrices. Proc. Am. Math. Soc. **128**, 119–123 (1999)
115. Koliha, J.J., Tran, T.D.: Semistable operators and singularly perturbed differential equations. J. Math. Anal. Appl. **231**, 446–458 (1999)
116. Koliha, J.J., Tran, T.D.: The Drazin inverse for closed linear operators. J. Oper. Theory **46**(2), 323–336 (2001)
117. Campbell, S.L.: The Drazin inverse and systems of second order linear differential equations. Linear Multilinear Algebra **14**, 195–198 (1983)
118. Cui, X., Wei, Y., Zhang, N.: Quotient convergence and multi-splitting methods for solving singular linear equations. Calcolo **44**, 21–31 (2007)
119. Hooker, J.W., Langenhop, C.E.: On regular systems of linear differential equations with constant coefficients. Rocky Mountain J. Math. **12**(4), 591–614 (1982)
120. Ascher, U.M., Petzold, L.R.: Computer Methods for Ordinary Differential Equations and Differential-Algebraic Equations. SIAM, Philadelphia (1998)
121. Brenan, K.E., Campbell, S.L., Petzold, L.R.: Numerical Solution of Initial-Value Problems in Differential-Algebraic Equations, Classics in Appl. Math. **14**, SIAM, Philadelphia (1995)

Chapter 2
Reverse Order Law

The problem of finding the generalized inverse of a product of matrices led to the investigation of the so-called "reverse order law". The reverse order law for many types of generalized inverses has been the subject of intensive research over the years. In the 1960s, Greville was the first to study it by considering the reverse order law for the Moore-Penrose inverse and gave a necessary and sufficient condition for the reverse order law

$$(AB)^\dagger = B^\dagger A^\dagger, \tag{2.1}$$

to hold for matrices A and B. This was followed by further research on this subject branching in several directions:

- Products with more than two matrices were considered;
- Different classes of generalized inverses were studied;
- Different settings were considered (operator algebras, C^*-algebras, rings etc.)

Over the years this topic has been the subject of interest in various investigations. In this chapter we will set as our primary goal a chronological and systematic presentation, thus taking into account both the time of publication and the level of generalization, of all the published results covering this topic and to point to some problems that are still open and the difficulties that one is faced with when attempting to solve them. Such an approach is intended to give the reader a clear picture of the current status of the research concerning this topic and also some guidelines for future research that they might be interested in doing.

We will discuss the reverse order laws for K-inverses when $K \in \{\{1\}, \{1, 2\}, \{1, 3\}, \{1, 2, 3\}, \{1, 3, 4\}\}$ and present all recently published results on this subject as well as some simple examples and open problems.

When we are talking about the reverse order law for the K-inverse, where $K \subseteq \{1, 2, 3, 4\}$, we are actually considering the following inclusions:

© Springer Nature Singapore Pte Ltd. 2017

D. Cvetković Ilić and Y. Wei, *Algebraic Properties of Generalized Inverses*,

Developments in Mathematics 52, DOI 10.1007/978-981-10-6349-7_2

$$BK \cdot AK \subseteq (AB)K,$$
$$(AB)K \subseteq BK \cdot AK,$$
$$(AB)K = BK \cdot AK.$$

The reverse order law problem for each of the above mentioned types of generalized inverses will receive special attention, and we will describe necessary and sufficient conditions in various settings such as that of matrices, algebras of bounded linear operators, C^*-algebras and, when possible, in general rings. Beside presenting to the reader some results that have large application, primarily in solving different types of matrix and operator equations, they will have the opportunity to familiarize themselves with the techniques that are used to generalize results obtained in the case of matrices to more general settings such as those of algebras of bounded linear operators, C^*-algebras or rings.

2.1   Reverse Order Laws for {1}-Inverses

In this section, we address the question of when the reverse order laws for {1}-inverses is valid. It is interesting that although the reverse order law has been considered for many types of generalized inverses and from various aspects too, there are only a few papers which are concerned with this problem for the {1}-inverse.

In his article, Rao [1] proves that if A and B are complex matrices such that AB is defined, and if either A is of full column rank or B is of full row rank, then

$$B\{1\}A\{1\} \subseteq (AB)\{1\}. \tag{2.2}$$

After this, Pringle and Rayner [2], state incorrectly that any of the two conditions from the Rao's result (i.e., if A is of full column rank or B is of full row rank) imply that

$$(AB)\{1\} = B\{1\}A\{1\} \tag{2.3}$$

which is noted in 1994 by Werner [3], who gives a simple counterexample to this assertion and proves that for given matrices A and B of appropriate sizes, (2.2) holds if and only if

$$\mathcal{N}(A) \subseteq \mathcal{R}(B) \text{ or } \mathcal{R}(B) \subseteq \mathcal{N}(A),$$

where $\mathcal{N}(A)$ and $\mathcal{R}(B)$ are the null space of A and the range of B, respectively. It can easily be seen that Werner's proof, when suitably modified, carries over to operators on Hilbert spaces.

Theorem 2.1 *Let $A \in \mathcal{B}(\mathcal{H}, \mathcal{K})$ and $B \in \mathcal{B}(\mathcal{L}, \mathcal{H})$ be regular operators such that the product AB is also regular. Then $B\{1\}A\{1\} \subseteq (AB)\{1\}$ if and only if*

$$\mathcal{N}(A) \subseteq \mathcal{R}(B) \text{ or } AB = 0. \tag{2.4}$$

Also he proves that (2.3) holds in particular in each of the following cases:

(i) A and B are both of full column rank
(ii) A and B are both of full row rank
(iii) A is nonsingular and/or B is nonsingular

but in general, the more difficult problem of finding equivalent descriptions of the condition (2.3) still remains open. The next paper on this topic was by M. Wei [4] where, using P-SVD of matrices A and B, some equivalents of (2.2) are derived and compared with the conditions given by Werner and finally certain necessary and sufficient conditions for (2.3) to hold are given.

Theorem 2.2 ([4]) *Let $A \in \mathbb{C}^{m \times n}$ and $B \in \mathbb{C}^{n \times p}$. The following conditions are equivalent:*

(i) $(AB)\{1\} = B\{1\}A\{1\}$
(ii) *One of the following conditions hold:*

    (a) $r(AB) = 0$, $n \geq \min\{m + r(B), p + r(A)\}$,
    (b) $r(A) + r(B) - r(AB) = n$ *and* $(r(A) = m$ *or* $r(B) = p)$

(iii) *One of the following conditions hold:*

    (a) $\mathcal{R}(B) \subseteq \mathcal{N}(A)$, $n \geq \min\{m + r(B), p + r(A)\}$,
    (b) $\mathcal{N}(A) \subseteq \mathcal{R}(B)$ *and* $(r(A) = m$ *or* $r(B) = p)$.

Let us now take a look at the following few examples.

Example 2.1 If $m = n$ and $A = I$, then for any $B \in \mathbb{C}^{n \times p}$ we have $(AB)\{1\} = B\{1\}$ and $A\{1\} = \{I\}$, so $(AB)\{1\} = B\{1\}A\{1\}$, which can also be concluded from Theorem 2.2,(iii)b.

Example 2.2 Let $A = \begin{bmatrix} 0 & 1 \\ 0 & 1 \end{bmatrix}$ and $B = \begin{bmatrix} 1 & 0 \\ 0 & 0 \end{bmatrix}$. Then evidently $AB = 0$ and from Theorem 2.1 (or using the fact that $(AB)\{1\} = \mathbb{C}^{2 \times 2}$) we have that $B\{1\}A\{1\} \subseteq (AB)\{1\}$. On the other hand, since

$$A\{1\} = \left\{ \begin{bmatrix} a_1 & a_2 \\ a_3 & 1 - a_3 \end{bmatrix} : a_1, a_2, a_3 \in \mathbb{C} \right\}$$

and

$$B\{1\} = \left\{ \begin{bmatrix} 1 & b_1 \\ b_2 & b_3 \end{bmatrix} : b_1, b_2, b_3 \in \mathbb{C} \right\},$$

we can check that $\begin{bmatrix} 0 & 0 \\ 1 & -1 \end{bmatrix} \in (AB)\{1\}$ can not be written as a product $B^{(1)}A^{(1)}$ for some $A^{(1)} \in A\{1\}$ and $B^{(1)} \in B\{1\}$. This means that $(AB)\{1\} \neq B\{1\}A\{1\}$ which can be checked also using Theorem 2.2.

Also, the inclusion (2.2) in the case of the product of more than two matrices was considered by M. Wei [4] by applying the multiple product singular value decomposition.

Theorem 2.3 ([5]) *Let* $A_i \in \mathbb{C}^{m_i \times m_{i+1}}$, $i = \overline{1, n}$, $n \geq 3$. *The following conditions are equivalent:*

(i) $A_n\{1\} \cdot A_{n-1}\{1\} \cdots A_1\{1\} \subseteq (A_1 A_2 \cdots A_n)\{1\}$

(ii) *One of the following conditions hold:*

   (a) $r(A_1 \cdots A_n) = 0$
   (b) $r(A_1 \cdots A_i) + r(A_{i+1}) - r(A_1 \cdots A_{i+1}) = m_{i+1}$, $i = \overline{1, n-1}$

(iii) *One of the following conditions hold:*

   (a) $\mathscr{R}(A_{i+1} \cdots A_n) \subseteq \mathscr{N}(A_1 \cdots A_i)$, *for some* $i \in \{1, \ldots, n-1\}$
   (b) $\mathscr{N}(A_1 \cdots A_i) \subseteq \mathscr{R}(A_{i+1})$, $i = \overline{1, n-1}$.

Recently, the previous result was generalized by Nikolov-Radenković [6] for bounded linear operators on Hilbert spaces. We will give a proof of this result.

Theorem 2.4 ([6]) *Let* $A_i \in \mathscr{B}(\mathscr{H}_{i+1}, \mathscr{H}_i)$, $i = \overline{1, n}$, *be such that* A_i, $i = \overline{1, n}$ *and* $A_1 A_2 \cdots A_j$, $j = \overline{2, n}$, *are regular operators. The following conditions are equivalent:*

(i) $A_n\{1\} \cdot A_{n-1}\{1\} \cdots A_1\{1\} \subseteq (A_1 A_2 \cdots A_n)\{1\}$

(ii) $A_1 A_2 \cdots A_n = 0$ *or* $\mathscr{N}(A_1 \cdots A_{j-1}) \subseteq \mathscr{R}(A_j)$, *for* $j = \overline{2, n}$

(iii) $A_1 A_2 \cdots A_n = 0$ *or* $A_k\{1\} \cdot A_{k-1}\{1\} \cdots A_1\{1\} \subseteq (A_1 A_2 \cdots A_k)\{1\}$, *for* $k = \overline{2, n}$.

Proof (ii) $\Rightarrow$ (iii) : If $A_1 A_2 \cdots A_n = 0$, it is evident that (iii) holds. Suppose that $A_1 A_2 \cdots A_n \neq 0$ and

$$\mathscr{N}(A_1 \cdots A_{j-1}) \subseteq \mathscr{R}(A_j), \text{ for } j = \overline{2, n}. \tag{2.5}$$

We will prove by induction on k that

$$A_k\{1\} \cdot A_{k-1}\{1\} \cdots A_1\{1\} \subseteq (A_1 A_2 \cdots A_k)\{1\} \tag{2.6}$$

holds for $k = \overline{2, n}$. From (2.5) it follows that $\mathscr{N}(A_1) \subseteq \mathscr{R}(A_2)$ which by (2.4) implies that (2.6) holds for $k = 2$. Suppose that (2.6) holds for $k = l - 1$, where $l \in \{2, 3, \ldots, n\}$, i.e.,

$$A_{l-1}\{1\} \cdot A_{l-2}\{1\} \cdots A_1\{1\} \subseteq (A_1 A_2 \cdots A_{l-1})\{1\}. \tag{2.7}$$

We prove that it must also hold for $k = l$. Since (2.5) holds for $j = l$, by (2.4) we have

$$A_l\{1\} \cdot (A_1 A_2 \cdots A_{l-1})\{1\} \subseteq (A_1 A_2 \cdots A_{l-1} A_l)\{1\}, \tag{2.8}$$

which by (2.7) implies that (2.6) holds for $k = l$. Hence, by induction it follows that (2.6) holds for $k = \overline{2, n}$.

(iii) $\Rightarrow$ (i) : This is evident.

(i) $\Rightarrow$ (ii) : Suppose that $A_1 A_2 \cdots A_n \neq 0$ and that $n > 2$ since the assertion in the case $n = 2$ follows by Theorem 2.1. Let $j \in \{3, 4, \ldots, n\}$ and $i \in \{1, 2, \ldots, j - 2\}$ be arbitrary. Then for arbitrary $A_i^{(1)} \in A_i\{1\}$ and $A_j^{(1)} \in A_j\{1\}$, we have that

$$
\begin{aligned}
& A_1 A_2 \cdots A_n A_n^{(1)} \cdots A_{j+1}^{(1)} \cdot (A_j^{(1)} + Y(I_{\mathcal{H}_j} - A_j A_j^{(1)})) A_{j-1}^{(1)} \cdots A_{i+1}^{(1)} \\
& (A_i^{(1)} + (I_{\mathcal{H}_{i+1}} - A_i^{(1)} A_i) X) \cdot A_{i-1}^{(1)} \cdots A_1^{(1)} A_1 \cdots A_n = A_1 \cdots A_n
\end{aligned}
\tag{2.9}
$$

holds for every $X \in \mathscr{B}(\mathcal{H}_i, \mathcal{H}_{i+1})$ and every $Y \in \mathscr{B}(\mathcal{H}_{j+1}, \mathcal{H}_j)$. Substituting $X = 0$ in (2.9), we get

$$
\begin{aligned}
& A_1 A_2 \cdots A_n A_n^{(1)} \cdots A_{j+1}^{(1)} \cdot (A_j^{(1)} + Y(I_{\mathcal{H}_j} - A_j A_j^{(1)})) A_{j-1}^{(1)} \cdots A_{i+1}^{(1)} A_i^{(1)} \cdot \\
& A_{i-1}^{(1)} \cdots A_1^{(1)} A_1 \cdots A_n = A_1 \cdots A_n.
\end{aligned}
\tag{2.10}
$$

Subtracting (2.10) from (2.9), we get that

$$
\begin{aligned}
& A_1 A_2 \cdots A_n A_n^{(1)} \cdots A_{j+1}^{(1)} \cdot (A_j^{(1)} + Y(I_{\mathcal{H}_j} - A_j A_j^{(1)})) A_{j-1}^{(1)} \cdot \\
& \cdots A_{i+1}^{(1)} (I_{\mathcal{H}_{i+1}} - A_i^{(1)} A_i) X A_{i-1}^{(1)} \cdots A_1^{(1)} A_1 \cdots A_n = 0.
\end{aligned}
\tag{2.11}
$$

Substituting $Y = 0$ in (2.11), we get

$$
\begin{aligned}
& A_1 A_2 \cdots A_n A_n^{(1)} \cdots A_{j+1}^{(1)} A_j^{(1)} A_{j-1}^{(1)} \cdots A_{i+1}^{(1)} (I_{\mathcal{H}_{i+1}} - A_i^{(1)} A_i) X \cdot \\
& A_{i-1}^{(1)} \cdots A_1^{(1)} A_1 \cdots A_n = 0.
\end{aligned}
\tag{2.12}
$$

Finally, from (2.12) and (2.11), we get that

$$
\begin{aligned}
& A_1 A_2 \cdots A_n A_n^{(1)} \cdots A_{j+1}^{(1)} \cdot Y(I_{\mathcal{H}_j} - A_j A_j^{(1)}) A_{j-1}^{(1)} \cdots A_{i+1}^{(1)} \cdot \\
& (I_{\mathcal{H}_{i+1}} - A_i^{(1)} A_i) X A_{i-1}^{(1)} \cdots A_1^{(1)} A_1 \cdots A_n = 0
\end{aligned}
\tag{2.13}
$$

holds for arbitrary $X \in \mathscr{B}(\mathcal{H}_i, \mathcal{H}_{i+1})$ and $Y \in \mathscr{B}(\mathcal{H}_{j+1}, \mathcal{H}_j)$.

Now, it follows that either

$$
A_1 A_2 \cdots A_n A_n^{(1)} \cdots A_{j+1}^{(1)} = 0
\tag{2.14}
$$

or

$$
(I_{\mathcal{H}_j} - A_j A_j^{(1)}) A_{j-1}^{(1)} \cdots A_{i+1}^{(1)} (I_{\mathcal{H}_{i+1}} - A_i^{(1)} A_i) = 0
\tag{2.15}
$$

or

$$
A_{i-1}^{(1)} \cdots A_1^{(1)} A_1 \cdots A_n = 0.
\tag{2.16}
$$

It is easy to see that (2.14) and (2.16) imply $A_1 A_2 \cdots A_n = 0$ which is not the case, so (2.15) must hold. Hence, for arbitrary $j \in \{3, 4, \ldots, n\}$ and $i \in \{1, 2, \ldots, j-2\}$, we have that

$$(I_{\mathcal{H}_j} - A_j A_j^{(1)}) A_{j-1}^{(1)} \cdots A_{i+1}^{(1)} A_i^{(1)} A_i = (I_{\mathcal{H}_j} - A_j A_j^{(1)}) A_{j-1}^{(1)} \cdots A_{i+1}^{(1)}. \quad (2.17)$$

Let $j \in \{2, 3, \ldots, n\}$, be arbitrary. Then by (i) it follows that

$$A_1 A_2 \cdots A_n A_n^{(1)} \cdots A_{j+1}^{(1)}.$$
$$(A_j^{(1)} + Y(I_{\mathcal{H}_j} - A_j A_j^{(1)}))(A_{j-1}^{(1)} + (I_{\mathcal{H}_j} - A_{j-1}^{(1)} A_{j-1}) X) \cdot \quad (2.18)$$
$$A_{j-2}^{(1)} \cdots A_1^{(1)} A_1 \cdots A_n = A_1 \cdots A_n$$

holds for arbitrary $X \in \mathcal{B}(\mathcal{H}_{j-1}, \mathcal{H}_j)$ and $Y \in \mathcal{B}(\mathcal{H}_j, \mathcal{H}_{j+1})$. Analogously to the previous part of the proof, we get that for arbitrary $j \in \{2, 3, \ldots, n\}$

$$(I_{\mathcal{H}_j} - A_j A_j^{(1)})(I_{\mathcal{H}_j} - A_{j-1}^{(1)} A_{j-1}) = 0. \quad (2.19)$$

Taking $j = 2$ in (2.19), we conclude that

$$\mathcal{N}(A_1) \subseteq \mathcal{R}(A_2). \quad (2.20)$$

Now, choose arbitrary $j \in \{3, 4, \ldots, n\}$. Using (2.17) and (2.19), we have

$$A_j A_j^{(1)} (I - A_{j-1}^{(1)} \cdots A_2^{(1)} A_1^{(1)} A_1 A_2 A_3 \cdots A_{j-1})$$
$$= I - A_{j-1}^{(1)} \cdots A_2^{(1)} A_1^{(1)} A_1 A_2 A_3 \cdots A_{j-1}, \quad (2.21)$$

which implies that $\mathcal{N}(A_1 A_2 \cdots A_{j-1}) \subseteq \mathcal{R}(A_j)$. $\square$

In [7] some necessary and sufficient conditions are given under which, in the matrix case, for some $(AB)^{(1)} \in (AB)\{1\}$ satisfying some special conditions there exist $A^{(1)} \in A\{1\}$ and $B^{(1)} \in B\{1\}$ such that $(AB)^{(1)} = B^{(1)} A^{(1)}$:

Theorem 2.5 ([7]) *Let* $A \in \mathbb{C}^{m \times n}$ *and* $B \in \mathbb{C}^{n \times p}$ *and let* $(AB)^{(1)} \in (AB)\{1\}$ *be arbitrarily given, and let*

$$t((AB)^{(1)}) = \dim (\mathcal{R}((AB)^{(1)}) \cap \mathcal{N}(B)) \text{ and}$$

$$v((AB)^{(1)}) = \dim (\mathcal{R}([(AB)^{(1)}]^*) \cap \mathcal{N}(A^*)).$$

Then $(AB)^{(1)} \in B\{1\}A\{1\}$ *if and only if:*

$$r((AB)^{(1)}) - t((AB)^{(1)}) - v((AB)^{(1)}) \geq r(A) + r(B) - n.$$

The reverse order law

$$(AB)\{1\} \subseteq B\{1\}A\{1\} \tag{2.22}$$

in the setting of matrices was completely solved in 1998 in [8], where using P-SVD of matrices A and B it was proved that (2.22) holds if and only if

$$\dim \mathcal{N}(A) - \dim (\mathcal{N}(A) \cap \mathcal{R}(B)) \geq \min \{\dim \mathcal{N}(A^*), \dim \mathcal{N}(B)\}.$$

In [9], some other necessary and sufficient conditions for (2.22) to hold were presented without using SVD or P-SVD of matrices A and B, potentially allowing for our purely algebraic proof to be generalized to more general settings:

Theorem 2.6 *Let $A \in \mathbb{C}^{m \times n}$ and $B \in \mathbb{C}^{n \times p}$. The following conditions are equivalent:*

(i) $(AB)\{1\} \subseteq B\{1\}A\{1\}$,
(ii) $r(A) + r(B) - n \leq r(AB) - \min\{m - r(A), p - r(B)\}$.

Proof Since (i) is equivalent with

$$(B^*A^*)\{1\} \subseteq A^*\{1\}B^*\{1\}, \tag{2.23}$$

without loss of generality we can suppose throughout the proof that

$$\min\{m - r(A), p - r(B)\} = m - r(A). \tag{2.24}$$

Indeed, if this is not the case we can have (2.24) by simply replacing A with B^* and B with A^*, given that $m - r(A) = \dim \mathcal{N}(A^*)$ and $p - r(B) = \dim \mathcal{N}(B)$. Now, assuming (2.24), we need to prove that (i) is equivalent with $m - r(AB) \leq n - r(B)$.

Evidently, (i) is equivalent with the fact that for any $(AB)^{(1)} \in (AB)\{1\}$ there exist $A^{(1)} \in A\{1\}$ and $B^{(1)} \in B\{1\}$ such that

$$(AB)^{(1)} = B^{(1)}A^{(1)}.$$

Using Lemma 1.1 from [9] (or the more general version of it—Lemma 3.5) as well as appropriate notations used therein, (i) is equivalent with the fact that for any $(A_1B_1)^{(1)} \in (A_1B_1)\{1\}$, $Z_2 \in \mathcal{B}(\mathcal{N}(A^*), \mathcal{R}(B^*))$, $Z_3 \in \mathcal{B}(\mathcal{R}(A), \mathcal{N}(B))$ and $Z_4 \in \mathcal{B}(\mathcal{N}(A^*), \mathcal{N}(B))$ there exist matrices $Y_2 \in \mathcal{B}(\mathcal{N}(B^*), \mathcal{R}(B^*))$, $Y_3 \in \mathcal{B}(\mathcal{R}(B), \mathcal{N}(B))$ and $Y_4 \in \mathcal{B}(\mathcal{N}(B^*), \mathcal{N}(B))$ and $X = \begin{bmatrix} X_1 & X_2 \\ X_3 & X_4 \end{bmatrix}$: $\begin{bmatrix} \mathcal{R}(A) \\ \mathcal{N}(A^*) \end{bmatrix} \rightarrow \begin{bmatrix} \mathcal{R}(B) \\ \mathcal{N}(B^*) \end{bmatrix}$ satisfying

$$A_1X_1 + A_2X_3 = I_{\mathcal{R}(A)}, \tag{2.25}$$

such that

$$\left[(A_1 B_1)^{(1)} \; Z_2 \right] = \left[B_1^{-1} \; Y_2 \right] X \tag{2.26}$$

$$\left[Z_3 \; Z_4 \right] = \left[Y_3 \; Y_4 \right] X. \tag{2.27}$$

In general for fixed Y_2 the Eq. (2.26) is solvable for X and we have that the set of solutions is given by

$$S = \left\{ \begin{bmatrix} B_1 \\ 0 \end{bmatrix} \left[(A_1 B_1)^{(1)} \; Z_2 \right] + \left(I - \begin{bmatrix} B_1 \\ 0 \end{bmatrix} \left[B_1^{-1} \; Y_2 \right] \right) W : \right.$$

$$W \in \mathbb{C}^{n \times m} \Big\}$$

$$= \left\{ \begin{bmatrix} B_1 (A_1 B_1)^{(1)} - B_1 Y_2 W_3 \; B_1 Z_2 - B_1 Y_2 W_4 \\ W_3 \qquad\qquad W_4 \end{bmatrix} : \right. \tag{2.28}$$

$$\left. \begin{bmatrix} W_1 \; W_2 \\ W_3 \; W_4 \end{bmatrix} : \begin{bmatrix} \mathscr{R}(A) \\ \mathscr{N}(A^*) \end{bmatrix} \to \begin{bmatrix} \mathscr{R}(B) \\ \mathscr{N}(B^*) \end{bmatrix} \right\},$$

given that obviously $\begin{bmatrix} B_1 \\ 0 \end{bmatrix} \in \left[B_1^{-1} \; Y_2 \right] \{1\}$.

Thus (i) is equivalent with the existence of at least one $X \in S \cap A\{1\}$ for which the Eq. (2.27) is solvable for $\left[Y_3 \; Y_4 \right]$. The solvability of Eq. (2.27) is equivalent with

$$\left[Z_3 \; Z_4 \right] (I - X^{(1)} X) = 0, \tag{2.29}$$

for some (any) $X^{(1)} \in X\{1\}$.

Hence (i) is equivalent with the existence of $X \in S \cap A\{1\}$ for which (2.29) holds. Write $X = \left[K_1 \; K_2 \right]$, where

$$K_1 = \begin{bmatrix} B_1 (A_1 B_1)^{(1)} - B_1 Y_2 W_3 \\ W_3 \end{bmatrix} \text{ and } K_2 = \begin{bmatrix} B_1 Z_2 - B_1 Y_2 W_4 \\ W_4 \end{bmatrix}.$$

Using Lemma 2.3 [10], we have that one inner inverse of X is given by $X^{(1)} = \begin{bmatrix} K_1^{(1)} - K_1^{(1)} K_2 T^{(1)} R_{K_1} \\ T^{(1)} R_{K_1} \end{bmatrix}$, where $T = R_{K_1} K_2$ and $R_{K_1} = I - K_1 K_1^{(1)}$. Thus (2.29) is equivalent with

$$(Z_4 - Z_3 K_1^{(1)} K_2)(I - T^{(1)} T) = 0, \; Z_3 (I - K_1^{(1)} K_1) = 0, \tag{2.30}$$

for some $T^{(1)} \in T\{1\}$.

(i) $\Rightarrow$ (ii): Taking $(A_1 B_1)^{(1)} = (A_1 B_1)^\dagger$, $Z_2 = 0$, $Z_3 = 0$ and a left invertible Z_4 (such Z_4 exists since $\dim \mathscr{N}(A^*) \le \dim \mathscr{N}(B)$), by (2.30) we get that T must be left invertible. Since $T = P_{\mathscr{N}(K_1 K_1^{(1)}), \mathscr{R}(K_1)} K_2$, we get that $\mathscr{N}(T) = \{0\}$ if and only if $\mathscr{N}(K_2) = \{0\}$ and $\mathscr{R}(K_1) \cap \mathscr{R}(K_2) = \{0\}$. The first condition, $\mathscr{N}(K_2) = \{0\}$ is satisfied if and only if $\mathscr{N}(W_4) = \{0\}$, which is possible only if $\dim \mathscr{N}(A^*) \le \dim \mathscr{N}(B^*)$. The second condition $\mathscr{R}(K_1) \cap \mathscr{R}(K_2) = \{0\}$ (in the case when K_1

and K_2 are left invertible) is equivalent with $\mathcal{N}(X) = \{0\}$, i.e.,

$$\mathcal{N}\left(\left[\, B_1(A_1 B_1)^\dagger \ 0\,\right]\right) \cap \mathcal{N}\left(\left[\, W_3 \ W_4\,\right]\right) = \{0\}. \tag{2.31}$$

Thus $\mathcal{N}((A_1 B_1)^\dagger) \cap \mathcal{N}(W_3) = \{0\}$, so the condition (2.31) is equivalent with $\mathcal{R}(W_3 \mid_{\mathcal{N}((A_1 B_1)^\dagger)}) \cap \mathcal{R}(W_4) = \{0\}$, which is possible only when $\dim \mathcal{N}((A_1 B_1)^\dagger) + \dim \mathcal{N}(A^*) \leq \dim \mathcal{N}(B^*)$. Since $\dim \mathcal{N}((A_1 B_1)^\dagger) = r(A) - r(AB)$, we get that $m - r(AB) \leq n - r(B)$.

(ii) $\Rightarrow$ (i): Suppose $(A_1 B_1)^{(1)} \in (A_1 B_1)\{1\}$, $Z_2 \in \mathcal{B}(\mathcal{N}(A^*), \mathcal{R}(B^*))$, $Z_3 \in \mathcal{B}(\mathcal{R}(A), \mathcal{N}(B))$ and $Z_4 \in \mathcal{B}(\mathcal{N}(A^*), \mathcal{N}(B))$ are given. We will show that there exists a left-invertible matrix $X \in S \cap A\{1\}$. Let $Y_2 = (A_1 B_1)^{(1)} A_2$ and $W_3 = X_3'$, where $\begin{bmatrix} X_1' \\ X_3' \end{bmatrix} : \mathcal{R}(A) \to \begin{bmatrix} \mathcal{R}(B^*) \\ \mathcal{N}(B) \end{bmatrix}$ is an arbitrary right inverse of $\left[\, A_1 \ A_2\,\right]$. Using Theorem 2.7 from [11], we will show that there is some W_4 such that

$$X' = \begin{bmatrix} B_1(A_1 B_1)^{(1)} & B_1 Z_2 \\ W_3 & W_4 \end{bmatrix} \tag{2.32}$$

is left invertible. It is easy to check that the first column of X' is left-invertible. Thus it remains to check the inequality

$$n(X_0) \leq d(W_3) + \dim\left(\mathcal{R}(W_3^*) \cap \mathcal{R}((B_1(A_1 B_1)^{(1)})^* \mid_{\mathcal{N}((B_1 Z_2)^*)}\right) \tag{2.33}$$

where $X_0 = \begin{bmatrix} B_1 Z_2 & B_1(A_1 B_1)^{(1)} \\ 0 & W_3 \end{bmatrix}$.

Note that

$$n(X_0^*) = n(W_3^*) + n((B_1(A_1 B_1)^{(1)})^* \mid_{\mathcal{N}((B_1 Z_2)^*)})$$
$$+ \dim\left(\mathcal{R}(W_3^*) \cap \mathcal{R}((B_1(A_1 B_1)^{(1)})^* \mid_{\mathcal{N}((B_1 Z_2)^*)})\right)$$

and since $n(X_0) = m - n + n(X_0^*)$, that (2.33) is equivalent with

$$m - n + n((B_1(A_1 B_1)^{(1)})^* \mid_{\mathcal{N}((B_1 Z_2)^*)}) \leq 0. \tag{2.34}$$

As $n((B_1(A_1 B_1)^{(1)})^*) \leq r(B) - r(AB)$ and $r(B) - r(AB) - n + m \leq 0$, we get that (2.34) holds for any $(A_1 B_1)^{(1)} \in (A_1 B_1)\{1\}$ and any $Z_2 \in \mathcal{B}(\mathcal{N}(A^*), \mathcal{R}(B^*))$.

Now by Theorem 2.7 from [11] there is some W_4 such that X' given by (2.32) is left invertible. It is easy to see that

$$X = \begin{bmatrix} B_1(A_1 B_1)^{(1)} - B_1 Y_2 W_3 & B_1 Z_2 - B_1 Y_2 W_4 \\ W_3 & W_4 \end{bmatrix}$$

is left invertible as well.   □

Let us take a look at the following examples.

Example 2.3 We will show that in the case when (ii) of Theorem 2.6 is not satisfied, which means that $(AB)\{1\} \not\subseteq B\{1\}A\{1\}$, we can find a general form of the inner inverses of AB, $(AB)^{(1)}$ for which there does not exist $A^{(1)}$ and $B^{(1)}$ such that $(AB)^{(1)} = B^{(1)}A^{(1)}$. Let $A = \begin{bmatrix} 1 & 0 \\ 0 & 0 \end{bmatrix}$ and $B = \begin{bmatrix} 0 & 0 \\ 1 & 0 \end{bmatrix}$. Then evidently $AB = 0$ and consequently $(AB)\{1\} = \mathbb{C}^{2 \times 2}$. Since

$$A\{1\} = \left\{ \begin{bmatrix} 1 & a_1 \\ a_2 & a_3 \end{bmatrix} : a_1, a_2, a_3 \in \mathbb{C} \right\}$$

and

$$B\{1\} = \left\{ \begin{bmatrix} b_1 & 1 \\ b_2 & b_3 \end{bmatrix} : b_1, b_2, b_3 \in \mathbb{C} \right\}$$

we can check that none of the matrices of the form $\begin{bmatrix} 0 & 0 \\ 0 & w \end{bmatrix}$, where $w \neq 0$, can be written as a product $B^{(1)}A^{(1)}$ for some $A^{(1)} \in A\{1\}$ and $B^{(1)} \in B\{1\}$.

Example 2.4 If $A \in \mathbb{C}^{m \times n}$ is left invertible and $B \in \mathbb{C}^{n \times p}$, then by Theorem 2.1 we have that $B\{1\}A\{1\} \subseteq (AB)\{1\}$. In this case $(AB)\{1\} = B\{1\}A\{1\}$ if and only if

$$r(B) \leq r(AB) - \min\{m - n, p - r(B)\}.$$

The last inequality is satisfied if and only if $m \leq n$ or $r(B) = p$. Since $A \in \mathbb{C}^{m \times n}$ is left invertible, we have $n \leq m$, so we can conclude that $(AB)\{1\} = B\{1\}A\{1\}$ if and only if A is invertible ($m = n$) or B is left invertible.

In spite of the many results obtained by various authors the problem of settling the reverse order law (2.22) for operators acting on separable Hilbert spaces remained open until 2015. This was finally completely resolved by Pavlović et al. [12] and this was using some radically new approaches involving some of the previous research on completions of operator matrices. These results will be presented in the Chap. 3.

2.2 Reverse Order Laws for {1, 2}-Inverses

In this section, we address all the known results so far results on the the reverse order laws for {1, 2}- generalized inverses. Shinozaki and Sibuya [7] proved that for matrices A, B such that the product AB is defined

$$(AB)\{1, 2\} \subseteq B\{1, 2\}A\{1, 2\} \tag{2.35}$$

always hold. To verify Shinozaki and Sibuya's result in the case of regular bounded linear operators on Hilbert spaces we will consider suitable representations of given regular operators $A \in \mathscr{B}(\mathscr{H}, \mathscr{K})$ and $B \in \mathscr{B}(\mathscr{L}, \mathscr{H})$ to first prove the lemma given below. More precisely if $A \in \mathscr{B}(\mathscr{H}, \mathscr{K})$ and $B \in \mathscr{B}(\mathscr{L}, \mathscr{H})$ are arbitrary regular operators, then using the following decompositions of the Hilbert spaces $\mathscr{L}$, $\mathscr{H}$ and $\mathscr{K}$,

$$\mathscr{L} = \mathscr{R}(B^*) \oplus \mathscr{N}(B), \quad \mathscr{H} = \mathscr{R}(B) \oplus \mathscr{N}(B^*), \quad \mathscr{K} = \mathscr{R}(A) \oplus \mathscr{N}(A^*),$$

we have that the corresponding decompositions of A and B are given by

$$
\begin{aligned}
A &= \begin{bmatrix} A_1 & A_2 \\ 0 & 0 \end{bmatrix} : \begin{bmatrix} \mathscr{R}(B) \\ \mathscr{N}(B^*) \end{bmatrix} \rightarrow \begin{bmatrix} \mathscr{R}(A) \\ \mathscr{N}(A^*) \end{bmatrix}, \\
B &= \begin{bmatrix} B_1 & 0 \\ 0 & 0 \end{bmatrix} : \begin{bmatrix} \mathscr{R}(B^*) \\ \mathscr{N}(B) \end{bmatrix} \rightarrow \begin{bmatrix} \mathscr{R}(B) \\ \mathscr{N}(B^*) \end{bmatrix},
\end{aligned}
\tag{2.36}
$$

where B_1 is an invertible operator and $\begin{bmatrix} A_1 & A_2 \end{bmatrix} : \begin{bmatrix} \mathscr{R}(B^*) \\ \mathscr{N}(B) \end{bmatrix} \rightarrow \mathscr{R}(A)$ is a right invertible operator. In that case AB is given by

$$AB = \begin{bmatrix} A_1 B_1 & 0 \\ 0 & 0 \end{bmatrix} : \begin{bmatrix} \mathscr{R}(B^*) \\ \mathscr{N}(B) \end{bmatrix} \rightarrow \begin{bmatrix} \mathscr{R}(A) \\ \mathscr{N}(A^*) \end{bmatrix}.$$

Now, using decompositions given above, we have the following result.

Lemma 2.1 *Let $A \in \mathscr{B}(\mathscr{H}, \mathscr{K})$ and $B \in \mathscr{B}(\mathscr{L}, \mathscr{H})$ be regular operators given by (2.36). Then*

(i) *an arbitrary {1, 2}-inverse of A is given by:*

$$A^{(1,2)} = \begin{bmatrix} X_1 & X_2 \\ X_3 & X_4 \end{bmatrix} : \begin{bmatrix} \mathscr{R}(A) \\ \mathscr{N}(A^*) \end{bmatrix} \rightarrow \begin{bmatrix} \mathscr{R}(B) \\ \mathscr{N}(B^*) \end{bmatrix},$$

where X_1 and X_3 satisfy

$$A_1 X_1 + A_2 X_3 = I_{\mathscr{R}(A)},$$

and X_2 and X_4 are of the form

$$
\begin{aligned}
X_2 &= X_1 A_1 Z_1 + X_1 A_2 Z_2, \\
X_4 &= X_3 A_1 Z_1 + X_3 A_2 Z_2,
\end{aligned}
$$

for some operators $Z_1 \in \mathscr{B}(\mathscr{N}(A^*), \mathscr{R}(B))$ and $Z_2 \in \mathscr{B}(\mathscr{N}(A^*), \mathscr{N}(B^*))$.
(ii) *an arbitrary {1, 2}-inverse of B is given by:*

$$B^{(1,2)} = \begin{bmatrix} B_1^{-1} & U \\ V & V B_1 U \end{bmatrix} : \begin{bmatrix} \mathscr{R}(B) \\ \mathscr{N}(B^*) \end{bmatrix} \rightarrow \begin{bmatrix} \mathscr{R}(B^*) \\ \mathscr{N}(B) \end{bmatrix},$$

where $U \in \mathscr{B}(\mathscr{N}(B^*), \mathscr{R}(B^*))$ and $V \in \mathscr{B}(\mathscr{R}(B), \mathscr{N}(B))$.

(iii) *if AB is regular, then an arbitrary $\{1, 2\}$-inverse of AB is given by:*

$$(AB)^{(1,2)} = \begin{bmatrix} (A_1 B_1)^{(1,2)} & Y_2 \\ Y_3 & Y_4 \end{bmatrix} : \begin{bmatrix} \mathscr{R}(A) \\ \mathscr{N}(A^*) \end{bmatrix} \rightarrow \begin{bmatrix} \mathscr{R}(B^*) \\ \mathscr{N}(B) \end{bmatrix},$$

where $(A_1 B_1)^{(1,2)} \in (A_1 B_1)\{1, 2\}$ and Y_i, $i = \overline{2, 4}$ satisfy the following system of the equations:

$$\begin{aligned} Y_2 &= (A_1 B_1)^{(1,2)} A_1 B_1 Y_2, \\ Y_3 &= Y_3 A_1 B_1 (A_1 B_1)^{(1,2)}, \\ Y_4 &= Y_3 A_1 B_1 Y_2. \end{aligned} \tag{2.37}$$

Proof (i) Without loss of generality, we can suppose that an arbitrary $\{1, 2\}$-inverse of A is given by:

$$A^{(1,2)} = \begin{bmatrix} X_1 & X_2 \\ X_3 & X_4 \end{bmatrix} : \begin{bmatrix} \mathscr{R}(A) \\ \mathscr{N}(A^*) \end{bmatrix} \rightarrow \begin{bmatrix} \mathscr{R}(B) \\ \mathscr{N}(B^*) \end{bmatrix}.$$

From $AXA = A$ and $XAX = X$, we get that $X \in A\{1, 2\}$ if and only if for X_i, $i = \overline{1, 4}$ the following equations

$$(A_1 X_1 + A_2 X_3) A_i = A_i, \quad i = 1, 2 \tag{2.38}$$
$$X_j (A_1 X_1 + A_2 X_3) = X_j, \quad j = 1, 3 \tag{2.39}$$
$$X_1 (A_1 X_2 + A_2 X_4) = X_2, \quad X_3 (A_1 X_2 + A_2 X_4) = X_4, \tag{2.40}$$

are satisfied. Since $S = \begin{bmatrix} A_1 & A_2 \end{bmatrix} : \begin{bmatrix} \mathscr{R}(B^*) \\ \mathscr{N}(B) \end{bmatrix} \rightarrow \mathscr{R}(A)$ is a right invertible operator, there exists $S_r^{-1} : \mathscr{R}(A) \rightarrow \begin{bmatrix} \mathscr{R}(B^*) \\ \mathscr{N}(B) \end{bmatrix}$ such that $\begin{bmatrix} A_1 & A_2 \end{bmatrix} S_r^{-1} = I_{\mathscr{R}(A)}$. Notice that (2.38) is equivalent to

$$\begin{bmatrix} A_1 & A_2 \end{bmatrix} \begin{bmatrix} X_1 \\ X_3 \end{bmatrix} \begin{bmatrix} A_1 & A_2 \end{bmatrix} = \begin{bmatrix} A_1 & A_2 \end{bmatrix}. \tag{2.41}$$

Multiplying (2.41) by S_r^{-1} from the right, we get that (2.41) is equivalent with $\begin{bmatrix} A_1 & A_2 \end{bmatrix} \begin{bmatrix} X_1 \\ X_3 \end{bmatrix} = I_{\mathscr{R}(A)}$, i.e.,

$$A_1 X_1 + A_2 X_3 = I_{\mathscr{R}(A)}. \tag{2.42}$$

Note, that for X_1 and X_3 which satisfy (2.42), (2.39) also holds. Condition (2.40) is equivalent to

$$\begin{bmatrix} X_1 \\ X_3 \end{bmatrix} \begin{bmatrix} A_1 & A_2 \end{bmatrix} \begin{bmatrix} X_2 \\ X_4 \end{bmatrix} = \begin{bmatrix} X_2 \\ X_4 \end{bmatrix}$$

i.e.,

$$(I - P)\begin{bmatrix} X_2 \\ X_4 \end{bmatrix} = 0,$$

where $P = \begin{bmatrix} X_1 \\ X_3 \end{bmatrix}\begin{bmatrix} A_1 & A_2 \end{bmatrix}$. Since P is a projection,

$$\begin{bmatrix} X_2 \\ X_4 \end{bmatrix} = P\begin{bmatrix} Z_1 \\ Z_2 \end{bmatrix},$$

i.e.,

$$\begin{bmatrix} X_2 \\ X_4 \end{bmatrix} = \begin{bmatrix} X_1 A_1 Z_1 + X_1 A_2 Z_2 \\ X_3 A_1 Z_1 + X_3 A_2 Z_2 \end{bmatrix},$$

where Z_1 and Z_2 are operators from appropriate spaces.

(ii) Suppose that an arbitrary {1, 2}-inverse of B is given by:

$$B^{(1,2)} = \begin{bmatrix} S & U \\ V & W \end{bmatrix} : \begin{bmatrix} \mathscr{R}(B) \\ \mathscr{N}(B^*) \end{bmatrix} \to \begin{bmatrix} \mathscr{R}(B^*) \\ \mathscr{N}(B) \end{bmatrix}.$$

From $BB^{(1,2)}B = B$ it follows that $B_1 S B_1 = B_1$ and since B_1 is invertible, $S = B_1^{-1}$. From $B^{(1,2)}BB^{(1,2)} = B^{(1,2)}$ we easily get $W = VB_1U$, where U and V are operators from appropriate spaces.

(iii) Let an arbitrary {1, 2}-inverse of AB be given by:

$$(AB)^{(1,2)} = \begin{bmatrix} Y_1 & Y_2 \\ Y_3 & Y_4 \end{bmatrix} : \begin{bmatrix} \mathscr{R}(A) \\ \mathscr{N}(A^*) \end{bmatrix} \to \begin{bmatrix} \mathscr{R}(B^*) \\ \mathscr{N}(B) \end{bmatrix}.$$

From $AB(AB)^{(1,2)}AB = AB$, we get

$$A_1 B_1 Y_1 A_1 B_1 = A_1 B_1, \tag{2.43}$$

and from $(AB)^{(1,2)}AB(AB)^{(1,2)} = (AB)^{(1,2)}$, we get

$$Y_1 A_1 B_1 Y_1 = Y_1, \tag{2.44}$$
$$Y_1 A_1 B_1 Y_2 = Y_2, \tag{2.45}$$
$$Y_3 A_1 B_1 Y_1 = Y_3, \tag{2.46}$$
$$Y_3 A_1 B_1 Y_2 = Y_4. \tag{2.47}$$

Now, by (2.43) and (2.44), we get that $Y_1 \in (A_1 B_1)\{1, 2\}$. Substituting $Y_1 = (A_1 B_1)^{(1,2)}$ in (2.45), (2.46) and (2.47), we get (2.37). □

Finally, we will give the proof of the result of Shinozaki and Sibuya in the case of regular bounded linear operators on Hilbert spaces the product of which is also regular.

Theorem 2.7 *Let $A \in \mathscr{B}(\mathscr{H}, \mathscr{K})$ and $B \in \mathscr{B}(\mathscr{L}, \mathscr{H})$ be regular operators such that AB is regular. Then*

$$(AB)\{1, 2\} \subseteq B\{1, 2\} \cdot A\{1, 2\}.$$

Proof Take an arbitrary $(AB)^{(1,2)} \in (AB)\{1, 2\}$. We will show that there exist $A^{(1,2)} \in A\{1, 2\}$ and $B^{(1,2)} \in B\{1, 2\}$ such that $(AB)^{(1,2)} = B^{(1,2)}A^{(1,2)}$. Without loss of generality, we can suppose that A and B are given by (2.36). By Lemma 2.1, we have that

$$(AB)^{(1,2)} = \begin{bmatrix} (A_1 B_1)^{(1,2)} & Y_2 \\ Y_3 & Y_4 \end{bmatrix} : \begin{bmatrix} \mathscr{R}(A) \\ \mathscr{N}(A^*) \end{bmatrix} \to \begin{bmatrix} \mathscr{R}(B^*) \\ \mathscr{N}(B) \end{bmatrix},$$

for $(A_1 B_1)^{(1,2)} \in (A_1 B_1)\{1, 2\}$ and some Y_i, $i = \overline{2, 4}$ which satisfy system (2.37). Since $\begin{bmatrix} A_1 & A_2 \end{bmatrix} : \begin{bmatrix} \mathscr{R}(B^*) \\ \mathscr{N}(B) \end{bmatrix} \to \mathscr{R}(A)$ is a right invertible operator, there exists (not unique in general) $\begin{bmatrix} X_1' \\ X_3' \end{bmatrix} : \mathscr{R}(A) \to \begin{bmatrix} \mathscr{R}(B^*) \\ \mathscr{N}(B) \end{bmatrix}$ such that $A_1 X_1' + A_2 X_3' = I_{\mathscr{R}(A)}$. Since B_1 is invertible, we have that $(A_1 B_1)(A_1 B_1)^{(1,2)} A_1 X_1' = A_1 X_1'$. Let $X_3 = X_3'$ and $X_1 = B_1(A_1 B_1)^{(1,2)} A_1 X_1'$. Obviously, $A_1 X_1 + A_2 X_3 = I_{\mathscr{R}(A)}$. Now, let

$$C = \begin{bmatrix} X_1 & X_1 A_1 B_1 Y_2 \\ X_3 & X_3 A_1 B_1 Y_2 \end{bmatrix} : \begin{bmatrix} \mathscr{R}(A) \\ \mathscr{N}(A^*) \end{bmatrix} \to \begin{bmatrix} \mathscr{R}(B) \\ \mathscr{N}(B^*) \end{bmatrix},$$

$$D = \begin{bmatrix} B_1^{-1} & U \\ Y_3 A_1 & Y_3 A_1 B_1 U \end{bmatrix} : \begin{bmatrix} \mathscr{R}(B) \\ \mathscr{N}(B^*) \end{bmatrix} \to \begin{bmatrix} \mathscr{R}(B^*) \\ \mathscr{N}(B) \end{bmatrix},$$

where $U = (A_1 B_1)^{(1,2)} A_2$. We will show that $C \in A\{1, 2\}$, $D \in B\{1, 2\}$ and that $(AB)^{(1,2)} = DC$. Using Lemma 2.1, we can check that $C \in A\{1, 2\}$ and $D \in B\{1, 2\}$. To prove that $(AB)^{(1,2)} = DC$, it suffices to show that the following system of the equations is satisfied:

$$\begin{aligned}
(A_1 B_1)^{(1,2)} &= B_1^{-1} X_1 + U X_3, \\
Y_2 &= B_1^{-1} X_1 A_1 B_1 Y_2 + U X_3 A_1 B_1 Y_2, \\
Y_3 &= Y_3 A_1 X_1 + Y_3 A_1 B_1 U X_3, \\
Y_4 &= Y_3 A_1 X_1 A_1 B_1 Y_2 + Y_3 A_1 B_1 U X_3 A_1 B_1 Y_2.
\end{aligned}$$

The first equation is satisfied, since $X_1 = B_1 (A_1 B_1)^{(1,2)} (I - A_2 X_3)$, while the other three equations are satisfied by virtue of (2.37). $\qquad\square$

The reverse inclusion of (2.35) was considered by De Pierro and M. Wei [8]. Using the product singular value decomposition of matrices they investigated when

$$B\{1, 2\}A\{1, 2\} \subseteq (AB)\{1, 2\} \tag{2.48}$$

is satisfied and proved the following:

Theorem 2.8 ([8]) *Let $A \in \mathbb{C}^{m\times n}$ and $B \in \mathbb{C}^{n\times p}$. The following conditions are equivalent:*

(i) $B\{1, 2\}A\{1, 2\} \subseteq (AB)\{1, 2\}$;
(ii) $A = 0$ *or* $B = 0$ *or* $r(B) = n$ *or* $r(A) = n$.

The following two examples are in order.

Example 2.5 In the case when (ii) of Theorem 2.8 is not satisfied, which means that $B\{1, 2\}A\{1, 2\} \nsubseteq (AB)\{1, 2\}$, we can find particular reflexive inverses of A and B, $A^{(1,2)}$ and $B^{(1,2)}$ such that $B^{(1,2)}A^{(1,2)} \in (AB)\{1, 2\}$. Let $A = \begin{bmatrix} -1 & 1 \\ -1 & 1 \end{bmatrix}$ and $B = \begin{bmatrix} 1 & 1 \\ 1 & 1 \end{bmatrix}$. Then evidently $AB = 0$ and consequently $(AB)\{1, 2\} = \{0\}$. But for $A^{(1,2)} = \begin{bmatrix} 0 & 0 \\ 1 & 0 \end{bmatrix}$ and $B^{(1,2)} = \begin{bmatrix} 1 & 0 \\ 0 & 0 \end{bmatrix}$, we have that $B^{(1,2)}A^{(1,2)} = 0 \in (AB)\{1, 2\}$.

Example 2.6 If $A \in \mathbb{C}^{m\times n}$ and $B \in \mathbb{C}^{n\times p}$, then $B\{1, 2\} \cdot A\{1, 2\} \subseteq (AB)\{1, 2\}$ implies $B\{1\} \cdot A\{1\} \subseteq (AB)\{1\}$ (see Theorems. 2.1 and 2.8).

Using a completely different approaches, Cvetković-Ilić and Nikolov [13] improved the results from [8] and verified Shinozaki and Sibuya's results in the case of regular bounded linear operators on Hilbert spaces the product of which is also regular. Notice that all the results stated in the sequel can be generalized to the C^*-algebra case.

Theorem 2.9 ([13]) *Let $A \in \mathcal{B}(\mathcal{H}, \mathcal{K})$ and $B \in \mathcal{B}(\mathcal{L}, \mathcal{H})$ be such that A, B and AB are regular operators. The following conditions are equivalent:*

(i) $B\{1, 2\} \cdot A\{1, 2\} \subseteq (AB)\{1, 2\}$,
(ii) $A = 0$ *or* $B = 0$ *or* $A \in \mathcal{B}_l^{-1}(\mathcal{H}, \mathcal{K})$ *or* $B \in \mathcal{B}_r^{-1}(\mathcal{H}, \mathcal{K})$.

Proof (i) $\Rightarrow$ (ii) : If (i) holds, then evidently $B^\dagger A^\dagger \in (AB)\{1, 2\}$, so

$$ABB^\dagger A^\dagger AB = AB \tag{2.49}$$

and

$$B^\dagger A^\dagger ABB^\dagger A^\dagger = B^\dagger A^\dagger. \tag{2.50}$$

Since, for any $X \in \mathcal{B}(\mathcal{K}, \mathcal{H})$, $A^\dagger + (I - A^\dagger A)XAA^\dagger \in A\{1, 2\}$, we get

$$ABB^\dagger(A^\dagger + (I - A^\dagger A)XAA^\dagger)AB = AB,$$

which using (2.49) further implies that

$$ABB^\dagger(I - A^\dagger A)XAB = 0, \tag{2.51}$$

for any $X \in \mathcal{B}(\mathcal{K}, \mathcal{H})$.
Similarly, for any $Y \in \mathcal{B}(\mathcal{H}, \mathcal{L})$, $B^\dagger + B^\dagger BY(I - BB^\dagger) \in B\{1, 2\}$, so

$$AB(B^\dagger + B^\dagger BY(I - BB^\dagger))A^\dagger AB = AB,$$

which using (2.49) implies that

$$ABY(I - BB^\dagger)A^\dagger AB = 0, \tag{2.52}$$

for any $Y \in \mathcal{B}(\mathcal{H}, \mathcal{L})$.
Since, for any $X \in \mathcal{B}(\mathcal{K}, \mathcal{H})$ and $Y \in \mathcal{B}(\mathcal{H}, \mathcal{L})$, we have that

$$AB(B^\dagger + B^\dagger BY(I - BB^\dagger))(A^\dagger + (I - A^\dagger A)XAA^\dagger)AB = AB, \tag{2.53}$$

using (2.49), (2.51) and (2.52), we get that

$$ABY(I - BB^\dagger)(I - A^\dagger A)XAB = 0,$$

for any $X \in \mathcal{B}(\mathcal{K}, \mathcal{H})$ and $Y \in \mathcal{B}(\mathcal{H}, \mathcal{L})$. Now,

$$AB = 0 \quad \text{or} \quad (I - BB^\dagger)(I - A^\dagger A) = 0. \tag{2.54}$$

Since, for any $X \in \mathcal{B}(\mathcal{K}, \mathcal{H})$,

$$B^\dagger(A^\dagger + (I - A^\dagger A)XAA^\dagger)ABB^\dagger(A^\dagger + (I - A^\dagger A)XAA^\dagger)$$
$$= B^\dagger(A^\dagger + (I - A^\dagger A)XAA^\dagger)$$

using (2.50) we get that

$$B^\dagger A^\dagger ABB^\dagger(I - A^\dagger A)XAA^\dagger + B^\dagger(I - A^\dagger A)XABB^\dagger A^\dagger$$
$$+ B^\dagger(I - A^\dagger A)XABB^\dagger(I - A^\dagger A)XAA^\dagger = B^\dagger(I - A^\dagger A)XAA^\dagger. \tag{2.55}$$

Now by (2.54), we get that the first and the third term on the left-hand side of (2.55) are zero, so

$$B^\dagger(I - A^\dagger A)X(ABB^\dagger A^\dagger - AA^\dagger) = 0,$$

for any $X \in \mathcal{B}(\mathcal{K}, \mathcal{H})$. Hence,

$$B^\dagger = B^\dagger A^\dagger A \quad \text{or} \quad ABB^\dagger A^\dagger = AA^\dagger.$$

Now, following (2.54) we have two cases:

Case 1. If $AB = 0$, then if $B^\dagger = B^\dagger A^\dagger A$ it follows that $B = 0$. If $ABB^\dagger A^\dagger = AA^\dagger$, we easily get that $A = 0$.

Case 2. If $(I - BB^\dagger)(I - A^\dagger A) = 0$, then if $B^\dagger = B^\dagger A^\dagger A$, it follows that $A^\dagger A = I$, i.e., A is left invertible. If $ABB^\dagger A^\dagger = AA^\dagger$, then multiplying $(I - BB^\dagger)(I - A^\dagger A) = 0$, by A from the left, we get

$$A = ABB^\dagger.$$

Now, given that $A^\dagger A$ and $BB^\dagger$ commute, we have that $BB^\dagger = I$, i.e., B is right invertible.

(ii) $\Rightarrow$ (i) : If A or B is zero, it is evident that (i) holds. Now, suppose that B is right invertible and let $B^{(1,2)} \in B\{1, 2\}$ be arbitrary. Evidently, $B^{(1,2)}$ is a right inverse of B, i.e., $BB^{(1,2)} = I$. Then, for arbitrary $A^{(1,2)} \in A\{1, 2\}$,

$$ABB^{(1,2)}A^{(1,2)}AB = AA^{(1,2)}AB = AB$$

and

$$B^{(1,2)}A^{(1,2)}ABB^{(1,2)}A^{(1,2)} = B^{(1,2)}A^{(1,2)}AA^{(1,2)} = B^{(1,2)}A^{(1,2)}.$$

If A is a left invertible operator, for any $A^{(1,2)} \in A\{1, 2\}$ we have that $A^{(1,2)}A = I$. Then, for arbitrary $A^{(1,2)} \in A\{1, 2\}$ and $B^{(1,2)} \in B\{1, 2\}$,

$$ABB^{(1,2)}A^{(1,2)}AB = ABB^{(1,2)}B = AB$$

and

$$B^{(1,2)}A^{(1,2)}ABB^{(1,2)}A^{(1,2)} = B^{(1,2)}BB^{(1,2)}A^{(1,2)} = B^{(1,2)}A^{(1,2)}.$$

$\square$

It is interesting to note that by the first part of the proof of Theorem 2.9, we can conclude that

$$B\{1, 2\} \cdot A\{1, 2\} \subseteq (AB)\{1\}$$

if and only if

$$AB = 0 \quad \text{or} \quad (I - BB^\dagger)(I - A^\dagger A) = 0$$

i.e.

$$AB = 0 \quad \text{or} \quad \mathscr{N}(A) \subseteq \mathscr{R}(B),$$

which is equivalent with $B\{1\} \cdot A\{1\} \subseteq (AB)\{1\}$ (see [3, 14]).

The proof that (2.48) is always satisfied in the case of a multiple product of regular operators is very similar to the one given in [5] (see [Theorem 4.1, [5]]) for the matrix case. We give it here for completeness' sake.

Theorem 2.10 *Let $A_i \in \mathscr{B}(\mathscr{H}_{i+1}, \mathscr{H}_i)$, be such that A_i, $i = \overline{1, n}$ and $A_1 A_2 \cdots A_j$, $j = \overline{2, n}$, are regular operators. Then*

$$(A_1 A_2 \cdots A_n)\{1, 2\} \subseteq A_n\{1, 2\} \cdot A_{n-1}\{1, 2\} \cdots A_1\{1, 2\}. \qquad (2.56)$$

Proof Suppose that $A_i \in \mathscr{B}(\mathscr{H}_{i+1}, \mathscr{H}_i)$, $i = \overline{1, n}$ and $A_1 A_2 \cdots A_j$, $j = \overline{2, n}$, are regular operators. We will prove that (2.56) holds by induction on n. For $n = 2$, the assertion holds by virtue of Theorem 2.7. Now suppose this is true for $2 \leq k \leq n$. For $k = n + 1$, let $A_1 A_2 \cdots A_k = B$. Using again Theorem 2.7, we obtain $(B A_{n+1})\{1, 2\} \subseteq A_{n+1}\{1, 2\} B\{1, 2\}$. From the induction hypothesis,

$$(A_1 \cdots A_n)\{1, 2\} \subseteq A_n\{1, 2\} \cdots A_1\{1, 2\},$$

so we get

$$(A_1 \cdots A_n A_{n+1})\{1, 2\} \subseteq A_{n+1}\{1, 2\}(A_1 \cdots A_n)\{1, 2\}$$
$$\subseteq A_{n+1}\{1, 2\} A_n\{1, 2\} \cdots A_1\{1, 2\}.$$

□

The reverse inclusion of (2.56) in the case of matrices was considered by M. Wei [4] who, applying the multiple product singular value decomposition (P-SVD), gave necessary and sufficient conditions for

$$A_n\{1, 2\} \cdot A_{n-1}\{1, 2\} \cdots A_1\{1, 2\} \subseteq (A_1 A_2 \cdots A_n)\{1, 2\}.$$

Theorem 2.11 ([4]) *Suppose that $A_i \in \mathbb{C}^{s_i \times s_{i+1}}$, $i = 1, 2, \ldots n$. Then the following conditions are equivalent:*

(i) $A_n\{1, 2\} \cdot A_{n-1}\{1, 2\} \cdots A_1\{1, 2\} \subseteq (A_1 A_2 \cdots A_n)\{1, 2\}$;
(ii) *One of the following conditions is satisfied:*

   (a) $r(A_1 \cdots A_n) > 0$ *and for each* $j \in \{1, \ldots, n - 1\}$, A_j *is of full column rank*
   (b) $r(A_1 \cdots A_n) > 0$ *and for each* $j \in \{2, \ldots, n\}$, A_j *is of full row rank*
   (c) $r(A_1 \cdots A_n) > 0$ *and there exists an integer* $q \in \{2, ..., n - 1\}$ *such that for each* $j \in \{1, \ldots, q - 1\}$, A_j *is of full column rank and for each* $j \in \{q, \ldots, n\}$, A_j *is of full row rank*
   (d) *There exists an integer* $q \in \{1, \ldots, n\}$, *such that* $r(A_q) = 0$.

The generalization of the previous result for the case of bounded regular operators on Hilbert spaces is given in [6] as follows:

Theorem 2.12 ([6]) *Let $A_i \in \mathscr{B}(\mathscr{H}_{i+1}, \mathscr{H}_i)$, be such that A_i, $i = \overline{1, n}$ and $A_1 A_2 \cdots A_j$, $j = \overline{2, n}$ are regular operators. The following conditions are equivalent:*

(i) $A_n\{1, 2\} \cdot A_{n-1}\{1, 2\} \cdots A_1\{1, 2\} = (A_1 A_2 \cdots A_n)\{1, 2\}$,

(ii) $A_n\{1, 2\} \cdot A_{n-1}\{1, 2\} \cdots A_1\{1, 2\} \subseteq (A_1 A_2 \cdots A_n)\{1, 2\}$,

(iii) *There exists an integer i, $1 \leq i \leq n$, such that $A_i = 0$,*

or

$A_1 A_2 \cdots A_n \neq 0$ and $A_i \in \mathscr{B}_r^{-1}(\mathscr{H}_{i+1}, \mathscr{H}_i)$, for $i = \overline{2, n}$,

or

$A_1 A_2 \cdots A_n \neq 0$ and $A_i \in \mathscr{B}_l^{-1}(\mathscr{H}_{i+1}, \mathscr{H}_i)$, for $i = \overline{1, n - 1}$,

or

$A_1 A_2 \cdots A_n \neq 0$ and there exists an integer k, $2 \leq k \leq n - 1$, such that $A_i \in \mathscr{B}_l^{-1}(\mathscr{H}_{i+1}, \mathscr{H}_i)$, for $i = \overline{1, k - 1}$, and $A_i \in \mathscr{B}_r^{-1}(\mathscr{H}_{i+1}, \mathscr{H}_i)$, for $i = \overline{k + 1, n}$.

Proof (i) $\Leftrightarrow$ (ii) : Follows from Theorem 2.10.

(ii) $\Rightarrow$ (iii) : We prove this by induction on n. For $n = 2$, this follows from Theorem 2.9. Assume that (ii) $\Rightarrow$ (iii) holds for $n = k - 1$; we will prove that the implication still holds for $n = k$. Suppose that

$$A_k\{1, 2\} \cdot A_{k-1}\{1, 2\} \cdots A_1\{1, 2\} \subseteq (A_1 A_2 \cdots A_k)\{1, 2\}. \tag{2.57}$$

By virtue of Theorem 2.10, we have

$$A_k\{1, 2\} \cdot (A_1 A_2 \cdots A_{k-1})\{1, 2\} \subseteq (A_1 A_2 \cdots A_k)\{1, 2\}. \tag{2.58}$$

which by Theorem 2.9 implies that at least one of the following cases must hold true:

$$A_1 A_2 \cdots A_{k-1} = 0 \text{ or } A_k = 0 \text{ or } A_1 A_2 \cdots A_{k-1} \in \mathscr{B}_l^{-1}(\mathscr{H}_k, \mathscr{H}_1)$$
$$\text{or } A_k \in \mathscr{B}_r^{-1}(\mathscr{H}_{k+1}, \mathscr{H}_k).$$

Now, we will consider all these cases:

Case 1. $A_1 A_2 \cdots A_{k-1} = 0$. Then $A_1 A_2 \cdots A_{k-1} A_k = 0$ which by (ii) implies

$$A_k\{1, 2\} \cdot A_{k-1}\{1, 2\} \cdots A_1\{1, 2\} = \{0\}. \tag{2.59}$$

Let $A_i^{(1,2)} \in A_i\{1, 2\}$, $i = \overline{1, k - 1}$ be arbitrary. Then from (2.59) we have

$$A_k^\dagger A_{k-1}^{(1,2)} \cdots A_1^{(1,2)} = 0. \tag{2.60}$$

Since for any $Z \in \mathscr{B}(\mathscr{H}_k, \mathscr{H}_{k+1})$, $A_k^\dagger + A_k^\dagger A_k Z(I_{\mathscr{H}_k} - A_k A_k^\dagger) \in A_k\{1, 2\}$, we get

$$(A_k^\dagger + A_k^\dagger A_k Z(I_{\mathscr{H}_k} - A_k A_k^\dagger)) A_{k-1}^{(1,2)} A_{k-2}^{(1,2)} \cdots A_1^{(1,2)} = 0,$$

which by (2.60) gives that $A_k^\dagger A_k Z A_{k-1}^{(1,2)} A_{k-2}^{(1,2)} \cdots A_1^{(1,2)} = 0$. Now,

$$A_k = 0 \text{ or } A_{k-1}^{(1,2)} \cdots A_1^{(1,2)} = 0. \tag{2.61}$$

If $A_k = 0$, then (iii) holds. Suppose that $A_k \neq 0$. Then $A_{k-1}^{(1,2)} A_{k-2}^{(1,2)} \cdots A_1^{(1,2)} = 0$ for arbitrary $A_i^{(1,2)} \in A_i\{1,2\}$, $i = \overline{1, k-1}$, implying

$$A_{k-1}\{1,2\}A_{k-2}\{1,2\}\cdots A_1\{1,2\} = \{0\} \subseteq (A_1 A_2 \cdots A_{k-2}A_{k-1})\{1,2\}. \quad (2.62)$$

By the induction hypothesis, from (2.62) it follows that at least one of the following conditions is satisfied:

(1) There exists $i \in \{1, 2, \ldots, k-1\}$ such that $A_i = 0$,
(2) $A_i \in \mathscr{B}_l^{-1}(\mathscr{H}_{i+1}, \mathscr{H}_i)$, $i = \overline{1, k-2}$,
(3) $A_i \in \mathscr{B}_r^{-1}(\mathscr{H}_{i+1}, \mathscr{H}_i)$, $i = \overline{2, k-1}$,
(4) There exists $i \in \{1, 2, \ldots, k-1\}$ such that $A_j \in \mathscr{B}_l^{-1}(\mathscr{H}_{j+1}, \mathscr{H}_j)$ for $j = \overline{1, i-1}$ and $A_j \in \mathscr{B}_r^{-1}(\mathscr{H}_{j+1}, \mathscr{H}_j)$, $j = \overline{i+1, k-1}$.

If (1) holds, then (iii) is satisfied. Suppose that (2) is true. Since $A_1 A_2 \cdots A_{k-1} = 0$ we get that $A_{k-1} = 0$ so (iii) holds. If (3) holds, then from $A_1 A_2 \cdots A_{k-1} = 0$ we get that $A_1 = 0$. Suppose that (4) holds. Multiplying $A_1 A_2 \cdots A_{k-1} = 0$ by $A_{i-1}^{\dagger} A_{i-2}^{\dagger} \cdots A_1^{\dagger}$ from the left, we get

$$A_i A_{i+1} \cdots A_{k-1} = 0. \quad (2.63)$$

Multiplying (2.63) by $A_{k-1}^{\dagger} A_{k-2}^{\dagger} \cdots A_{i+1}^{\dagger}$ from the right we get $A_i = 0$. Hence, (iii) is satisfied.

Case 2. If $A_k = 0$, then (iii) obviously holds.

Case 3. Suppose that $A_1 A_2 \cdots A_{k-1} \in \mathscr{B}_l^{-1}(\mathscr{H}_k, \mathscr{H}_1)$. Then $A_{k-1} \in \mathscr{B}_l^{-1}(\mathscr{H}_k, \mathscr{H}_{k-1})$. From Theorem 2.9, we have

$$(A_{k-1} A_k)\{1,2\} \subseteq A_k\{1,2\}A_{k-1}\{1,2\},$$

so it follows that

$$(A_{k-1} A_k)\{1,2\} \cdot A_{k-2}\{1,2\} \cdots A_1\{1,2\}$$
$$\subseteq A_k\{1,2\} \cdot A_{k-1}\{1,2\} \cdots A_1\{1,2\} \subseteq (A_1 A_2 \cdots A_k)\{1,2\}. \quad (2.64)$$

By the induction hypothesis, from (2.64) it follows that at least one of the following conditions is true:

(1') There exists $i \in \{1, 2, \ldots, k-2\}$ such that $A_i = 0$ or $A_{k-1} A_k = 0$,
(2') $A_i \in \mathscr{B}_l^{-1}(\mathscr{H}_{i+1}, \mathscr{H}_i)$, $i = \overline{1, k-2}$,
(3') $A_i \in \mathscr{B}_r^{-1}(\mathscr{H}_{i+1}, \mathscr{H}_i)$, $i = \overline{2, k-2}$, and $A_{k-1} A_k \in \mathscr{B}_r^{-1}(\mathscr{H}_{k+1}, \mathscr{H}_{k-1})$,
(4') There exists $i \in \{1, 2, \ldots, k-1\}$ such that $A_j \in \mathscr{B}_l^{-1}(\mathscr{H}_{j+1}, \mathscr{H}_j)$ for $j = \overline{1, i-1}$ and $A_j \in \mathscr{B}_r^{-1}(\mathscr{H}_{j+1}, \mathscr{H}_j)$, $j = \overline{i+1, k-2}$ and $A_{k-1} A_k \in \mathscr{B}_r^{-1}(\mathscr{H}_{k+1}, \mathscr{H}_{k-1})$.

As before, we can check that in all these cases, (iii) is satisfied.

Case 4. Suppose that $A_k \in \mathscr{B}_r^{-1}(\mathscr{H}_{k+1}, \mathscr{H}_k)$. Then $A_k A_k^{(1,2)} = I_{\mathscr{H}_k}$ for arbitrary $A_k^{(1,2)} \in A_k\{1, 2\}$. Let $A_i^{(1,2)} \in A_i\{1, 2\}$, $i = \overline{1, k}$ be arbitrary. Then

$$A_1 A_2 \cdots A_k A_k^{(1,2)} A_{k-1}^{(1,2)} \cdots A_1^{(1,2)} A_1 A_2 \cdots A_k = A_1 A_2 \cdots A_k \qquad (2.65)$$

and

$$A_k^{(1,2)} A_{k-1}^{(1,2)} \cdots A_1^{(1,2)} A_1 A_2 \cdots A_k A_k^{(1,2)} A_{k-1}^{(1,2)} \cdots A_1^{(1,2)}$$
$$= A_k^{(1,2)} A_{k-1}^{(1,2)} \cdots A_1^{(1,2)}. \qquad (2.66)$$

Multiplying (2.65) by $A_k^{(1,2)}$ from the right and (2.66) by A_k from the left, we get

$$A_1 A_2 \cdots A_{k-1} A_{k-1}^{(1,2)} A_{k-2}^{(1,2)} \cdots A_1^{(1,2)} A_1 A_2 \cdots A_{k-1} = A_1 A_2 \cdots A_{k-1} \qquad (2.67)$$

and

$$A_{k-1}^{(1,2)} A_{k-2}^{(1,2)} \cdots A_1^{(1,2)} A_1 A_2 \cdots A_{k-1} A_{k-1}^{(1,2)} A_{k-2}^{(1,2)} \cdots A_1^{(1,2)}$$
$$= A_{k-1}^{(1,2)} A_{k-2}^{(1,2)} \cdots A_1^{(1,2)}. \qquad (2.68)$$

Evidently,

$$A_{k-1}\{1, 2\} \cdot A_{k-2}\{1, 2\} \cdots A_1\{1, 2\} \subseteq (A_1 A_2 \cdots A_{k-1})\{1, 2\}. \qquad (2.69)$$

By the induction hypothesis, from (2.69) it follows that at least one of the following conditions is true:

(1″) There exists $i \in \{1, 2, \ldots, k - 1\}$ such that $A_i = 0$,
(2″) $A_i \in \mathscr{B}_l^{-1}(\mathscr{H}_{i+1}, \mathscr{H}_i)$, $i = \overline{1, k - 2}$,
(3″) $A_i \in \mathscr{B}_r^{-1}(\mathscr{H}_{i+1}, \mathscr{H}_i)$, $i = \overline{2, k - 1}$,
(4″) There exists $i \in \{1, 2, \ldots, k - 1\}$ such that $A_j \in \mathscr{B}_l^{-1}(\mathscr{H}_{j+1}, \mathscr{H}_j)$ for $j = \overline{1, i - 1}$ and $A_j \in \mathscr{B}_r^{-1}(\mathscr{H}_{j+1}, \mathscr{H}_j)$, $j = \overline{i + 1, k - 1}$.

It is easy to check that in all these four cases (iii) is satisfied.

(iii) $\Rightarrow$ (ii) : If $A_1 A_2 \cdots A_n = 0$, then it is evident that (ii) holds. Suppose that $A_1 A_2 \cdots A_n \neq 0$ and let $A_i^{(1,2)} \in A_i\{1, 2\}$, $i = \overline{1, n}$ be arbitrary. If $A_i \in \mathscr{B}_r^{-1}(\mathscr{H}_{i+1}, \mathscr{H}_i)$ for $i = \overline{2, n}$, then $A_i A_i^{(1,2)} = I_{\mathscr{H}_i}$ for $i = \overline{2, n}$. Now,

$$A_1 A_2 \cdots A_{n-1} A_n A_n^{(1,2)} A_{n-1}^{(1,2)} \cdots A_1^{(1,2)}$$
$$= A_1 A_2 \cdots A_{n-2} A_{n-1} A_{n-1}^{(1,2)} A_{n-2}^{(1,2)} \cdots A_1^{(1,2)}$$

$$\vdots \tag{2.70}$$

$$= A_1 A_2 A_2^{(1,2)} A_1^{(1,2)} = A_1 A_1^{(1,2)}.$$

From (2.70), it follows

$$A_1 A_2 \cdots A_n A_n^{(1,2)} A_{n-1}^{(1,2)} \cdots A_1^{(1,2)} A_1 A_2 \cdots A_n = A_1 A_2 \cdots A_n$$

and

$$A_n^{(1,2)} A_{n-1}^{(1,2)} \cdots A_1^{(1,2)} A_1 A_2 \cdots A_n A_n^{(1,2)} A_{n-1}^{(1,2)} = A_n^{(1,2)} A_{n-1}^{(1,2)} \cdots A_1^{(1,2)},$$

so $A_n^{(1,2)} A_{n-1}^{(1,2)} \cdots A_1^{(1,2)} \in (A_1 A_2 \cdots A_n)\{1, 2\}$. Hence (ii) holds.

Analogously, if $A_i \in \mathcal{B}_l^{-1}(\mathcal{H}_{i+1}, \mathcal{H}_i)$ for $i = \overline{1, n-1}$ or if there exists $k \in \{2, ..., n-1\}$ such that $A_i \in \mathcal{B}_l^{-1}(\mathcal{H}_{i+1}, \mathcal{H}_i)$ for $i = \overline{1, k-1}$, and $A_i \in \mathcal{B}_r^{-1}(\mathcal{H}_{i+1}, \mathcal{H}_i)$ for $i = \overline{k+1, n}$, we can prove that $A_n^{(1,2)} A_{n-1}^{(1,2)} \cdots A_1^{(1,2)} \in (A_1 A_2 \cdots A_n)\{1, 2\}$. Thus (ii) holds. $\square$

Since the proof of the previous result is algebraic, it can easily be generalized to C^*-algebras and rings. It is thus safe to say that the reverse order law for $\{1, 2\}$-inverses has been completely solved.

2.3 Reverse Order Laws for $\{1, 3\}$ and $\{1, 4\}$-Inverses

The reverse order laws for $\{1, 3\}$ and $\{1, 4\}$-inverses were for the first time considered by M. Wei and Guo [15] in the matrix case. They presented some equivalent conditions for

$$B\{1, 3\}A\{1, 3\} \subseteq (AB)\{1, 3\} \tag{2.71}$$

and

$$(AB)\{1, 3\} \subseteq B\{1, 3\}A\{1, 3\} \tag{2.72}$$

obtained by applying the product singular value decomposition (P-SVD) of matrices. Namely, in [15] they proved that for $A \in \mathbb{C}^{m \times n}$ and $B \in \mathbb{C}^{n \times p}$ one has $B\{1, 3\} \cdot A\{1, 3\} \subseteq (AB)\{1, 3\}$ if and only if

$$Z_{12} = 0 \text{ and } Z_{14} = 0$$

and that that $(AB)\{1, 3\} \subseteq B\{1, 3\} \cdot A\{1, 3\}$ holds if and only if

$$\dim(\mathcal{R}(Z_{14})) = \dim(\mathcal{R}(Z_{12}, Z_{14})), \text{ and}$$
$$0 \le \min\{p - r_2, m - r_1\} \le n - r_1 - r_2^2 - \mathrm{r}(Z_{14}),$$

where the submatrices Z_{12}, Z_{14} and the constants r_1, r_2, r_2^2 are described in the P-SVD of matrices A and B given in Theorem 1.1 and Corollary 1.1 of [15].

Evidently a disadvantage of the results presented in [15] lies in the fact that the necessary and sufficient conditions for (2.71) and (2.72) to be satisfied contain information about the subblocks produced by P-SVD. In other words, they are dependent on P-SVD. In order to overcome this shortcoming, two methods are employed. One of the methods use certain operator matrix representations (see [16]) and the other one is based on some maximal and minimal ranks of matrix expressions (see [17]). Using these two different methods, in both of the papers [16, 17] it is proved that

$$B\{1, 3\}A\{1, 3\} \subseteq (AB)\{1, 3\} \Leftrightarrow \mathcal{R}(A^*AB) \subseteq \mathcal{R}(B)$$

but in the first one in the case of regular operators and in the second one in the setting of matrices. These results are more elegant because they require no information on the P-SVD. Note that in the matrix case $\mathcal{R}(A^*AB) \subseteq \mathcal{R}(B)$ is equivalent to $\mathrm{r}(B, A^*AB) = \mathrm{r}(B)$.

All these results were generalized in the paper of Cvetković-Ilić and Harte [18] where purely algebraic necessary and sufficient conditions for (2.71) in C^*-algebras are offered, extending rank conditions for matrices and range conditions for Hilbert space operators. To present the result form [18] and its proof, first we will introduce some notations and give some preliminaries.

Let $\mathscr{A}$ be a complex unital C*-algebra. Then we have the following characterization of $a\{1, 3\}$, where $a \in \mathscr{A}$ is regular:

Lemma 2.2 *Let* $a \in \mathscr{A}$ *be regular and* $b \in \mathscr{A}$. *Then* $b \in a\{1, 3\}$ *if and only if* $a^\dagger ab = a^\dagger$.

Lemma 2.2 can be expressed by saying

$$a\{1, 3\} = \{a^\dagger + (1 - a^\dagger a)y : y \in \mathscr{A}\}. \tag{2.73}$$

Theorem 2.13 ([18]) *If* $a, b \in \mathscr{A}$ *are such that* a, b, ab *and* $a(1 - bb^\dagger)$ *are regular, then the following conditions are equivalent:*

(1′) $bb^\dagger a^*ab = a^*ab$,
(2′) $b\{1, 3\} \cdot a\{1, 3\} \subseteq (ab)\{1, 3\}$,
(3′) $b^\dagger a^\dagger \in (ab)\{1, 3\}$,
(4′) $b^\dagger a^\dagger \in (ab)\{1, 2, 3\}$.

Proof With $p = bb^\dagger$, $q = b^\dagger b$ and $r = aa^\dagger$, we have that $b = \begin{bmatrix} b & 0 \\ 0 & 0 \end{bmatrix}_{p,q}$ and $a = \begin{bmatrix} a_1 & a_2 \\ 0 & 0 \end{bmatrix}_{r,p}$. Using Lemma 2.2 and (2.73), we see that arbitrary $b^{(1,3)} \in b\{1, 3\}$ can be represented as $b^{(1,3)} = \begin{bmatrix} b^\dagger & 0 \\ u & v \end{bmatrix}_{q,p}$, for some $u \in (1-q)\mathscr{A}p$ and $v \in (1-q)\mathscr{A}(1-p)$,

as well as that $a^\dagger = a^*(aa^*)^\dagger = \begin{bmatrix} a_1^*d^\dagger & 0 \\ a_2^*d^\dagger & 0 \end{bmatrix}_{p,r}$, where $d = a_1a_1^* + a_2a_2^*$. Remark

that $d \in r\mathscr{A}r$ is invertible in that subalgebra, $(d)_{r\mathscr{A}r}^{-1} = d^\dagger$ and $dd^\dagger = d^\dagger d = r$.
Also, again by (2.73), any $a^{(1,3)}$ has the form $a^{(1,3)} = a^\dagger + (1 - a^\dagger a)x$, for some

$x = \begin{bmatrix} x_1 & x_2 \\ x_3 & x_4 \end{bmatrix}_{p,r}$ i.e., $a^{(1,3)} = \begin{bmatrix} z_1 & z_2 \\ z_3 & z_4 \end{bmatrix}_{p,r}$, where

$$z_1 = a_1^*d^\dagger + (1 - a_1^*d^\dagger a_1)x_1 - a_1^*d^\dagger a_2x_3,$$
$$z_2 = (1 - a_1^*d^\dagger a_1)x_2 - a_1^*d^\dagger a_2x_4,$$
$$z_3 = a_2^*d^\dagger - a_2^*d^\dagger a_1x_1 + (1 - a_2^*d^\dagger a_2)x_3,$$
$$z_4 = -a_2^*d^\dagger a_1x_2 + (1 - a_2^*d^\dagger a_2)x_4.$$

With these preliminaries, we turn to the four conditions of the statement; we will
show $(1') \Rightarrow (2') \Rightarrow (3') \Rightarrow (1')$ and then $(1') \Rightarrow (4') \Rightarrow (3')$.

$(1') \Rightarrow (2')$: Suppose that $bb^\dagger a^*ab = a^*ab$ which is equivalent to $a_2^*a_1 = 0$, i.e.,
$a_1^*a_2 = 0$. For arbitrary $a^{(1,3)}$ and $b^{(1,3)}$ we have that

$$abb^{(1,3)}a^{(1,3)}ab = \begin{bmatrix} a_1z_1a_1b & 0 \\ 0 & 0 \end{bmatrix}_{r,q}.$$

Let $s = a_1a_1^\dagger$. Since $d \in s\mathscr{A}s + (1 - s)\mathscr{A}(1 - s)$, we have that $d^\dagger \in s\mathscr{A}s + (1 -
s)\mathscr{A}(1 - s)$. Now, $a_1^*d^\dagger a_2 \in \mathscr{A} \cdot (s\mathscr{A}s + (1 - s)\mathscr{A}(1 - s)) \cdot (1 - s)\mathscr{A} = \{0\}$.
Hence, $a_1^*d^\dagger a_2 = 0$, i.e., $a_2^*d^\dagger a_1 = 0$.
Since,

$$a_1z_1a_1 = a_1a_1^*d^\dagger a_1 + a_1(1 - a_1^*d^\dagger a_1)x_1a_1$$
$$= (d - a_2a_2^*)d^\dagger a_1 + (a_1 - (d - a_2a_2^*)d^\dagger a_1)x_1a_1$$
$$= a_1,$$

it follows that $abb^{(1,3)}a^{(1,3)}ab = ab$. To prove that $abb^{(1,3)}a^{(1,3)}$ is Hermitian it is
sufficient to prove that a_1z_1 is Hermitian and $a_1z_2 = 0$. By computation, we get that
$a_1z_1 = a_1a_1^*d^\dagger = a_1a_1^*(a_1a_1^*)^\dagger$ which is Hermitian. Also,

$$a_1z_2 = (a_1 - a_1a_1^*d^\dagger a_1)x_2 - a_1a_1^*d^\dagger a_2x_4$$
$$= (a_1 - (d - a_2a_2^*)d^\dagger a_1)x_2$$
$$= a_2a_2^*d^\dagger a_1x_2$$
$$= 0.$$

(2′) ⇒ (3′): This is evident.

(3′) ⇒ (1′): From $abb^\dagger a^\dagger ab = ab$ it follows that $a_1 a_1^* d^\dagger a_1 = a_1$, i.e., $a_2 a_2^* d^\dagger a_1 = 0$. Similarly, from $(abb^\dagger a^\dagger)^* = abb^\dagger a^\dagger$, we get that $a_1 a_1^* d^\dagger$ is Hermitian. Now, $d^\dagger a_1 a_1^* a_1 = a_1$, i.e., $a_2 a_2^* a_1 = 0$. Multiplying the last equality by $a_2^\dagger$ from the left side, we get $a_2^* a_1 = 0$ which is equivalent to the statement (1′).

(4′) ⇒ (3′): This is obvious.

(1′) ⇒ (4′): We need to prove that $b^\dagger a^\dagger abb^\dagger a^\dagger = b^\dagger a^\dagger$ which is equivalent to $b^\dagger a_1^* d^\dagger a_1 a_1^* d^\dagger = b^\dagger a_1^* d^\dagger$. The last equality follows from the fact that $d^\dagger a_1 a_1^* = s$. □

Example 2.7 Let $b \in \mathscr{A}$ be right invertible. Then for any $a \in \mathscr{A}$ such that a, ab are regular, we have $b\{1, 3\} \cdot a\{1, 3\} \subseteq (ab)\{1, 3\}$.

Example 2.8 Let $p, q \in \mathscr{A}$ be orthogonal projections. Then $q\{1, 3\} \cdot p\{1, 3\} \subseteq (pq)\{1, 3\}$ if and only if $qpq = pq$, which is equivalent with the fact that p and q commute, which is in turn equivalent with the fact that pg is an orthogonal projection.

A similar result in the case $K = \{1, 4\}$ follows from Theorem 2.13 by reversal of products:

Theorem 2.14 *If $a, b \in \mathscr{A}$ are such that a, b, ba and $(1 - a^\dagger a)b$ are regular, then the following conditions are equivalent:*

(1″) $abb^* a^\dagger a = abb^*$,
(2″) $b\{1, 4\} \cdot a\{1, 4\} \subseteq (ab)\{1, 4\}$,
(3″) $b^\dagger a^\dagger \in (ab)\{1, 4\}$,
(4″) $b^\dagger a^\dagger \in (ab)\{1, 2, 4\}$.

Example 2.9 Let $a \in \mathscr{A}$ be left invertible. Then for any $b \in \mathscr{A}$ such that b and ba are regular, we have $b\{1, 4\} \cdot a\{1, 4\} \subseteq (ab)\{1, 4\}$.

Example 2.10 Let $p, q \in \mathscr{A}$ be orthogonal projections such that pq, $(1 - p)q$ and $p(1 - q)$ are regular. Then $q\{1, 3\} \cdot p\{1, 3\} \subseteq (pq)\{1, 3\}$ if and only if $q\{1, 4\} \cdot p\{1, 4\} \subseteq (pq)\{1, 4\}$ if and only if pq is an orthogonal projection.

The inclusion $(AB)\{1, 3\} \subseteq B\{1, 3\}A\{1, 3\}$ was considered by Liu and Yang [19] in the matrix case.

Theorem 2.15 ([19]) *Let $A \in \mathbb{C}^{m \times n}$, $B \in \mathbb{C}^{n \times k}$. Then $(AB)\{1, 3\} \subseteq B\{1, 3\}A\{1, 3\}$ if and only if*

$$r(A^* AB \ \ B) + r(A) = r(AB) + \min\{r(A^* \ \ B), \max\{n + r(A) - m, n + r(B) - k\}\}.$$

For (2.72), some equivalent conditions with the one given in [15, 19] can be found in [20]:

Theorem 2.16 ([20]) *Let $A \in \mathbb{C}^{n \times m}$ and $B \in \mathbb{C}^{m \times k}$. Then the following conditions are equivalent:*

(i) $(AB)\{1,3\} \subseteq B\{1,3\} \cdot A\{1,3\}$,

(ii) $(I - SS^\dagger)((AB)^\dagger - B^\dagger A^\dagger) = 0$ and $r(C) \geq \min\{n - r(A), k - r(B)\}$,

where $S = B^\dagger(I - A^\dagger A)$ and $C = I - A^\dagger A - S^\dagger S$.

Notice that $C = P_{\mathscr{N}(A) \cap \mathscr{N}(B^*)}$, so $r(C) = \dim(\mathscr{N}(A) \cap \mathscr{N}(B^*))$.

Example 2.11 Let $A \in \mathbb{C}^{n \times m}$ be left invertible and $B \in \mathbb{C}^{m \times k}$. Then $(AB)\{1,3\} \subseteq B\{1,3\} \cdot A\{1,3\}$ if and only if $(AB)^\dagger = B^\dagger A^\dagger$ and either A is invertible or B is left-invertible.

Corollary 2.1 *Let $A \in \mathbb{C}^{n \times m}$ and $B \in \mathbb{C}^{m \times k}$. Then the following conditions are equivalent:*

(i*) $(AB)\{1,3\} \subseteq B\{1,3\} \cdot A\{1,3\}$,

(ii*) $(I - SS^\dagger)((AB)^\dagger - B^\dagger A^\dagger) = 0$, *and at least one of the two conditions below holds:*

(a) $r(C) \geq k - r(B)$, $k - r(B) < n - r(A)$,

(b) $r(C) \geq n - r(A)$, $k - r(B) \geq n - r(A)$,

   where $S = B^\dagger(I - A^\dagger A)$ and $C = I - A^\dagger A - S^\dagger S$.

Example 2.12 Let $A \in \mathbb{C}^{n \times m}$ and $B \in \mathbb{C}^{m \times k}$. If $m < n$ and $m < k$ then

$$(AB)\{1,3\} \nsubseteq B\{1,3\} \cdot A\{1,3\}.$$

Open question: As we can see, in contrast to the case of inclusion (2.71) of which Theorem 2.13 provides a purely algebraic characterization, none of the results we have presented do so for the inclusion (2.72). To our knowledge no such results can be found in literature so far, which leaves the formulated problem still unsolved.

   The reverse order law problem for $\{1,3\}$ and $\{1,4\}$-inverses, in the matrix setting, was considered by M. Wei [4]. He obtained necessary and sufficient conditions for the following inclusions to hold:

$$A_n\{1,3\} \cdot A_{n-1}\{1,3\} \cdots A_1\{1,3\} \subseteq (A_1 A_2 \cdots A_n)\{1,3\}$$

and

$$A_n\{1,4\} \cdot A_{n-1}\{1,4\} \cdots A_1\{1,4\} \subseteq (A_1 A_2 \cdots A_n)\{1,4\}$$

by applying the multiple product singular value decomposition (P-SVD).

Using the next lemma, his results were generalized in [21] in the case of regular bounded linear operators on Hilbert spaces and new simple conditions which involve only ranges of operators were presented.

Lemma 2.3 *Let $A \in \mathscr{B}(\mathscr{H}, \mathscr{K})$ be regular. Then*

$$X \in A\{1,3\} \Leftrightarrow A^* AX = A^*.$$

Theorem 2.17 *Let* $A_i \in \mathcal{B}(\mathcal{H}_{i+1}, \mathcal{H}_i)$ *be regular operators such that* $A_1 A_2 \cdots A_n$ *is regular. The following are equivalent:*

(i) $A_n\{1, 3\} \cdot A_{n-1}\{1, 3\} \cdots A_1\{1, 3\} \subseteq (A_1 A_2 \cdots A_n)\{1, 3\},$
(ii) $\mathcal{R}(A_k^* A_{k-1}^* \cdots A_1^* A_1 A_2 \cdots A_n) \subseteq \mathcal{R}(A_{k+1}),$ *for* $k = \overline{1, n-1}.$

Proof (i) $\Rightarrow$ (ii) : If $A_1 A_2 \cdots A_n = 0$, then

$$\mathcal{R}(A_k^* A_{k-1}^* \cdots A_1^* A_1 A_2 \cdots A_n) = \{0\} \subseteq \mathcal{R}(A_{k+1}),$$

for $k = \overline{1, n-1}$, so (ii) holds.
Assume now that $A_1 A_2 \cdots A_n \neq 0$. Let $A_i^{(1,3)} \in A_i\{1, 3\}, i = \overline{1, n}$ be arbitrary. By Lemma 2.3 it follows that

$$(A_1 A_2 \cdots A_n)^* A_1 A_2 \cdots A_n A_n^{(1,3)} A_{n-1}^{(1,3)} \cdots A_1^{(1,3)} = (A_1 A_2 \cdots A_n)^*. \quad (2.74)$$

Let $i \in \{1, 2, \dots, n-1\}$ be arbitrary. Since, for arbitrary $X \in \mathcal{B}(\mathcal{H}_i, \mathcal{H}_{i+1})$, we have that $A_i^{(1,3)} + (I_{\mathcal{H}_{i+1}} - A_i^{(1,3)} A_i)X \in A_i\{1, 3\}$, by Lemma 2.3 we have

$$
\begin{aligned}
&(A_1 A_2 \cdots A_n)^* A_1 A_2 \cdots A_n \cdot \\
&A_n^{(1,3)} \cdots A_{i+1}^{(1,3)}(A_i^{(1,3)} + (I_{\mathcal{H}_{i+1}} - A_i^{(1,3)} A_i)X)A_{i-1}^{(1,3)} \cdots A_1^{(1,3)} \\
&= (A_1 A_2 \cdots A_n)^*.
\end{aligned}
\quad (2.75)
$$

Substracting (2.74) from (2.75), we get that

$$
\begin{aligned}
&(A_1 A_2 \cdots A_n)^* A_1 A_2 \cdots A_n \cdot \\
&A_n^{(1,3)} \cdots A_{i+1}^{(1,3)}(I_{\mathcal{H}_{i+1}} - A_i^{(1,3)} A_i)X A_{i-1}^{(1,3)} \cdots A_1^{(1,3)} = 0.
\end{aligned}
\quad (2.76)
$$

From (2.76) it follows that
$$A_{i-1}^{(1,3)} \cdots A_1^{(1,3)} = 0 \quad (2.77)$$

or

$$(A_1 A_2 \cdots A_n)^* A_1 A_2 \cdots A_n A_n^{(1,3)} \cdots A_{i+1}^{(1,3)}(I_{\mathcal{H}_{i+1}} - A_i^{(1,3)} A_i) = 0. \quad (2.78)$$

If (2.77) holds, then from (2.74) it follows that $A_1 A_2 \cdots A_n = 0$, which is a contradiction, so (2.78) must hold for arbitrary $i \in \{1, 2, \dots, n-1\}$.
  Condition (ii) is equivalent to

$$
\begin{aligned}
&(A_1 A_2 \cdots A_n)^* A_1 A_2 \cdots A_{n-k} A_{n-k+1} A_{n-k+1}^{(1,3)} \\
&= (A_1 A_2 \cdots A_n)^* A_1 A_2 \cdots A_{n-k}, \quad k = \overline{1, n-1}.
\end{aligned}
\quad (2.79)
$$

Now, we will prove (2.79) by induction on k.

From (2.78) and (2.74) it follows that

$$
(A_1 A_2 \cdots A_n)^* A_1 A_2 \cdots A_{n-1} A_n A_n^{(1,3)}
$$
$$
= (A_1 A_2 \cdots A_n)^* A_1 A_2 \cdots \cdots A_n A_n^{(1,3)} A_{n-1}^{(1,3)} A_{n-1}
$$
$$
= (A_1 A_2 \cdots A_n)^* A_1 A_2 \cdots \cdots A_n A_n^{(1,3)} A_{n-1}^{(1,3)} A_{n-2}^{(1,3)} A_{n-2} A_{n-1}
$$
$$
\vdots
$$
$$
= (A_1 A_2 \cdots A_n)^* A_1 A_2 \cdots \cdots A_n A_n^{(1,3)} A_{n-1}^{(1,3)} \cdots A_1^{(1,3)} A_1 A_2 \cdots A_{n-2} A_{n-1}
$$
$$
= (A_1 A_2 \cdots A_n)^* A_1 A_2 \cdots A_{n-2} A_{n-1},
$$

so (2.79) holds for $k = 1$.

Assume now that (2.79) holds for $k < l \leq n$, i.e.,

$$
(A_1 A_2 \cdots A_n)^* A_1 A_2 \cdots A_{n-k} A_{n-k+1} A_{n-k+1}^{(1,3)}
$$
$$
= (A_1 A_2 \cdots A_n)^* A_1 A_2 \cdots A_{n-k}, \quad k = 1, 2, \ldots, l - 1 \tag{2.80}
$$

and prove that (2.79) is true for $k = l$. Using (2.80), we have

$$
(A_1 A_2 \cdots A_n)^* A_1 A_2 \cdots A_{n-l} A_{n-l+1} A_{n-l+1}^{(1,3)}
$$
$$
= (A_1 A_2 \cdots A_n)^* A_1 A_2 \cdots \cdots A_n A_n^{(1,3)} A_{n-1}^{(1,3)} \cdots A_{n-l+1}^{(1,3)}. \tag{2.81}
$$

Now, using (2.81) and (2.74), we get

$$
(A_1 A_2 \cdots A_n)^* A_1 A_2 \cdots A_{n-l} A_{n-l+1} A_{n-l+1}^{(1,3)}
$$
$$
= (A_1 A_2 \cdots A_n)^* A_1 A_2 \cdots A_{n-l},
$$

so (2.79) holds for $k = l$.

(ii) $\Rightarrow$ (i) : Let $A_i^{(1,3)} \in A_i\{1, 3\}$, $i = \overline{1, n}$ be arbitrary. Condition (ii) is equivalent to

$$
(A_1 A_2 \cdots A_n)^* A_1 A_2 \cdots A_{n-k} A_{n-k+1} A_{n-k+1}^{(1,3)}
$$
$$
= (A_1 A_2 \cdots A_n)^* A_1 A_2 \cdots A_{n-k}, \quad k = 1, 2, \ldots, n - 1. \tag{2.82}
$$

Now, from (2.82) it follows

$$(A_1 A_2 \cdots A_n)^* A_1 A_2 \cdots A_{n-1} A_n A_n^{(1,3)} A_{n-1}^{(1,3)} \cdots A_1^{(1,3)}$$
$$= (A_1 A_2 \cdots A_n)^* A_1 A_2 \cdots A_{n-2} A_{n-1} A_{n-1}^{(1,3)} A_{n-2}^{(1,3)} \cdots A_1^{(1,3)}$$
$$= (A_1 A_2 \cdots A_n)^* A_1 A_2 \cdots A_{n-2} A_{n-2}^{(1,3)} \cdots A_1^{(1,3)}$$
$$\vdots$$
$$= (A_1 A_2 \cdots A_n)^* A_1 A_1^{(1,3)}$$
$$= (A_1 A_2 \cdots A_n)^*.$$

Hence by Lemma 2.3 it follows that

$$A_n^{(1,3)} A_{n-1}^{(1,3)} \cdots A_1^{(1,3)} \in (A_1 A_2 \cdots A_n)\{1, 3\}.$$

$\square$

The next result follows from Theorem 2.17 by taking adjoints:

Theorem 2.18 *Let $A_i \in \mathscr{B}(\mathscr{H}_{i+1}, \mathscr{H}_i)$ be regular operators such that $A_1 A_2 \cdots A_n$ is regular. The following are equivalent:*

(i) $A_n\{1, 4\} \cdot A_{n-1}\{1, 4\} \cdots A_1\{1, 4\} \subseteq (A_1 A_2 \cdots A_n)\{1, 4\},$

(ii) $\mathscr{R}(A_{k+1} A_{k+2} \cdots A_n A_n^* A_{n-1}^* \cdots A_1^*) \subseteq \mathscr{R}(A_k^*)$ *for $k = \overline{1, n-1}$.*

2.4 Reverse Order Laws for {1, 2, 3} and {1, 2, 4}-Inverses

The reverse order law for $\{1, 2, 3\}$-inverses for the matrix case was considered by Xiong and Zheng [22]. They presented necessary and sufficient conditions under which

$$B\{1, 2, 3\}A\{1, 2, 3\} \subseteq (AB)\{1, 2, 3\} \tag{2.83}$$

is satisfied. The method of the proof of their result, which will be stated below, involved expressions for maximal and minimal ranks of the generalized Schur complement.

Theorem 2.19 ([22]) *Let $A \in \mathbb{C}^{m \times n}$ and $B \in \mathbb{C}^{n \times k}$. Then the following statements are equivalent:*

(i) $B\{1, 2, 3\}A\{1, 2, 3\} \subseteq (AB)\{1, 2, 3\};$

(ii) $r(B, A^* AB) = r(B)$ *and* $r(AB) = \min\{r(A), r(B)\} = r(A) + r(B) - r\begin{pmatrix} A \\ B^* \end{pmatrix}.$

This result was generalized to the C^*-algebra case by Cvetković-Ilić and Harte [18] using the following characterization of the set $a\{1, 2, 3\}$:

Lemma 2.4 *Let $a \in \mathscr{A}$ be regular and $b \in \mathscr{A}$. Then $b \in a\{1, 2, 3\}$ if and only if $a^* ab = a^*$ and $baa^\dagger = b$.*

Theorem 2.20 ([18] *If* $a, b \in \mathscr{A}$ *are such that* a, b, ab *and* $a - abb^{\dagger}$ *are regular, then the following conditions are equivalent:*

(i) $b\{1, 2, 3\}a\{1, 2, 3\} \subseteq (ab)\{1, 2, 3\}$,
(ii) $bb^{\dagger}a^*ab = a^*ab$ *and* $(bb^{\dagger} - (abb^{\dagger})^{\dagger}abb^{\dagger})\mathscr{A}(aa^{\dagger} - (ab)(ab)^{\dagger}) = \{0\}$.

Proof Let $p = bb^{\dagger}, q = b^{\dagger}b$ and $r = aa^{\dagger}$. Then $b = \begin{bmatrix} b & 0 \\ 0 & 0 \end{bmatrix}_{p,q}$ and $a = \begin{bmatrix} a_1 & a_2 \\ 0 & 0 \end{bmatrix}_{r,p}$.

We have that $b\{1, 2, 3\} = \left\{ \begin{bmatrix} b^{\dagger} & 0 \\ u & 0 \end{bmatrix}_{q,p} : u \in (1 - q)\mathscr{A}p \right\}$ and $a^{\dagger} = a^*(aa^*)^{\dagger} =$

$\begin{bmatrix} a_1^*d^{\dagger} & 0 \\ a_2^*d^{\dagger} & 0 \end{bmatrix}_{p,r}$, where $d = a_1a_1^* + a_2a_2^*$. Using Lemma 2.4, $a\{1, 2, 3\} = \left\{ \begin{bmatrix} z_1 & 0 \\ z_3 & 0 \end{bmatrix}_{p,r} : \right.$

$a_1^*a_1z_1 + a_1^*a_2z_3 = a_1^*, a_2^*a_1z_1 + a_2^*a_2z_3 = a_2^*, z_1 \in p\mathscr{A}r, z_3 \in (1 - p)\mathscr{A}r\}$.

Hence $x \in b\{1, 2, 3\} \cdot a\{1, 2, 3\}$ if and only if $x = \begin{bmatrix} b^{\dagger}z_1 & 0 \\ uz_1 & 0 \end{bmatrix}_{q,r}$ for some $u \in$

$(1 - q)\mathscr{A}p$ and some $z_1 \in p\mathscr{A}r$ such that for some $z_3 \in (1 - p)\mathscr{A}r$ the following hold:

$$a_1^*a_1z_1 + a_1^*a_2z_3 = a_1^*, \quad a_2^*a_1z_1 + a_2^*a_2z_3 = a_2^*. \tag{2.84}$$

By Lemma 2.4, $b\{1, 2, 3\}a\{1, 2, 3\} \subseteq (ab)\{1, 2, 3\}$ if and only if

$$(ab)^*(ab)b^{(1,2,3)} \cdot a^{(1,2,3)} = (ab)^*,$$
$$b^{(1,2,3)} \cdot a^{(1,2,3)}(ab)(ab)^{\dagger} = b^{(1,2,3)} \cdot a^{(1,2,3)} \tag{2.85}$$

hold for any $a^{(1,2,3)}$ and $b^{(1,2,3)}$.

Now, using the matrix forms introduced above, we find that (2.85) is equivalent to the following equalities:

$$(ab)^*(ab)b^{\dagger}z_1 = (ab)^*,$$
$$z_1(ab)(ab)^{\dagger} = z_1, \tag{2.86}$$

for any $z_1 \in p\mathscr{A}r$ which satisfies (2.84).

(ii) $\Rightarrow$ (i) : Suppose that (ii) holds. Since $bb^{\dagger}a^*ab = a^*ab$, is equivalent to $a_2^*a_1 = 0$, i.e., $a_1^*a_2 = 0$, we have that (2.84) is equivalent to

$$a_1^*a_1z_1 = a_1^*, \quad a_2^*a_2z_3 = a_2^*.$$

Now, to prove that $b\{1, 2, 3\}a\{1, 2, 3\} \subseteq (ab)\{1, 2, 3\}$ it is sufficient to prove that (2.86) holds for every $z_1 \in p\mathscr{A}r$ which satisfies the equation $a_1z_1 = a_1a_1^{\dagger}$. Denote the set of all such z_1 by Z. Note that $z_1 = bb^{\dagger}zaa^{\dagger}$ for some $z \in \mathscr{A}$ which is a solution of the equation $abb^\dagger zaa^\dagger = abb^\dagger(abb^\dagger)^\dagger$. So, $Z = \{(abb^\dagger)^\dagger aa^\dagger + bb^\dagger yaa^\dagger - (abb^\dagger)^\dagger abb^\dagger yaa^\dagger : y \in \mathscr{A}\}$.

The first equality from (2.86) is satisfied because for every $z_1 \in Z$:

$$(ab)^*(ab)b^\dagger z_1 = (a_1 b)^* a_1 bb^\dagger z_1 = b^* a_1^* a_1 z_1 = b^* a_1^* = (a_1 b)^* = (ab)^*.$$

Now, the second equality from (2.86) is equivalent to

$$\begin{aligned}
(abb^\dagger)^\dagger aa^\dagger + bb^\dagger yaa^\dagger - (abb^\dagger)^\dagger abb^\dagger yaa^\dagger &= (abb^\dagger)^\dagger(ab)(ab)^\dagger \\
+ bb^\dagger y(ab)(ab)^\dagger - (abb^\dagger)^\dagger abb^\dagger y(ab)(ab)^\dagger. &
\end{aligned} \tag{2.87}$$

Since, $(ab)(ab)^\dagger = abb^\dagger(abb^\dagger)^\dagger$, we get that $(abb^\dagger)^\dagger(ab)(ab)^\dagger = (abb^\dagger)^\dagger = (abb^\dagger)^\dagger aa^\dagger$, so (2.87) is equivalent to

$$\left(bb^\dagger - (abb^\dagger)^\dagger(abb^\dagger)\right)y\left(aa^\dagger - (ab)(ab)^\dagger\right) = 0$$

which holds since $(bb^\dagger - (abb^\dagger)^\dagger abb^\dagger)\mathscr{A}(aa^\dagger - (ab)(ab)^\dagger) = \{0\}$.

(i) $\Rightarrow$ (ii) : If (i) holds, then $b^\dagger a^\dagger \in (ab)\{1, 2, 3\}$. Now, from $abb^\dagger a^\dagger ab = ab$, it follows that $a_1 a_1^* d^\dagger a_1 = a_1$, i.e., $a_2 a_2^* d^\dagger a_1 = 0$. Similarly, from $(abb^\dagger a^\dagger)^* = abb^\dagger a^\dagger$, we get that $a_1 a_1^* d^\dagger$ is Hermitian. Now, $d^\dagger a_1 a_1^* a_1 = a_1$, i.e., $a_2 a_2^* a_1 = 0$. Multiplying the last equality by $a_2^\dagger$ from the left side, we get $a_2^* a_1 = 0$ which is equivalent to $bb^\dagger a^* ab = a^* ab$. Now, (2.86) holds for every $z_1 \in p\mathscr{A}r$ which satisfies the equation $a_1 z_1 = a_1 a_1^\dagger$. Hence, $\left(bb^\dagger - (abb^\dagger)^\dagger(abb^\dagger)\right)y\left(aa^\dagger - (ab)(ab)^\dagger\right) = 0$, for every $y \in \mathscr{A}$, i.e., $(bb^\dagger - (abb^\dagger)^\dagger abb^\dagger)\mathscr{A}(aa^\dagger - (ab)(ab)^\dagger) = \{0\}$. $\qquad \square$

Note that if the algebra $\mathscr{A}$ is prime, in the sense that

$$a\mathscr{A}b = \{0\} \Longrightarrow 0 \in \{a, b\},$$

then the second half of condition (ii) of Theorem 2.20 is equivalent to

$$bb^\dagger - (abb^\dagger)^\dagger abb^\dagger = 0 \text{ or } aa^\dagger - (ab)(ab)^\dagger = 0.$$

The C*-algebra $\mathscr{A} = \mathscr{B}(\mathscr{H})$ of operators on Hilbert space is prime, in particular (Lemma 3 [23]) the matrix algebra. We thus have the following results.

Corollary 2.2 *Let $A \in \mathbb{C}^{m \times n}$ and $B \in \mathbb{C}^{n \times k}$. Then the following statements are equivalent:*

(i) $B\{1, 2, 3\}A\{1, 2, 3\} \subseteq (AB)\{1, 2, 3\}$,

(ii) $BB^\dagger A^* AB = A^* AB$ and $\left((ABB^\dagger)^\dagger ABB^\dagger = BB^\dagger \text{ or } (AB)(AB)^\dagger = AA^\dagger\right)$.

Example 2.13 We will show that in the case when $B\{1, 2, 3\}A\{1, 2, 3\} \not\subseteq (AB)\{1, 2, 3\}$ we can find particular $A^{(1,2,3)}$ and $B^{(1,2,3)}$ such that $B^{(1,2,3)}A^{(1,2,3)} \in$

$(AB)\{1, 2, 3\}$: Let $A = \begin{bmatrix} 1 & 0 \\ 0 & 0 \end{bmatrix}$ and $B = \begin{bmatrix} 0 & 0 \\ 1 & 0 \end{bmatrix}$. Then evidently $AB = 0$, which implies $(AB)\{1, 2, 3\} = \{0\}$ and by Corollary 2.2 that $B\{1, 2, 3\}A\{1, 2, 3\} \not\subseteq (AB)\{1, 2, 3\}$. But for $A^{(1,2,3)} = \begin{bmatrix} 1 & 0 \\ 0 & 0 \end{bmatrix}$ and $B^{(1,2,3)} = \begin{bmatrix} 0 & 1 \\ 0 & d \end{bmatrix}$, where d is any complex number, we have that $B^{(1,2,3)} A^{(1,2,3)} = 0 \in (AB)\{1, 2, 3\}$.

Example 2.14 Let $b \in \mathscr{A}$ be right invertible and $a \in \mathscr{A}$ be regular such that ab is regular. Then $b\{1, 2, 3\} \cdot a\{1, 2, 3\} \subseteq (ab)\{1, 2, 3\}$ if and only if either a is left invertible or $aa^\dagger = (ab)(ab)^\dagger$.

Also, in [24] using some block-operator matrix techniques, the authors presented some different conditions than the one presented in Corollary 2.2 for (2.83) to hold in the case of linear bounded operators on Hilbert spaces:

Theorem 2.21 ([24]) *Let* $A \in \mathscr{B}(\mathscr{H}, \mathscr{K})$ *and* $B \in \mathscr{B}(\mathscr{L}, \mathscr{H})$ *be such that* A, B, AB *are regular operators and* $AB \neq 0$. *The following conditions are equivalent:*

(i) $B\{1, 2, 3\}A\{1, 2, 3\} \subseteq (AB)\{1, 2, 3\}$;
(ii) $\mathscr{R}(B) = \mathscr{R}(A^*AB) \oplus^\perp [\mathscr{R}(B) \cap \mathscr{N}(A)]$, $\mathscr{R}(AB) = \mathscr{R}(A)$.

In the matrix case, in [25], it is shown that $B\{1, 2, 3\} \cdot A\{1, 2, 3\} \subseteq (AB)\{1, 2, 3\}$ implies $B\{1, 2, 3\} \cdot A\{1, 2, 3\} = (AB)\{1, 2, 3\}$. Here we will present a completely different proof which can easily be adapted to more general cases.

Theorem 2.22 *Let* $A \in \mathbb{C}^{n \times m}$ *and* $B \in \mathbb{C}^{m \times k}$. *The following conditions are equivalent:*

(i) $B\{1, 2, 3\} \cdot A\{1, 2, 3\} \subseteq (AB)\{1, 2, 3\}$,
(ii) $BB^\dagger A^*AB = A^*AB$ and $\left((ABB^\dagger)^\dagger ABB^\dagger = BB^\dagger \text{ or } (AB)(AB)^\dagger = AA^\dagger \right)$,
(iii) $B\{1, 2, 3\} \cdot A\{1, 2, 3\} = (AB)\{1, 2, 3\}$.

Proof (i) $\Leftrightarrow$ (ii): Follows from Corollary 2.2.

(i) $\Rightarrow$ (iii): Let $P = BB^\dagger$, $Q = B^\dagger B$ and $R = AA^\dagger$. We have that $A = A_1 + A_2$, where $A_1 = AP$ and $A_2 = A(I - P)$. To prove (iii), take arbitrary $X \in (AB)\{1, 2, 3\}$. We will show that there exist $Y \in B\{1, 2, 3\}$ and $Z \in A\{1, 2, 3\}$ such that $X = YZ$. Since $X \in (AB)\{1, 2, 3\}$, it is of the form $X = QX_1R + (I - Q)X_3R$, for some $X_1 \in \mathbb{C}^{k \times n}$ and $X_3 \in \mathbb{C}^{k \times n}$ such that $QX_1R \in (A_1B)\{1, 2, 3\}$ and $(I - Q)X_3A_1BX_1R = (I - Q)X_3R$.
Let $Z = BX_1R + A_2^\dagger$ and $Y = B^\dagger + (I - Q)X_3A_1$. We have $B\{1, 2, 3\} = \{B^\dagger + (I - Q)UP : U \in \mathbb{C}^{k \times m}\}$, so $Y \in B\{1, 2, 3\}$. To prove that $Z \in A\{1, 2, 3\}$, we can check that the first three Penrose equations are satisfied using that $A_2^*A_1 = 0$, which follows from the condition $BB^\dagger A^*AB = A^*AB$. Since $X = YZ$, it follows that $B\{1, 2, 3\} \cdot A\{1, 2, 3\} = (AB)\{1, 2, 3\}$.

(iii) $\Rightarrow$ (i): This is evident. $\square$

The opposite reverse order law

$$(AB)\{1, 2, 3\} \subseteq B\{1, 2, 3\} \cdot A\{1, 2, 3\} \tag{2.88}$$

on the set of matrices is considered in [25] where purely algebraic necessary and sufficient conditions for (2.88) to hold are offered.

Theorem 2.23 ([25]) *Let $A \in \mathbb{C}^{n \times m}$ and $B \in \mathbb{C}^{m \times k}$. The following conditions are equivalent:*

(i) $(AB)\{1, 2, 3\} \subseteq B\{1, 2, 3\} \cdot A\{1, 2, 3\}$,
(ii) $(I - B^\dagger(B^\dagger(I - A^\dagger A))^\dagger)((AB)^\dagger - B^\dagger A^\dagger) = 0$

It is very important to remark that the results from Theorems 2.23 and 2.22 can be generalized to the setting of bounded linear operators on Hilbert spaces and to the C^*-algebra setting by imposing the additional condition of regularity of suitable elements.

Example 2.15 We will show that $(AB)\{1, 2, 3\} \subseteq B\{1, 2, 3\}A\{1, 2, 3\}$ doesn't imply that $B\{1, 2, 3\}A\{1, 2, 3\} \subseteq (AB)\{1, 2, 3\}$, although the reverse implication is always true. Let $A = \begin{bmatrix} 1 & 0 \\ 0 & 0 \end{bmatrix}$ and $B = \begin{bmatrix} 0 & 0 \\ 1 & 0 \end{bmatrix}$. Then evidently $A^\dagger = \begin{bmatrix} 1 & 0 \\ 0 & 0 \end{bmatrix}$ and $B^\dagger = \begin{bmatrix} 0 & 1 \\ 0 & 0 \end{bmatrix}$. So, we can check that $(AB)^\dagger = B^\dagger A^\dagger$, which implies by Theorem 2.23 that $(AB)\{1, 2, 3\} \subseteq B\{1, 2, 3\} \cdot A\{1, 2, 3\}$. On the other hand, neither of the conditions $(ABB^\dagger)^\dagger ABB^\dagger = BB^\dagger$ and $(AB)(AB)^\dagger = AA^\dagger$ is satisfied, so from Theorem 2.22 it follows that $B\{1, 2, 3\}A\{1, 2, 3\} \not\subseteq (AB)\{1, 2, 3\}$

By taking adjoints, we obtain analogous results for $\{1, 2, 4\}-$ generalized inverses.

Theorem 2.24 *Let $a, b \in \mathscr{A}$ be such that a, b, ab and $(1 - a^\dagger a)b$ are regular. Then the following conditions are equivalent:*

(i') $b\{1, 2, 4\}a\{1, 2, 4\} \subseteq (ab)\{1, 2, 4\}$
(ii') $a^\dagger abb^*a^* = bb^*a^*$ and $(a^\dagger a - a^\dagger ab(a^\dagger ab)^\dagger)\mathscr{A}(b^\dagger b - (ab)^\dagger(ab)) = \{0\}$.

Theorem 2.25 *Let $A \in \mathbb{C}^{m \times n}$ and $B \in \mathbb{C}^{n \times k}$. Then the following statements are equivalent:*

(i) $B\{1, 2, 4\}A\{1, 2, 4\} \subseteq (AB)\{1, 2, 4\}$,
(ii) $ABB^*A^\dagger A = ABB^*$ and $((A^\dagger AB)(A^\dagger AB)^\dagger = A^\dagger A$ or $(AB)^\dagger(AB) = B^\dagger B)$,
(iii) $B\{1, 2, 4\}A\{1, 2, 4\} = (AB)\{1, 2, 4\}$,

Theorem 2.26 ([24]) *Let $\mathscr{H}, \mathscr{K}$ and $\mathscr{L}$ be Hilbert spaces and let $A \in \mathscr{B}(\mathscr{H}, \mathscr{K})$, $B \in \mathscr{B}(\mathscr{L}, \mathscr{H})$ be such that $\mathscr{R}(A), \mathscr{R}(B)$ and $\mathscr{R}(AB)$ are closed and $AB \neq 0$. Then the following statements are equivalent:*

(i) $B\{1, 2, 4\}A\{1, 2, 4\} \subseteq (AB)\{1, 2, 4\}$.

(ii) $\mathscr{R}(A^*) = \mathscr{R}(BB^*A^*) \oplus^\perp [\mathscr{R}(A^*) \cap \mathscr{N}(B^*)]$, $\mathscr{N}(AB) = \mathscr{N}(B)$

Theorem 2.27 *Let $A \in \mathbb{C}^{n \times m}$ and $B \in \mathbb{C}^{m \times k}$. The following conditions are equivalent:*

(i) $(AB)\{1, 2, 4\} \subseteq B\{1, 2, 4\} \cdot A\{1, 2, 4\}$,
(ii) $((AB)^\dagger - B^\dagger A^\dagger)(I - ((I - BB^\dagger)A^\dagger)^\dagger A^\dagger) = 0$

2.5   Reverse Order Laws for {1, 3, 4}-Generalized Inverses

Reverse order laws for $\{1, 3, 4\}$-generalized inverses of matrices $A \in \mathbb{C}^{n \times m}$ and $B \in \mathbb{C}^{m \times k}$ were considered by Liu and Yang [26], who gave certain necessary and sufficient conditions for

$$B\{1, 3, 4\} \cdot A\{1, 3, 4\} \subseteq (AB)\{1, 3, 4\} \tag{2.89}$$

and

$$(AB)\{1, 3, 4\} \subseteq B\{1, 3, 4\} \cdot A\{1, 3, 4\}. \tag{2.90}$$

Let

$$n_1 = \mathrm{r}\begin{bmatrix} B^* \\ B^*A^*A \end{bmatrix}, \; n_2 = \mathrm{r}\begin{bmatrix} B^*B & B^*A^* \\ B^*A^*AB & B^*A^*AA^* \end{bmatrix}, \; n_3 = \mathrm{r}\begin{bmatrix} A \\ ABB^* \end{bmatrix}$$

and

$$n_4 = \mathrm{r}\begin{bmatrix} AA^* & ABB^*A^* \\ B^*A^* & B^*BB^*A^* \end{bmatrix}.$$

With this notations the following is proved:

Theorem 2.28 ([26]) *Let $A \in \mathbb{C}^{n \times m}$ and $B \in \mathbb{C}^{m \times k}$. Then the following conditions are equivalent:*

(i) $B\{1, 3, 4\} \cdot A\{1, 3, 4\} \subseteq (AB)\{1, 3, 4\}$,
(ii) $\mathrm{r}(A) = \min\{n_3, n_4 + k - \mathrm{r}(B)\}$, $\mathrm{r}(B) = \min\{n_1, n_2 + m - \mathrm{r}(A)\}$.

Theorem 2.29 ([26]) *Let $A \in \mathbb{C}^{n \times m}$ and $B \in \mathbb{C}^{m \times k}$. Then the following conditions are equivalent:*

(i) $(AB)\{1, 3, 4\} \subseteq B\{1, 3, 4\} \cdot A\{1, 3, 4\}$,
(ii) $\mathrm{r}(AB) = \max\{\min\{n_2, n_4\}, \min\{k - \mathrm{r}(B), n - \mathrm{r}(A)\} - m + n_1 + n_3\}$.

What the authors of [26] did not realize was that $(AB)\{1, 3, 4\} \subseteq B\{1, 3, 4\} \cdot A\{1, 3, 4\}$ is actually equivalent to $(AB)\{1, 3, 4\} = B\{1, 3, 4\} \cdot A\{1, 3, 4\}$, which was shown in [27] in the matrix case and later generalized in [28] to the C^*-algebra setting. In the following theorem the proof of this fact in C^*-algebras will be presented together with some necessary and sufficient conditions for (2.90) to hold.

Theorem 2.30 ([28]) *Let $a, b \in \mathscr{A}$ be such that $a, b, ab, a(1-bb^\dagger)$ and $(1-a^\dagger a)b$ are generalized invertible. Then the following conditions are equivalent:*

(i) $(ab)\{1, 3, 4\} \subseteq b\{1, 3, 4\} \cdot a\{1, 3, 4\},$
(ii) $(ab)\{1, 3, 4\} = b\{1, 3, 4\} \cdot a\{1, 3, 4\},$
(iii) $(ab)^\dagger = b^\dagger a^\dagger, (b^\dagger b - (ab)^\dagger (ab))\mathscr{A}(aa^\dagger - (ab)(ab)^\dagger) = \{0\}$ *and the equation*

$$(1 - b^\dagger b)z(1 - aa^\dagger) = (1 - b^\dagger b)x(1 - s_2 s_2^\dagger) f_b e_a (1 - s_1^\dagger s_1)y(1 - aa^\dagger), \quad (2.91)$$

is solvable for any $z \in \mathscr{A}$, where $s_1 = 1 - (ab)^\dagger(ab)$, $s_2 = (1 - bb^\dagger)a^\dagger$, $e_a = 1 - a^\dagger a$ and $f_b = 1 - bb^\dagger$.

To give a proof of Theorem 2.30, we will need some auxiliary results given as follows:

Lemma 2.5 *Let $a \in \mathscr{A}$ be generalized invertible and $b \in \mathscr{A}$. Then the following statements are equivalent:*

(1) $b \in a\{1, 3, 4\}$
(2) $a^*ab = a^*$ and $baa^* = a^*$.
(3) *There exists $y \in \mathscr{A}$ such that $b = a^\dagger + (1 - a^\dagger a)y(1 - aa^\dagger)$.*

Proof (1) $\Rightarrow$ (2) If $b \in a\{1, 3, 4\}$, then

$$a^*ab = a^*(ab)^* = (aba)^* = a^* \quad \text{and} \quad baa^* = (aba)^* = a^*.$$

(2) $\Rightarrow$ (1) If $a^*ab = a^*$ and $baa^* = a^*$, then

$$aba = aa^\dagger aba = (a^\dagger)^* a^* aba = (a^\dagger)^* a^* a = a,$$
$$ab = b^* a^* ab = (ab)^*(ab),$$
$$ba = baa^* b^* = (ba)(ba)^*,$$

so, $b \in a\{1, 3, 4\}$.

(1) $\Rightarrow$ (3) If $b \in a\{1, 3, 4\}$, then $b \in a\{1, 3\}$, so $b = a^\dagger + (1 - a^\dagger a)t$, for some $t \in \mathscr{A}$. Then $ba = a^\dagger a$, so $(1 - a^\dagger a)ta = 0$. Put $z = (1 - a^\dagger a)t$. We have that z is a solution of the equation $za = 0$, so $z = y(1 - aa^\dagger)$, for some $y \in \mathscr{A}$. Now, $b = a^\dagger + (1 - a^\dagger a)y(1 - aa^\dagger)$.

(3) $\Rightarrow$ (1) This is evident. □

Now, let $p = bb^\dagger$, $q = b^\dagger b$ and $r = aa^\dagger$.

Remark 2.1 By Lemma 2.5, we get that $b\{1, 3, 4\} = \left\{ \begin{bmatrix} b^\dagger & 0 \\ 0 & u \end{bmatrix}_{q,p} : u \in (1-q)\mathscr{A}(1-p) \right\}$

and

$$a\{1,3,4\} = \left\{ \begin{bmatrix} a_1^* d^\dagger z_2 - a_1^* d^\dagger a_1 z_2 - a_1^* d^\dagger a_2 z_4 \\ a_2^* d^\dagger z_4 - a_2^* d^\dagger a_1 z_2 - a_2^* d^\dagger a_2 z_4 \end{bmatrix}_{p,r} : z_2 \in p\mathscr{A}(1-r), \right.$$

$$\left. z_4 \in (1-p)\mathscr{A}(1-r) \right\}.$$

Lemma 2.6 *Let $a, b \in \mathscr{A}$ be such that $a, b, ab \in \mathscr{A}^\dagger$ and $(ab)^\dagger = b^\dagger a^\dagger$. Then*

$$(abb^\dagger)^\dagger = b(ab)^\dagger. \tag{2.92}$$

Proof of Theorem 2.30: (i) $\Leftrightarrow$ (iii) : The fact that $(ab)\{1,3,4\} \subseteq b\{1,3,4\} \cdot a\{1,3,4\}$ is equivalent to the fact that for every $(ab)^{(1,3,4)}$ there exist $a^{(1,3,4)}$ and $b^{(1,3,4)}$ such that $(ab)^{(1,3,4)} = b^{(1,3,4)} \cdot a^{(1,3,4)}$.

Since, $ab = \begin{bmatrix} a_1 b & 0 \\ 0 & 0 \end{bmatrix}_{r,q}$, by Lemma 2.5,

$$(ab)\{1,3,4\} = \left\{ \begin{bmatrix} s & (b^\dagger b - (a_1 b)^\dagger (a_1 b)) y_2 \\ y_3 (aa^\dagger - (ab)(ab)^\dagger) & y_4 \end{bmatrix}_{q,r} : \right.$$

$$\left. y = \begin{bmatrix} y_1 & y_2 \\ y_3 & y_4 \end{bmatrix}_{q,r} \in \mathscr{A} \right\},$$

where $s = (a_1 b)^\dagger + \left(1 - (a_1 b)^\dagger (a_1 b)\right) y_1 \left(1 - (a_1 b)(a_1 b)^\dagger\right)$.

Now, using (ii) of Remark 2.1, we may conclude that $(ab)\{1,3,4\} \subseteq b\{1,3,4\} \cdot a\{1,3,4\}$ holds if and only if for arbitrary $y = \begin{bmatrix} y_1 & y_2 \\ y_3 & y_4 \end{bmatrix}_{q,r} \in \mathscr{A}$, there exist $u, z \in \mathscr{A}$ such that

$$(a_1 b)^\dagger + \left(1 - (a_1 b)^\dagger (a_1 b)\right) y_1 \left(1 - (a_1 b)(a_1 b)^\dagger\right) = b^\dagger a_1^* d^\dagger, \tag{2.93}$$

$$(b^\dagger b - (a_1 b)^\dagger (a_1 b)) y_2 = b^\dagger (z_2 - a_1^* d^\dagger a_1 z_2 - a_1^* d^\dagger a_2 z_4), \tag{2.94}$$

$$y_3 (aa^\dagger - (ab)(ab)^\dagger) = (1 - b^\dagger b) u a_2^* d^\dagger, \tag{2.95}$$

$$y_4 = (1 - b^\dagger b) u (z_4 - a_2^* d^\dagger a_1 z_2 - a_2^* d^\dagger a_2 z_4), \tag{2.96}$$

where $z_2 = pz(1-r)$ and $z_4 = (1-p)z(1-r)$.
The fact that the equation (2.93) holds for every $y_1 \in q\mathscr{A}r$ is equivalent to

$$(ab)^\dagger = b^\dagger a^* d^\dagger = b^\dagger a^\dagger, \ (b^\dagger b - (ab)^\dagger (ab))\mathscr{A}(aa^\dagger - (ab)(ab)^\dagger) = \{0\}. \tag{2.97}$$

Now, using Lemma 2.6 and the fact that $(ab)^\dagger = b^\dagger a^\dagger \Rightarrow a_1^* d^\dagger a_2 = 0$, we have that the equations (2.94), (2.95) and (2.96), respectively have the forms:

$$\left(1 - (ab)^\dagger(ab)\right)b^\dagger by(1 - aa^\dagger) = \left(1 - (ab)^\dagger(ab)\right)b^\dagger z(1 - aa^\dagger), \quad (2.98)$$

$$\left(1 - b^\dagger b\right)ya(1 - bb^\dagger)a^\dagger = \left(1 - b^\dagger b\right)u(1 - bb^\dagger)a^\dagger, \quad (2.99)$$

$$\left(1 - b^\dagger b\right)y(1 - aa^\dagger) = \left(1 - b^\dagger b\right)uf_be_az(1 - aa^\dagger). \quad (2.100)$$

It is evident that the equations (2.98) and (2.99) are solvable for any $y \in \mathscr{A}$ and that solutions of these equations are respectively, given by

$$z = by + t_1 - s_1^\dagger s_1 t_1(1 - aa^\dagger), \quad u = ya + t_2 - (1 - b^\dagger b)t_2 s_2 s_2^\dagger,$$

for $t_1, t_2 \in \mathscr{A}$. Notice that $s_1 \in \mathscr{A}^\dagger$ because it is a projection and $s_2 \in \mathscr{A}^\dagger$ since $a(1 - bb^\dagger) \in s_2\{1\}$.

Since $abb^\dagger = ab(ab)^\dagger a$, we have that

$$a(1 - bb^\dagger)(1 - a^\dagger a)(1 - bb^\dagger) = 0. \quad (2.101)$$

Also, from $a_2^* d^\dagger a_1 = 0$, we get that

$$(1 - bb^\dagger)(1 - a^\dagger a)(1 - bb^\dagger) = (1 - bb^\dagger)(1 - a^\dagger a). \quad (2.102)$$

Now, by (2.101) and (2.102), it follows that for z and u satisfying the equations (2.98) and (2.99), respectively, the equation (2.100) is equivalent to

$$\left(1 - b^\dagger b\right)y(1 - aa^\dagger) = \left(1 - b^\dagger b\right)t_2(1 - s_2 s_2^\dagger)f_b e_a(1 - s_1^\dagger s_1)t_1(1 - aa^\dagger). \quad (2.103)$$

Now, (i) holds if and only if (2.97) holds and for arbitrary $y \in \mathscr{A}$ there exist $t_1, t_2 \in \mathscr{A}$ such that the equation (2.103) is satisfied.

(i) $\Rightarrow$ (ii) : If (i) holds, then there exist $a^{(1,3,4)}$ and $b^{(1,3,4)}$ such that $(ab)^\dagger = b^{(1,3,4)}a^{(1,3,4)}$. Now, if we multiply the last equality by $b^\dagger b$ from the left and by $aa^\dagger$ from the right, we get that $(ab)^\dagger = b^\dagger bb^{(1,3,4)}a^{(1,3,4)}aa^\dagger = b^\dagger bb^\dagger a^\dagger aa^\dagger = b^\dagger a^\dagger$. Now, by Theorem 2.32, we get that (ii) holds.

(ii) $\Rightarrow$ (i) : This is evident.   □

For matrices in the special case when $k = n$ it was proved in [27] that the solvability of the equation (2.91) is equivalent to the condition $n \leq m$. Hence, when $n > m$ then

$$(AB)\{1, 3, 4\} \nsubseteq B\{1, 3, 4\} \cdot A\{1, 3, 4\}. \quad (2.104)$$

The case $n \leq m$ is treated in the next result.

Theorem 2.31 ([27]) *Let $A \in \mathbb{C}^{n \times m}$ and $B \in \mathbb{C}^{m \times n}$ and $n \leq m$. The following conditions are equivalent:*

(i) $(AB)\{1, 3, 4\} \subseteq B\{1, 3, 4\} \cdot A\{1, 3, 4\}$,

(ii) $(AB)^\dagger = B^\dagger A^\dagger$ *and* $\left(B = A^\dagger AB \text{ or } A = ABB^\dagger\right)$.

Example 2.16 We will show that $(AB)\{1,3,4\} \subseteq B\{1,3,4\} \cdot A\{1,3,4\}$ is not equivalent with $B\{1,3,4\} \cdot A\{1,3,4\} \subseteq (AB)\{1,3,4\}$. Also, in the case when $(AB)\{1,3,4\} \not\subseteq B\{1,3,4\} \cdot A\{1,3,4\}$, for some $(AB)^{(1,3,4)}$, we will find particular $A^{(1,3,4)}$ and $B^{(1,3,4)}$ such that $(AB)^{(1,3,4)} = B^{(1,3,4)}A^{(1,3,4)} \in (AB)\{1,2,3\}$: Let $A = \begin{bmatrix} 1 & 0 \\ 0 & 0 \end{bmatrix}$ and $B = \begin{bmatrix} 0 & 0 \\ 1 & 0 \end{bmatrix}$. Then evidently $AB = 0$, which implies by Theorem 2.31 that $(AB)\{1,3,4\} \not\subseteq B\{1,3,4\} \cdot A\{1,3,4\}$, and also that $(AB)\{1,3,4\} = \mathbb{C}^{2\times 2}$. This further implies that $B\{1,3,4\} \cdot A\{1,3,4\} \subseteq (AB)\{1,3,4\}$. Since $A\{1,3,4\} = \left\{ \begin{bmatrix} 1 & 0 \\ 0 & c \end{bmatrix} : c \in \mathbb{C} \right\}$, $B\{1,3,4\} = \left\{ \begin{bmatrix} 0 & 1 \\ d & 0 \end{bmatrix} : d \in \mathbb{C} \right\}$ then for each bidiagonal matrix of the form $\begin{bmatrix} 0 & x \\ y & 0 \end{bmatrix} \in (AB)\{1,3,4\}$, there exist $A^{(1,3,4)}$ and $B^{(1,3,4)}$ such that $\begin{bmatrix} 0 & x \\ y & 0 \end{bmatrix} = B^{(1,3,4)}A^{(1,3,4)}$.

Example 2.17 It can be shown that for $A \in \mathbb{C}^{n\times m}$ and $B \in \mathbb{C}^{m\times k}$, the condition $\min\{n,k\} \leq m$ is necessary for $(AB)\{1,3,4\} \subseteq B\{1,3,4\} \cdot A\{1,3,4\}$ to hold.

Recently, some algebraic conditions for $B\{1,3,4\} \cdot A\{1,3,4\} \subseteq (AB)\{1,3,4\}$ to hold in the matrix case were given in [27] and their generalization in the case of C^*-algebras in [28]. Here, we will present only the more general version in the C^*-algebra case.

Theorem 2.32 ([28]) *Let $a, b \in \mathscr{A}$ be such that $a, b, ab, a(1-bb^\dagger)$ and $(1-a^\dagger a)b$ are generalized invertible. Then the following conditions are equivalent:*

  (i) $b\{1,3,4\} \cdot a\{1,3,4\} \subseteq (ab)\{1,3,4\}$,
  (ii) $bb^\dagger a^* ab = a^* ab$ *and* $abb^* a^\dagger a = abb^*$,
  (iii) $b^\dagger a^\dagger = (ab)^\dagger$.

Proof (i) $\Rightarrow$ (ii) Since $b^\dagger a^\dagger \in (ab)\{1,3,4\}$, from $abb^\dagger a^\dagger ab = ab$ it follows that $a_1 a_1^* d^\dagger a_1 = a_1$ i.e., $a_2 a_2^* d^\dagger a_1 = 0$. Similarly, from $(abb^\dagger a^\dagger)^* = abb^\dagger a^\dagger$, we get that $a_1 a_1^* d^\dagger$ is Hermitian. Now, $d^\dagger a_1 a_1^* a_1 = a_1$, i.e., $a_2 a_2^* a_1 = 0$. Multiplying the last equality by $a_2^\dagger$ from the left side, we get $a_2^* a_1 = 0$ which is equivalent to $bb^\dagger a^* ab = a^* ab$. Similarly, we get that $abb^* a^\dagger a = abb^*$.

  (ii) $\Rightarrow$ (i) The condition $bb^\dagger a^* ab = a^* ab$ is equivalent to $a_2^* a_1 = 0$, i.e., $a_1^* a_2 = 0$. Let $s = a_1 a_1^\dagger$. Since $d \in s\mathscr{A}s + (1-s)\mathscr{A}(1-s)$, we have that $d^\dagger \in s\mathscr{A}s + (1-s)\mathscr{A}(1-s)$. Now, $a_1^* d^\dagger a_2 \in \mathscr{A}s \cdot (s\mathscr{A}s + (1-s)\mathscr{A}(1-s)) \cdot (1-s)\mathscr{A} = \{0\}$. Hence, $a_1^* d^\dagger a_2 = 0$, i.e., $a_2^* d^\dagger a_1 = 0$. For arbitrary $a^{(1,3,4)}$ and $b^{(1,3,4)}$ we have that

$$abb^{(1,3,4)}a^{(1,3,4)}ab = \begin{bmatrix} a_1 a_1^* d^\dagger a_1 b & 0 \\ 0 & 0 \end{bmatrix}_{r,q}.$$

Since

$$a_1 a_1^* d^\dagger a_1 = (d - a_2 a_2^*) d^\dagger a_1 = a_1,$$

it follows that $abb^{(1,3,4)} a^{(1,3,4)} ab = ab$. To prove that $abb^{(1,3)} a^{(1,3)}$ is Hermitian it is sufficient to prove that $a_1 a_1^* d^\dagger$ is Hermitian and $a_1(z_2 - a_1^* d^\dagger a_1 z_2) = 0$. By computation, we get that $a_1 z_1 = a_1 a_1^* d^\dagger = a_1 a_1^* (a_1 a_1^*)^\dagger$ which is Hermitian. Also,

$$a_1 z_2 - a_1 a_1^* d^\dagger a_1 z_2 = (a_1 - (d - a_2 a_2^*) d^\dagger a_1) z_2 = 0.$$

Hence, $b\{1, 3, 4\} \cdot a\{1, 3, 4\} \subseteq (ab)\{1, 3\}$. Similarly, the condition $abb^* a^\dagger a = abb^*$ implies that $b\{1, 3, 4\} \cdot a\{1, 3, 4\} \subseteq (ab)\{1, 4\}$, so (i) holds.

(i) $\Rightarrow$ (iii) : It is sufficient to prove that $b^\dagger a^\dagger$ is an outer inverse of ab. That is equivalent to $b^\dagger a_1^* d^\dagger a_1 a_1^* d^\dagger = b^\dagger a_1^* d^\dagger$ which holds since $a_1^* d^\dagger a_1 a_1^* = a_1^*$.

(iii) $\Rightarrow$ (ii) : The proof of this part follows directly from the proof of the part (i) $\Rightarrow$ (ii). $\qquad\square$

References

1. Rao, C.R.: A note on a generalized inverse of a matrix with applications to problems in mathematical statistics, J. R. Stat. Soc. Ser. B Stat. Methodol. **24**(1), 152–158 (1962)
2. Pringle, R.M., Rayner, A.A.: Generalized Inverse Matrices with Applications to Statistics. Griffin, London (1971)
3. Werner, H.J.: When is $B^- A^-$ a generalized inverse of AB. Linear Algebra Appl. **210**, 255–263 (1994)
4. Wei, M.: Reverse order laws for generalized inverses of multiple matrix products. Linear Algebra Appl. **293**(1–3), 273–288 (1999)
5. Wei, Y.: The Drazin inverse of a modified matrix. Appl. Math. Comput. **125**, 295–301 (2002)
6. Radenković, J.N.: Reverse order laws for generalized inverses of multiple operator products. Linear Multilinear Algebra **64**, 1266–1282 (2016)
7. Shinozaki, N., Sibuya, M.: Further results on the reverse order law. Linear Algebra Appl. **27**, 9–16 (1979)
8. De Pierro, A.R., Wei, M.: Reverse order laws for reflexive generalized inverse of products of matrices. Linear Algebra Appl. **277**, 299–311 (1998)
9. Cvetković-Ilić, D.S., Djikić, M.: Various solutions to reverse order law problems. Linear Multilinear Algebra **64**, 1207–1219 (2016)
10. Wang, Q.W.: A system of matrix equations and a linear matrix equation over arbitrary regular rings with identity. Linear Algebra Appl. **384**, 43–54 (2004)
11. Hai, G., Chen, A.: On the right(left) invertible completions for operator matrices. Integral Equations Operator Theory **67**, 79–93 (2010)
12. Pavlović, V., Cvetković-Ilić, D.S.: Applications of completions of operator matrices to reverse order law for {1}-inverses of operators on Hilbert spaces. Linear Algebra Appl. **484**, 219–236 (2015)
13. Cvetković-Ilić, D.S., Nikolov, J.: Reverse order laws for reflexive generalized inverse of operators. Linear Multilinear Algebra **63**(6), 1167–175 (2015)
14. Wei, M.: Equivalent conditions for generalized inverses of products. Linear Algebra Appl. **266**, 347–363 (1997)
15. Wei, M., Guo, W.: Reverse order laws for least squares g-inverses and minimum norm g-inverses of products of two matrices. Linear Algebra Appl. **342**, 117–132 (2002)

16. Djordjević, D.S., Stanimirović, P.S.: On the generalized Drazin inverse and generalized resolvent. Czechoslovak Math. J. **51**(126), 617–634 (2001)
17. Takane, Y., Tian, Y., Yanai, H.: On reverse-order laws for least-squares g-inverses and minimum norm g-inverses of a matrix product. Aequationes Math. **73**, 56–70 (2007)
18. Cvetković-Ilić, D.S., Harte, R.: Reverse order laws in C*-algebras. Linear Algebra Appl. **434**(5), 1388–1394 (2011)
19. Liu, D., Yang, H.: Further results on the reverse order law for {1, 3}-inverse and {1, 4}-inverse of a matrix product, J. Inequal. Appl. Article ID 312767, 13 pages (2010)
20. Cvetković-Ilić, D.S.: New conditions for the reverse order laws for {1, 3} and {1, 4}-generalized inverses. Electron. J. Linear Algebra **23**, 231–234 (2012)
21. Benítez, J.: Moore-Penrose inverses and commuting elements of C*-algebras. J. Math. Anal. Appl. **345**, 766–770 (2008)
22. Xiong, Z., Zheng, B.: The reverse order law for {1, 2, 3}- and {1, 2, 4}-inverses of a two-matrix product. Appl. Math. Lett. **21**, 649–655 (2008)
23. Baksalary, J.K., Baksalary, O.M.: Nonsingularity of linear combinations of idempotent matrices. Linear Algebra Appl. **388**, 25–29 (2004)
24. Liu, X., Wu, S., Cvetković-Ilić, D.S.: New results on reverse order law for {1, 2, 3} and {1, 2, 4}-inverses of bounded operators. Math. Comput. **82**(283), 1597–1607 (2013)
25. Cvetković-Ilić, D.S., Nikolov, J.: Reverse order laws for {1, 2, 3}-generalized inverses. Appl. Math. Comput. **234**, 114–117 (2014)
26. Liu, D., Yang, H.: The reverse order law for {1, 3, 4}- inverse of the product of two matrices. Appl. Math. Comput. **215**, 4293–4303 (2010)
27. Cvetković-Ilić, D.S., Pavlović, V.: A comment on some recent results concerning the reverse order law for {1, 3, 4}-inverses. Appl. Math. Comput. **217**, 105–109 (2010)
28. Cvetković-Ilić, D.S.: Reverse order laws for {1, 3, 4}-generalized inverses in C*-algebras. Appl. Math. Lett. **24**(2), 210–213 (2011)

Chapter 3
Completions of Operator Matrices and Generalized Inverses

In this section we will discuss various problems of completions of operator matrices and present an application of such results to some problems concerning generalized inverses and to that of invertibility of linear combinations of operators. It is worth mentioning that this very intensively studied topic of Operator theory finds large application in the theory of generalized inverses.

Although the reverse order law for {1}-generalized inverses of matrices was completely resolved already by 1998, the corresponding problem for the operators on separable Hilbert spaces was only solved in 2015. Namely, the reverse order law for {1}-generalized inverses for the operators on separable Hilbert spaces was completely solved in the paper of Pavlović et al. [1]. One of the objective of this chapter is to present the approach taken in resolving the reverse order law for {1}-generalized inverses for the operators on separable Hilbert spaces which involves some of the previous research on completions of operator matrices to left and right invertibility.

We will first go over some characteristic results on the problem of completions of operator matrices, with a special emphasis on some instructive examples, and then demonstrate usability of results of that type by showing how they can be applied to one of the topics in generalized inverses of operators that has seen a great interest over the years. Also, we will consider the existence of Drazin invertible completions of an upper triangular operator matrix and applications of results on completions of operator matrices to the problem of invertibility of a linear combination of operators with the special emphasis on the set of projectors and orthogonal projectors.

© Springer Nature Singapore Pte Ltd. 2017
D. Cvetković Ilić and Y. Wei, *Algebraic Properties of Generalized Inverses*,
Developments in Mathematics 52, DOI 10.1007/978-981-10-6349-7_3

3.1 Some Specific Problems of Completions of Operator Matrices

Various aspects of operator matrices and their properties have long motivated researchers in operator theory. Completion of partially given operator matrices to operators of fixed prescribed type is an extensively studied area of operator theory, which is a topic of many various currently undergoing investigations. In this section we will consider only some specific problems from that field which will be usefull later in finding necessary and sufficient conditions for the reverse order law for {1}-generalized inverses for the operators on separable Hilbert spaces to hold. When we talk about completion problems we usually consider the following three types of operator matrices

$$M_C = \begin{bmatrix} A & C \\ 0 & B \end{bmatrix} : \begin{bmatrix} \mathcal{H} \\ \mathcal{K} \end{bmatrix} \to \begin{bmatrix} \mathcal{H} \\ \mathcal{K} \end{bmatrix},$$

$$M_X = \begin{bmatrix} A & C \\ X & B \end{bmatrix} : \begin{bmatrix} \mathcal{H} \\ \mathcal{K} \end{bmatrix} \to \begin{bmatrix} \mathcal{H} \\ \mathcal{K} \end{bmatrix}$$

and

$$M_{(X\ Y)} = \begin{bmatrix} A & B \\ X & Y \end{bmatrix} : \begin{bmatrix} \mathcal{H} \\ \mathcal{K} \end{bmatrix} \to \begin{bmatrix} \mathcal{H} \\ \mathcal{K} \end{bmatrix}$$

and for which the following three questions frequently arise:

Question 1: Is there an operator $C \in \mathcal{B}(\mathcal{Y}, \mathcal{X})$ (resp. X and X, Y) such that M_C (resp. M_X and $M_{(X\ Y)}$) is invertible (right invertible, left invertible, regular...) ?

Question 2: $\bigcap_{C \in \mathcal{B}(\mathcal{Y}, \mathcal{X})} \sigma_*(M_C) =$? where σ_* is any type of spectrum such as the point, continuous, residual, defect, left, right, essential, Weyl spectrum etc.

Question 3: For given operators $A \in \mathcal{B}(\mathcal{X})$ and $B \in \mathcal{B}(\mathcal{Y})$, is there an operator $C' \in \mathcal{B}(\mathcal{Y}, \mathcal{X})$ such that

$$\sigma_*(M_{C'}) = \bigcap_{C \in \mathcal{B}(\mathcal{Y}, \mathcal{X})} \sigma_*(M_C),$$

where again σ_* is any type of spectrum?

In the case of the operator matrix M_C it is clear that $\sigma(M_C) \subseteq \sigma(A) \cup \sigma(B)$. In the following two examples we will show that this inclusion is sometimes actually an equality, but that also it can be a proper one:

Example 3.1 If $\{g_i\}_{i=1}^{\infty}$ is an orthonormal basis of $\mathcal{K}$, define an operator B_0 by

$$\begin{cases} B_0 g_1 = 0, \\ B_0 g_i = g_{i-1}, \ i = 2, 3, \ldots \end{cases}$$

If $\{f_i\}_{i=1}^{\infty}$ is an orthonormal basis of $\mathscr{H}$, define an operator A_0 by $A_0 f_i = f_{i+1}$, $i = 1, 2, \ldots$, and an operator C_0 by $C_0 = (\cdot, g_1) f_1$ from $\mathscr{K}$ into $\mathscr{H}$. Then it is easy to see that $\sigma(A_0) = \sigma(B_0) = \{\lambda : |\lambda| \leq 1\}$. But, in this case, M_{C_0} is a unitary operator, $\sigma(M_{C_0}) \subseteq \{\lambda : |\lambda| = 1\}$, so we have the inclusion $\sigma(M_{C_0}) \subset \sigma(A) \cup \sigma(B)$.

Example 3.2 If $A \in \mathscr{B}(\mathscr{H})$ and $B \in \mathscr{B}(\mathscr{K})$ are normal operators, then for any $C \in \mathscr{B}(\mathscr{K}, \mathscr{H})$, $\sigma(M_C) = \sigma(A) \cup \sigma(B)$ (see Theorem 5 from [2]).

Also, the following example shows that the inclusion $\sigma_{gD}(M_C) \subseteq \sigma_{gD}(A) \cup \sigma_{gD}(B)$ may be strict in the case of the generalized Drazin spectrum:

Example 3.3 Define operators A, B_1, $C_1 \in \mathscr{B}(l_2)$ by

$$A(x_1, x_2, x_3, \ldots) = (0, x_1, x_2, x_3, \ldots),$$
$$B_1(x_1, x_2, x_3, \ldots) = (x_2, x_3, x_4, \ldots),$$
$$C_1(x_1, x_2, x_3, \ldots) = (x_1, 0, 0, \ldots).$$

Consider the operator

$$M_C = \begin{pmatrix} A & C \\ 0 & B \end{pmatrix} : l_2 \oplus (l_2 \oplus l_2) \to l_2 \oplus (l_2 \oplus l_2)$$

$$= \begin{pmatrix} A & C_1 & 0 \\ 0 & B_1 & 0 \\ 0 & 0 & 0 \end{pmatrix} : l_2 \oplus l_2 \oplus l_2 \to l_2 \oplus l_2 \oplus l_2,$$

where $B = \begin{pmatrix} B_1 & 0 \\ 0 & 0 \end{pmatrix} : l_2 \oplus l_2 \to l_2 \oplus l_2$, $C = (C_1, 0) : l_2 \oplus l_2 \to l_2$.
A direct calculation shows that

(i) $\sigma(M_C) = \{\lambda \in \mathbb{C} : |\lambda| = 1\} \cup \{0\}$, $\sigma(A) = \sigma(B) = \{\lambda \in \mathbb{C} : |\lambda| \leq 1\}$;
(ii) $\sigma_{gD}(M_C) = \{\lambda \in \mathbb{C} : |\lambda| = 1\} \cup \{0\}$, $\sigma_{gD}(A) = \sigma_{gD}(B) = \{\lambda \in \mathbb{C} : |\lambda| \leq 1\}$.

On the other hand the inclusion $\sigma_c(M_C) \subseteq \sigma_c(A) \cup \sigma_c(B)$ is not true in general, in the case of the continuous spectrum which will be shown in the next example:

Example 3.4 Let $\mathscr{H} = \mathscr{K} = l_2$. Define the operators A, B, C by

$$A(x_1, x_2, x_3, x_4, \ldots) = (0, 0, x_1, x_2, x_3, x_4, \ldots)$$
$$B(x_1, x_2, x_3, x_4, \ldots) = (x_3, \frac{x_4}{\sqrt{4}}, \frac{x_5}{\sqrt{5}}, \frac{x_6}{\sqrt{6}}, \ldots)$$
$$C(x_1, x_2, x_3, x_4, \ldots) = (x_1, x_2, 0, 0, \ldots),$$

for any $x = (x_n)_{n=1}^{\infty} \in l_2$. Consider $M_C = \begin{pmatrix} A & C \\ 0 & B \end{pmatrix} : l_2 \oplus l_2 \to l_2 \oplus l_2$. A direct calculation shows that $0 \in \sigma_c(M_c)$, but $0 \notin \sigma_C(A) \cup \sigma_C(B)$ which implies $\sigma_C(M_C) \nsubseteq \sigma_C(A) \cup \sigma_C(B)$.

Given operators $A \in \mathscr{B}(\mathscr{X})$ and $B \in \mathscr{B}(\mathscr{Y})$, the question of existence of an operator $C \in \mathscr{B}(\mathscr{K}, \mathscr{H})$ such that the operator matrix M_C is invertible was considered for the first time in [2] in the case when $\mathscr{X}$ and $\mathscr{Y}$ are separable Hilbert spaces. The results from [2] are generalized in [3] in the case of Banach spaces. In [4], the same problem is considered in the case of Banach spaces and the set of all $C \in \mathscr{B}(\mathscr{Y}, \mathscr{X})$ for which M_C is invertible is completely described and additionally the set of all $C \in \mathscr{B}(\mathscr{Y}, \mathscr{X})$ such that M_C is invertible, denoted by $S(A, B)$, is completely described (in the case when $\mathscr{X}$ and $\mathscr{Y}$ are Banach spaces).

Theorem 3.1 ([4]) *Let $A \in \mathscr{B}(\mathscr{X})$ and $B \in \mathscr{B}(\mathscr{Y})$ be given operators. The operator matrix M_C is invertible for some $C \in \mathscr{B}(\mathscr{Y}, \mathscr{X})$ if and only if*

(i) *A is left invertible,*
(ii) *B is right invertible,*
(iii) *$\mathscr{N}(B) \cong \mathscr{X}/\mathscr{R}(A)$.*

If conditions (i)−(iii) are satisfied, the set of all $C \in \mathscr{B}(\mathscr{Y}, \mathscr{X})$ such that M_C is invertible is given by

$$S(A, B) = \{C \in \mathscr{B}(\mathscr{Y}, \mathscr{X}) : C = \begin{bmatrix} C_1 & 0 \\ 0 & C_4 \end{bmatrix} : \begin{bmatrix} \mathscr{P} \\ \mathscr{N}(B) \end{bmatrix} \rightarrow \begin{bmatrix} \mathscr{R}(A) \\ \mathscr{S} \end{bmatrix},$$
$$C_4 \text{ is invertible}, \ \mathscr{X} = \mathscr{R}(A) \oplus \mathscr{S} \text{ and } \mathscr{Y} = \mathscr{P} \oplus \mathscr{N}(B)\}. \quad (3.1)$$

In Remark 2.5 in [4], it is proved that if we take arbitrary but fixed decompositions of $\mathscr{X}$ and $\mathscr{Y}$, $\mathscr{X} = \mathscr{R}(A) \oplus \mathscr{S}$ and $\mathscr{Y} = \mathscr{P} \oplus \mathscr{N}(B)$, then

$$S(A, B) = \{C \in \mathscr{B}(\mathscr{Y}, \mathscr{X}) : C = \begin{bmatrix} C_1 & C_2 \\ C_3 & C_4 \end{bmatrix} : \begin{bmatrix} \mathscr{P} \\ \mathscr{N}(B) \end{bmatrix} \rightarrow \begin{bmatrix} \mathscr{R}(A) \\ \mathscr{S} \end{bmatrix},$$
$$C_4 \text{ is invertible}\}. \quad (3.2)$$

Based on the above results and using the fact that the invertibility of $C_4 \in \mathscr{B}(\mathscr{N}(B), \mathscr{S})$ simply means that $P_{\mathscr{S},\mathscr{R}(A)}C|_{\mathscr{N}(B)}$ is an injective operator with range $\mathscr{S}$, in the case of separable Hilbert spaces we have the following characterization of invertibility of an upper triangular operator matrix:

Theorem 3.2 *Let $\mathscr{H}$ and $\mathscr{K}$ be separable Hilbert spaces and let $A \in \mathscr{B}(\mathscr{H})$, $B \in \mathscr{B}(\mathscr{K})$ and $C \in \mathscr{B}(\mathscr{K}, \mathscr{H})$ be given operators. The operator matrix $\begin{bmatrix} A & C \\ 0 & B \end{bmatrix}$ is invertible if and only if A is left invertible, B is right invertible and $P_{\mathscr{S},\mathscr{R}(A)}C|_{\mathscr{N}(B)}$ is an injective operator with (closed) range $\mathscr{S}$, where $\mathscr{H} = \mathscr{R}(A) \oplus \mathscr{S}$.*

Aside from the existence of invertible completions of the aforementioned operator matrix, the problems of existence of completions of the operator matrix M_C that are Fredholm, semi-Fredholm, Kato, Browder etc. have subsequently been studied in literature. In Sect. 3.4, we will consider such problem in the case of Drazin invertible completions.

Moving on, in [5] the problem is considered of completing an operator matrix

$$M_{(X,Y)} = \begin{bmatrix} A & C \\ X & Y \end{bmatrix} : \begin{bmatrix} \mathcal{H}_1 \\ \mathcal{H}_2 \end{bmatrix} \to \begin{bmatrix} \mathcal{H}_1 \\ \mathcal{H}_2 \end{bmatrix} \tag{3.3}$$

to left (right) invertibility in the case when $A \in \mathcal{B}(\mathcal{H}_1)$ and $C \in \mathcal{B}(\mathcal{H}_2, \mathcal{H}_1)$ are given and $\mathcal{H}_1$, $\mathcal{H}_2$ are separable Hilbert spaces.

Theorem 3.3 ([5]) *Let $M_{(X,Y)}$ be given by (3.3).*

(i) *If* $\dim\mathcal{H}_2 = \infty$, *then there exist* $X \in \mathcal{B}(\mathcal{H}_1, \mathcal{H}_2)$ *and* $Y \in \mathcal{B}(\mathcal{H}_2)$ *such that* $M_{(X,Y)}$ *is left invertible.*

(ii) *If* $\dim\mathcal{H}_2 < \infty$, *then* $M_{(X,Y)}$ *is left invertible for some operators* $X \in \mathcal{B}(\mathcal{H}_1, \mathcal{H}_2)$ *and* $Y \in \mathcal{B}(\mathcal{H}_2)$ *if and only if* $\dim\mathcal{N}\left(\begin{bmatrix} A & C \end{bmatrix}\right) \leq \dim\mathcal{H}_2$ *and* $\mathcal{R}(A)$ *is closed.*

Here, we will present a result of this type in the case when

$$M_{(X,Y)} = \begin{bmatrix} A & C \\ X & Y \end{bmatrix} : \begin{bmatrix} \mathcal{H}_1 \\ \mathcal{H}_2 \end{bmatrix} \to \begin{bmatrix} \mathcal{H}_3 \\ \mathcal{H}_4 \end{bmatrix} \tag{3.4}$$

and $\mathcal{H}_i$, $i = \overline{1,4}$ are separable Hilbert spaces. So, we will give a modification of Theorem 3.3 for the operator matrix $M_{(X,Y)}$ given by (3.4).

Theorem 3.4 *Let $M_{(X,Y)}$ be given by (3.4).*

(i) *If* $\dim\mathcal{H}_4 = \infty$, *then there exist* $X \in \mathcal{B}(\mathcal{H}_1, \mathcal{H}_4)$ *and* $Y \in \mathcal{B}(\mathcal{H}_2, \mathcal{H}_4)$ *such that* $M_{(X,Y)}$ *is left invertible.*

(ii) *If* $\dim\mathcal{H}_4 < \infty$, *then* $M_{(X,Y)}$ *is left invertible for some operators* $X \in \mathcal{B}(\mathcal{H}_1, \mathcal{H}_4)$ *and* $Y \in \mathcal{B}(\mathcal{H}_2, \mathcal{H}_4)$ *if and only if* $\dim\mathcal{N}\left(\begin{bmatrix} A & C \end{bmatrix}\right) \leq \dim\mathcal{H}_4$ *and* $\mathcal{R}(A) + \mathcal{R}(C)$ *is closed.*

Proof (i) If $\dim\mathcal{H}_4 = \infty$, then there exists a closed infinite dimensional subspace of $\mathcal{H}_4$, $\mathcal{M}$ such that $\dim\mathcal{M}^\perp = \dim\mathcal{H}_1$. Now, there exist left invertible operators $J_1 : \mathcal{H}_1 \to \mathcal{M}^\perp$ and $J_2 : \mathcal{H}_2 \to \mathcal{M}$. Let

$$X = \begin{bmatrix} 0 \\ J_1 \end{bmatrix} : \mathcal{H}_1 \to \begin{bmatrix} \mathcal{M} \\ \mathcal{M}^\perp \end{bmatrix}, \quad Y = \begin{bmatrix} J_2 \\ 0 \end{bmatrix} : \mathcal{H}_2 \to \begin{bmatrix} \mathcal{M} \\ \mathcal{M}^\perp \end{bmatrix}.$$

Let

$$X^- = \begin{bmatrix} 0 & (J_1)_l^{-1} \end{bmatrix} : \begin{bmatrix} \mathcal{M} \\ \mathcal{M}^\perp \end{bmatrix} \to \mathcal{H}_1, \, Y^- = \begin{bmatrix} (J_2)_l^{-1} & 0 \end{bmatrix} : \begin{bmatrix} \mathcal{M} \\ \mathcal{M}^\perp \end{bmatrix} \to \mathcal{H}_2.$$

Now,

$$\begin{bmatrix} 0 & X^- \\ 0 & Y^- \end{bmatrix} \begin{bmatrix} A & C \\ X & Y \end{bmatrix} = \begin{bmatrix} I & 0 \\ 0 & I \end{bmatrix},$$

i.e., $M_{(X,Y)}$ is left-invertible.

(ii) Suppose that $\dim \mathcal{H}_4 < \infty$. If there exist regular $X \in \mathcal{B}(\mathcal{H}_1, \mathcal{H}_4)$ and $Y \in \mathcal{B}(\mathcal{H}_2, \mathcal{H}_4)$ such that $M_{(X,Y)}$ is left invertible, from the fact that $\mathcal{R}(M_{(X,Y)})$ is closed we get that

$$\mathcal{R}\left(\begin{bmatrix} A^* & X^* \\ C^* & Y^* \end{bmatrix}\right) = \mathcal{R}\left(\begin{bmatrix} A^* & 0 \\ C^* & 0 \end{bmatrix}\right) + \mathcal{R}\left(\begin{bmatrix} 0 & X^* \\ 0 & Y^* \end{bmatrix}\right)$$

is closed. It follows that $\mathcal{R}\left(\begin{bmatrix} A & C \\ 0 & 0 \end{bmatrix}\right)$ is closed, i.e., $\mathcal{R}(A) + \mathcal{R}(C)$ is closed since $\mathcal{R}\left(\begin{bmatrix} 0 & X^* \\ 0 & Y^* \end{bmatrix}\right)$ is a finite dimensional subspace. From the injectivity of $M_{(X,Y)}$, it follows that $\mathcal{N}\left(\begin{bmatrix} A & C \end{bmatrix}\right) \cap \mathcal{N}\left(\begin{bmatrix} X & Y \end{bmatrix}\right) = \{0\}$ which implies that

$$\dim \mathcal{N}\left(\begin{bmatrix} A & C \end{bmatrix}\right) \leq \dim \mathcal{N}\left(\begin{bmatrix} X & Y \end{bmatrix}\right)^\perp \leq \dim \mathcal{H}_4.$$

For the converse, suppose that $\dim \mathcal{N}\left(\begin{bmatrix} A & C \end{bmatrix}\right) \leq \dim \mathcal{H}_4$ and $\mathcal{R}(A) + \mathcal{R}(C)$ is closed. Since $\mathcal{N}\left(\begin{bmatrix} A & C \end{bmatrix}\right) = \mathcal{K}_1 \oplus \mathcal{K}_2 \oplus \mathcal{K}_3$, where $\mathcal{K}_1 = \left\{ \begin{bmatrix} x \\ 0 \end{bmatrix} : x \in \mathcal{N}(A) \right\}$, $\mathcal{K}_2 = \left\{ \begin{bmatrix} 0 \\ y \end{bmatrix} : y \in \mathcal{N}(C) \right\}$ and $\mathcal{K}_3 = \left\{ \begin{bmatrix} x \\ y \end{bmatrix} : x \in \mathcal{N}(A)^\perp, y \in \mathcal{N}(C)^\perp, Ax + Cy = 0 \right\}$, there exists a subspace $\mathcal{M}$ of $\mathcal{H}_4$ such that $\dim \mathcal{M} = \dim \mathcal{K}_1$. Then $\dim \mathcal{M}^\perp \geq \dim \mathcal{K}_2 + \dim \mathcal{K}_3$. Now, there exist left invertible operators $J_1 : \mathcal{N}(A) \to \mathcal{M}$ and $J_2 : P_{\mathcal{H}_2} \mathcal{N}\left(\begin{bmatrix} A & C \end{bmatrix}\right) \to \mathcal{M}^\perp$. Let

$$X = \begin{bmatrix} J_1 & 0 \\ 0 & 0 \end{bmatrix} : \begin{bmatrix} \mathcal{N}(A) \\ \mathcal{N}(A)^\perp \end{bmatrix} \to \begin{bmatrix} \mathcal{M} \\ \mathcal{M}^\perp \end{bmatrix}$$

and

$$Y = \begin{bmatrix} 0 & 0 \\ 0 & J_2 \end{bmatrix} : \begin{bmatrix} \left(P_{\mathcal{H}_2} \mathcal{N}\left(\begin{bmatrix} A & C \end{bmatrix}\right)\right)^\perp \\ P_{\mathcal{H}_2} \mathcal{N}\left(\begin{bmatrix} A & C \end{bmatrix}\right) \end{bmatrix} \to \begin{bmatrix} \mathcal{M} \\ \mathcal{M}^\perp \end{bmatrix}.$$

Now, as in Theorem 2.1 [5], we can check that $M_{(X,Y)}$ is left-invertible, i.e., we will prove that $\mathcal{R}(M_{(X,Y)})$ is closed and $M_{(X,Y)}$ is injective. From the fact that $\dim \mathcal{H}_4 < \infty$ and Kato's lemma we have that $\mathcal{R}(M_{(X,Y)})$ is closed. On the other hand, let

$$\begin{bmatrix} A & C \\ X & Y \end{bmatrix} \begin{bmatrix} x \\ y \end{bmatrix} = \begin{bmatrix} 0 \\ 0 \end{bmatrix},$$

which is equivalent to

$$Ax + Cy = 0, \quad Xx + Yy = 0.$$

Then it follows that $y \in P_{\mathcal{H}_2} \mathcal{N}([A \; C])$. Also, we have that $Xx = Yy = 0$ which implies that $y = 0$. Thus, $Ax = 0$ which by definition of X implies that $x = 0$. This proves that $M_{(X,Y)}$ is injective. $\qquad \square$

As for completions of an operator matrix

$$M_X = \begin{bmatrix} A & C \\ X & B \end{bmatrix} : \begin{bmatrix} \mathcal{H}_1 \\ \mathcal{H}_2 \end{bmatrix} \rightarrow \begin{bmatrix} \mathcal{H}_1 \\ \mathcal{H}_2 \end{bmatrix}, \tag{3.5}$$

where $A \in \mathcal{B}(\mathcal{H}_1)$, $B \in \mathcal{B}(\mathcal{H}_2)$ and $C \in \mathcal{B}(\mathcal{H}_2, \mathcal{H}_1)$ are given, the first to ever address any kind of questions (for separable Hilbert spaces not necessarily of finite dimension) related to it was Takahashi. More specifically, in his paper [6] he gave necessary and sufficient conditions for the existence of $X \in \mathcal{B}(\mathcal{H}_1)$ such that M_X is invertible.

Although Takahashi's paper was published in 1995, there have only been several papers since, namely [7, 9–14], which deal with various completions of the operator matrix of the form M_X. Actually in [13] exactly the same problem was considered as in [6] but using methods of geometrical structure of operators and in it some necessary and sufficient conditions were given different than those from [6]. In [9] the authors considered the problem of completions of M_X given by (3.5) to right (left) invertibility in the case when $A \in \mathcal{B}(\mathcal{H}_1)$, $B \in \mathcal{B}(\mathcal{H}_2)$ and $C \in \mathcal{B}(\mathcal{H}_2, \mathcal{H}_1)$ are given.

Theorem 3.5 ([9]) *Let* $A \in \mathcal{B}(\mathcal{H}_1)$, $B \in \mathcal{B}(\mathcal{H}_2)$ *and* $C \in \mathcal{B}(\mathcal{H}_2, \mathcal{H}_1)$ *be given. Then* M_X *is right invertible for some* $X \in \mathcal{B}(\mathcal{H}_1, \mathcal{H}_2)$ *if and only if* $\mathcal{R}(A) + \mathcal{R}(C) = \mathcal{H}_1$ *and one of the following conditions holds:*

(1) $\mathcal{N}(A \mid C; \mathcal{H}_2)$ *contains a non-compact operator,*

(2) $M_0 = \begin{bmatrix} A & C \\ 0 & B \end{bmatrix}$ *is a right semi-Fredholm operator and*

$$d(M_0) \leq n(A) + \dim\left(\mathcal{R}(A) \cap \mathcal{R}(C|_{\mathcal{N}(B)})\right),$$

where $\mathcal{N}(A \mid C; \mathcal{H}_2) = \{G \in \mathcal{B}(\mathcal{H}_2, \mathcal{H}_1) : \mathcal{R}(AG) \subseteq \mathcal{R}(C)\}$.
Here we will present a result of this type in the case when

$$M_X = \begin{bmatrix} A & C \\ X & B \end{bmatrix} : \begin{bmatrix} \mathcal{H}_1 \\ \mathcal{H}_2 \end{bmatrix} \rightarrow \begin{bmatrix} \mathcal{H}_3 \\ \mathcal{H}_4 \end{bmatrix} \tag{3.6}$$

and give a modification of Theorem 3.5 which shortens significantly one implication of the original one. Since for the proof we need some auxiliary results, we begin by stating these.

Lemma 3.1 ([15]) *If* $\mathcal{H}$ *is an infinite dimensional Hilbert space, then* $T \in \mathcal{B}(\mathcal{H})$ *is compact if and only if* $\mathcal{R}(T)$ *contains no closed infinite dimensional subspaces.*

Lemma 3.2 *Let $\mathcal{H}_1$ and $\mathcal{H}_2$ be separable Hilbert spaces. If $U \subseteq \mathcal{H}_1$ and $V \subseteq \mathcal{H}_2$ are closed subspaces with $\dim U = \dim V$, then there exists $T \in \mathcal{B}(\mathcal{H}_1, \mathcal{H}_2)$ such that $\mathcal{N}(T) = U^{\perp}$, $\mathcal{R}(T) = V$ and $T|_U : U \to V$ is unitary. In particular, if $U = \mathcal{H}_1$, then T is left invertible; if $V = \mathcal{H}_2$, then T is right invertible.*

Lemma 3.3 ([9]) *Let $S \in \mathcal{B}(\mathcal{H}_1, \mathcal{H}_2)$, and let T be a closed linear operator from $\mathcal{H}_2$ into $\mathcal{H}_3$. If $\mathcal{R}(S) \subseteq \mathcal{D}(T)$, then $TS \in \mathcal{B}(\mathcal{H}_1, \mathcal{H}_3)$.*

For Hilbert spaces $\mathcal{H}_i$, $i = \overline{1,4}$ and operators $A \in \mathcal{B}(\mathcal{H}_1, \mathcal{H}_3)$ and $C \in \mathcal{B}(\mathcal{H}_2, \mathcal{H}_3)$, let

$$\mathcal{N}(A \mid C; \mathcal{H}_4) = \{G \in \mathcal{B}(\mathcal{H}_4, \mathcal{H}_1) : \mathcal{R}(AG) \subseteq \mathcal{R}(C)\}.$$

It is well known that $G \in \mathcal{B}(\mathcal{H}_4, \mathcal{H}_1)$ belongs to $\mathcal{N}(A \mid C; \mathcal{H}_4)$ if and only if there exists $H \in \mathcal{B}(\mathcal{H}_4, \mathcal{H}_2)$ such that $AG = CH$.

Lemma 3.4 ([9]) *Let $A \in \mathcal{B}(\mathcal{H}_1, \mathcal{H}_3)$, $B \in \mathcal{B}(\mathcal{H}_2, \mathcal{H}_4)$ and $C \in \mathcal{B}(\mathcal{H}_2, \mathcal{H}_3)$ be given operators. Assume that*

$$M_0 = M(A, B, C; 0) = \begin{bmatrix} A & C \\ 0 & B \end{bmatrix}$$

is a right Fredholm operator on $\mathcal{H}_1 \oplus \mathcal{H}_2$. Then B is a right Fredholm operator, $\mathcal{R}(A) + \mathcal{R}(C|_{\mathcal{N}(B)})$ is a closed subspace, and

$$d(M_0) = \dim(\mathcal{R}(A) + \mathcal{R}(C|_{\mathcal{N}(B)}))^{\perp} + d(B),$$
$$n(M_0) = n(A) + n(C|_{\mathcal{N}(B)}) + \dim(\mathcal{R}(A) \cap \mathcal{R}(C|_{\mathcal{N}(B)})).$$

Finally, we will give a a modification of Theorem 3.5:

Theorem 3.6 *Let $A \in \mathcal{B}(\mathcal{H}_1, \mathcal{H}_3)$, $B \in \mathcal{B}(\mathcal{H}_2, \mathcal{H}_4)$ and $C \in \mathcal{B}(\mathcal{H}_2, \mathcal{H}_3)$ be given. Then M_X is right invertible for some $X \in \mathcal{B}(\mathcal{H}_1, \mathcal{H}_4)$ if and only if $\mathcal{R}(A) + \mathcal{R}(C) = \mathcal{H}_3$ and one of the following conditions holds:*

(1) *$\mathcal{N}(A \mid C; \mathcal{H}_4)$ contains a non-compact operator,*

(2) *$M_0 = \begin{bmatrix} A & C \\ 0 & B \end{bmatrix}$ is a right semi-Fredholm operator and*

$$d(M_0) \leq n(A) + \dim\left(\mathcal{R}(A) \cap \mathcal{R}(C|_{\mathcal{N}(B)})\right).$$

Proof Suppose M_X given by (3.6) is right invertible for some $X \in \mathcal{B}(\mathcal{H}_1, \mathcal{H}_4)$. This implies that $\begin{bmatrix} A & C \end{bmatrix}$ is right invertible and so $\mathcal{R}(A) + \mathcal{R}(C) = \mathcal{H}_3$. Let $\mathcal{H}_2' = (\mathcal{N}(C) \cap \mathcal{N}(B))^{\perp}$. Then

$$M_X = \begin{bmatrix} A & 0 & C' \\ X & 0 & B' \end{bmatrix} : \begin{bmatrix} \mathcal{H}_1 \\ (\mathcal{H}_2')^{\perp} \\ \mathcal{H}_2' \end{bmatrix} \to \begin{bmatrix} \mathcal{H}_3 \\ \mathcal{H}_4 \end{bmatrix},$$

where $\mathcal{N}(C') \cap \mathcal{N}(B') = \{0\}$. Clearly

$$M'_X = \begin{bmatrix} A & C' \\ X & B' \end{bmatrix} : \begin{bmatrix} \mathcal{H}_1 \\ \mathcal{H}'_2 \end{bmatrix} \to \begin{bmatrix} \mathcal{H}_3 \\ \mathcal{H}_4 \end{bmatrix}$$

is right invertible. Thus there is a bounded linear operator

$$\begin{bmatrix} E & G \\ F & H \end{bmatrix} : \begin{bmatrix} \mathcal{H}_3 \\ \mathcal{H}_4 \end{bmatrix} \to \begin{bmatrix} \mathcal{H}_1 \\ \mathcal{H}'_2 \end{bmatrix}$$

such that

$$\begin{bmatrix} A & C' \\ X & B' \end{bmatrix} \begin{bmatrix} E & G \\ F & H \end{bmatrix} = \begin{bmatrix} I_{\mathcal{H}_3} & 0 \\ 0 & I_{\mathcal{H}_4} \end{bmatrix}.$$

From $AG + C'H = 0$ it follows that $\mathcal{R}(AG) \subseteq \mathcal{R}(C') = \mathcal{R}(C)$ so, if G is a non-compact operator then (1) holds. If on the other hand G is compact, then from $XG + B'H = I$, we see that $B'H$ is a Fredholm operator and $d(B'H) = n(B'H)$. Since

$$\begin{bmatrix} I & 0 \\ -B'F & I \end{bmatrix} \begin{bmatrix} A & C' \\ 0 & B' \end{bmatrix} \begin{bmatrix} E & G \\ F & H \end{bmatrix} = \begin{bmatrix} I_{\mathcal{H}_3} & 0 \\ 0 & B'H \end{bmatrix}$$

and $B'H$ is a Fredholm operator, it follows that $M'_0 = \begin{bmatrix} A & C' \\ 0 & B' \end{bmatrix}$ is a right Fredholm operator. As $\mathcal{R}(M_0) = \mathcal{R}(M'_0)$ the operator M_0 is right Fredholm. Also

$$d(M_0) = d(M'_0) \le d\left(\begin{bmatrix} I_{\mathcal{H}_3} & 0 \\ 0 & B'H \end{bmatrix}\right) = d(B'H) = n(B'H)$$

$$\le n(M'_0) = n(A) + n\left(C'|_{\mathcal{N}(B')}\right) + \dim\left(\mathcal{R}\left(C'|_{\mathcal{N}(B')}\right) \cap \mathcal{R}(A)\right)$$

$$= n(A) + \dim\left(\mathcal{R}(A) \cap \mathcal{R}(C|_{\mathcal{N}(B)})\right).$$

For the converse implication: If $\mathcal{N}(A \mid C; \mathcal{H}_4)$ contains a non-compact operator, then $\mathcal{H}_1$ and $\mathcal{H}_4$ are infinite dimensional. By Lemma 3.1, there exists a closed subspace $\mathcal{M} \subseteq \mathcal{H}_1$ with $\dim \mathcal{M} = \dim \mathcal{H}_4 = \infty$ such that $\mathcal{R}(A|_{\mathcal{M}}) \subseteq \mathcal{R}(C)$, and hence $\mathcal{R}(AP_{\mathcal{M}}) = \mathcal{R}(A|_{\mathcal{M}}) \subseteq \mathcal{R}(C) \subseteq$, where $C^+ : \mathcal{R}(C) \oplus \mathcal{R}(C)^\perp \to \mathcal{H}_2$ is defined to be 0 on $\mathcal{R}(C)^\perp$ and $(C|_{\mathcal{N}(C)^\perp})^{-1}$ on $\mathcal{R}(C)$. This, together with $AP_{\mathcal{M}} \in \mathcal{B}(\mathcal{H}_1, \mathcal{H}_3)$ and Lemma 3.3, shows that $C^+AP_{\mathcal{M}} \in \mathcal{B}(\mathcal{H}_2, \mathcal{H}_3)$. On the other hand, it follows from Lemma 3.2 that there exists a right invertible operator $T \in \mathcal{B}(\mathcal{H}_1, \mathcal{H}_4)$ such that $\mathcal{N}(T) = \mathcal{M}^\perp$. Define an operator $X \in \mathcal{B}(\mathcal{H}_1, \mathcal{H}_4)$ by

$$X = T + BC^+AP_{\mathcal{M}}.$$

Then M_X is a right invertible operator. To prove that let $u \in \mathcal{H}_3$ and $v \in \mathcal{H}_4$ be arbitrary. Since $\mathcal{R}(A) + \mathcal{R}(C) = \mathcal{H}_3$ and $\mathcal{R}(A|_{\mathcal{M}}) \subseteq \mathcal{R}(C)$, there exist $x_1 \in \mathcal{M}^\perp$

and $y_1 \in \mathcal{H}_2$ such that $Ax_1 + Cy_1 = u$. Also, by right invertibility of T, there exists $x_2 \in \mathcal{M}$ such that $Tx_2 = v - By_1$. Let $x_0 = x_1 + x_2$ and $y_0 = y_1 - C^+Ax_2$. Then

$$\begin{bmatrix} A & C \\ X & B \end{bmatrix} \begin{bmatrix} x_0 \\ y_0 \end{bmatrix} = \begin{bmatrix} u \\ v \end{bmatrix}.$$

This establishes right invertibility of M_X.

If (2) holds, put $E = \mathcal{R}(A) + \mathcal{R}(C|_{\mathcal{N}(B)})$. From Lemma 3.4 and the right Fredholmness of M_0 we can infer that B is a right Fredholm operator, E is closed and $\dim E^{\perp} = d(M_0) - d(B) < \infty$. From $\mathcal{R}(A) + \mathcal{R}(C) = \mathcal{H}_3$ it follows that $\mathcal{R}(P_{E^{\perp}}C) = E^{\perp}$. Let $G = (P_{E^{\perp}}C)^+ E^{\perp}$ and $S = BG \oplus \mathcal{R}(B)^{\perp}$. Then clearly $G \subseteq \mathcal{N}(B)^{\perp}$ and so $\dim E^{\perp} = \dim G = \dim BG$. Therefore $\dim S = d(M_0)$. On the other hand, since $d(M_0) \leq n(A) + \dim(\mathcal{R}(A) \cap \mathcal{R}(C|_{\mathcal{N}(B)}))$, there exists a subspace $\mathcal{M} \subseteq \mathcal{H}_1$ with $\dim \mathcal{M} = d(M_0)$ such that $\mathcal{R}(A|_{\mathcal{M}}) \subseteq \mathcal{R}(C|_{\mathcal{N}(B)})$. From $\dim \mathcal{M} = \dim S = d(M_0) < \infty$ and Lemma 3.2, there exists an operator $J : \mathcal{H}_1 \to S$ such that $\mathcal{N}(J) = \mathcal{M}^{\perp}$ and $J|_{\mathcal{M}} : \mathcal{M} \to S$ is unitary. Define $X \in \mathcal{B}(\mathcal{H}_1, \mathcal{H}_2)$ by

$$X = \begin{bmatrix} J \\ 0 \end{bmatrix} : \mathcal{H}_1 \to S \oplus S^{\perp}.$$

Then M_X as an operator from $\mathcal{H}_1 \oplus \mathcal{N}(B) \oplus G \oplus (\mathcal{N}(B)^{\perp} \ominus G)$ into $E \oplus E^{\perp} \oplus S \oplus S^{\perp}$ has the following operator matrix:

$$M_X = \begin{bmatrix} A_1 & C_1 & C_2 & C_3 \\ 0 & 0 & C_4 & 0 \\ J & 0 & B_1 & B_3 \\ 0 & 0 & 0 & B_2 \end{bmatrix},$$

where $\mathcal{N}(B)^{\perp} \ominus G = \{y \in \mathcal{N}(B)^{\perp} : y \in G^{\perp}\}$. Obviously, C_4 is invertible. From the right Fredholmness of B we can infer that B_2 is invertible. Thus there is an invertible operator $U \in \mathcal{B}(\mathcal{H}_1, \mathcal{H}_2)$ such that

$$UM_X = U \begin{bmatrix} A_1 & C_1 & C_2 & C_3 \\ 0 & 0 & C_4 & 0 \\ J & 0 & B_1 & B_3 \\ 0 & 0 & 0 & B_2 \end{bmatrix} = \begin{bmatrix} A_1 & C_1 & 0 & 0 \\ 0 & 0 & C_4 & 0 \\ J & 0 & 0 & 0 \\ 0 & 0 & 0 & B_2 \end{bmatrix}.$$

It follows that M_X is a right invertible operator if and only if

$$\begin{bmatrix} A_1 & C_1 \\ J & 0 \end{bmatrix} : \mathcal{H}_1 \oplus \mathcal{N}(B) \to E \oplus S,$$

is a right invertible operator.

For any $u \in E$ and $v \in S$, it follows from $E = \mathcal{R}(A) + \mathcal{R}(C|_{\mathcal{N}(B)})$, $\mathcal{R}(A|_{\mathcal{M}}) \subseteq \mathcal{R}(C|_{\mathcal{N}(B)})$ and the definition of J that there exist $x_1 \in \mathcal{M}$, $x_2 \in \mathcal{M}^{\perp}$ and $y_1 \in$

$\mathcal{N}(B)$ such that

$$J x_1 = v, \quad A x_2 + C y_1 = u.$$

Since $\mathcal{R}(A|_{\mathcal{M}}) \subseteq \mathcal{R}(C|_{\mathcal{N}(B)})$, there exists $y_2 \in \mathcal{N}(B)$ with

$$A x_1 + C y_2 = 0.$$

Note that $A_1 = A : \mathcal{H}_1 \to E$, $C_1 = C|_{\mathcal{N}(B)} : \mathcal{N}(B) \to E$ and $\mathcal{N}(J) = \mathcal{M}^\perp$, and hence

$$\begin{bmatrix} A_1 & C_1 \\ J & 0 \end{bmatrix} \begin{bmatrix} x_1 + x_2 \\ y_1 + y_2 \end{bmatrix} = \begin{bmatrix} u \\ v \end{bmatrix}.$$

From the argument above we get that M_X is a right invertible operator. $\qquad\square$

Remark 3.1 The condition (1) from the previous theorem is equivalent to the existence of a closed infinite dimensional subspace $\mathcal{M}$ of $\mathcal{H}_1$ such that $\mathcal{R}(A|_{\mathcal{M}}) \subseteq \mathcal{R}(C)$.

As a corollary of Theorem 3.6 we have the following result concerning completions to left invertibility, that parallels Theorem 2.7 [9].

Corollary 3.1 *Let* $A \in \mathcal{B}(\mathcal{H}_1, \mathcal{H}_3)$, $B \in \mathcal{B}(\mathcal{H}_2, \mathcal{H}_4)$ *and* $C \in \mathcal{B}(\mathcal{H}_2, \mathcal{H}_3)$ *be given. Then* M_X *is left invertible for some* $X \in \mathcal{B}(\mathcal{H}_1, \mathcal{H}_4)$ *if and only if* $\mathcal{R}(B^*) + \mathcal{R}(C^*) = \mathcal{H}_2$ *and one of the following conditions holds:*

(1) $\mathcal{N}(B^* \mid C^*; \mathcal{H}_1)$ *contains a non-compact operator,*

(2) $M_0 = \begin{bmatrix} A & C \\ 0 & B \end{bmatrix}$ *is a left semi-Fredholm operator and*

$$n(M_0) \leq d(B) + \dim \left(\mathcal{R}(B^*) \cap \mathcal{R}\left(C^*|_{\mathcal{N}(A^*)}\right) \right).$$

3.2 Applications of Completions of Operator Matrices to Reverse Order Law for {1}-Inverses of Operators on Hilbert Spaces

The reverse order law problem for {1}-inverses for operators acting on separable Hilbert spaces was completely resolved in the paper [1] and this was done using a radically new approach than in the recent papers on this subject, one that involves some of the previous research on completions of operator matrices to left and right invertibility. More exactly, the solution of this problem relies heavily on the results on completions of operator matrices presented in Sect. 3.1, so that the results of the present section can in a way be regarded as an interesting application of the research related to the topic of completions of operator matrices.

First, we will need the following observations.

Let $A \in \mathcal{B}(\mathcal{H}, \mathcal{K})$ and $B \in \mathcal{B}(\mathcal{L}, \mathcal{H})$ be arbitrary regular operators. Using the following decompositions of the spaces $\mathcal{H}$, $\mathcal{K}$ and $\mathcal{L}$,

$$\mathcal{L} = \mathcal{R}(B^*) \oplus \mathcal{N}(B), \quad \mathcal{H} = \mathcal{R}(B) \oplus \mathcal{N}(B^*), \quad \mathcal{K} = \mathcal{R}(A) \oplus \mathcal{N}(A^*),$$

we have that the corresponding representations of operators A and B are given by

$$
\begin{aligned}
A &= \begin{bmatrix} A_1 & A_2 \\ 0 & 0 \end{bmatrix} : \begin{bmatrix} \mathcal{R}(B) \\ \mathcal{N}(B^*) \end{bmatrix} \to \begin{bmatrix} \mathcal{R}(A) \\ \mathcal{N}(A^*) \end{bmatrix}, \\
B &= \begin{bmatrix} B_1 & 0 \\ 0 & 0 \end{bmatrix} : \begin{bmatrix} \mathcal{R}(B^*) \\ \mathcal{N}(B) \end{bmatrix} \to \begin{bmatrix} \mathcal{R}(B) \\ \mathcal{N}(B^*) \end{bmatrix},
\end{aligned}
\tag{3.7}
$$

where B_1 is invertible and $\begin{bmatrix} A_1 & A_2 \end{bmatrix} : \begin{bmatrix} \mathcal{R}(B^*) \\ \mathcal{N}(B) \end{bmatrix} \to \mathcal{R}(A)$ is right invertible. In that case the operator AB is given by

$$AB = \begin{bmatrix} A_1 B_1 & 0 \\ 0 & 0 \end{bmatrix} : \begin{bmatrix} \mathcal{R}(B^*) \\ \mathcal{N}(B) \end{bmatrix} \to \begin{bmatrix} \mathcal{R}(A) \\ \mathcal{N}(A^*) \end{bmatrix}. \tag{3.8}$$

The following lemma gives a description of all the $\{1\}$-inverses of A, B and AB in terms of their representations corresponding to appropriate decompositions of spaces.

Lemma 3.5 *Let $A \in \mathcal{B}(\mathcal{H}, \mathcal{K})$ and $B \in \mathcal{B}(\mathcal{L}, \mathcal{H})$ be regular operators given by (3.7). Then*

(i) *an arbitrary $\{1\}$-inverse of A is given by:*

$$A^{(1)} = \begin{bmatrix} X_1 & X_2 \\ X_3 & X_4 \end{bmatrix} : \begin{bmatrix} \mathcal{R}(A) \\ \mathcal{N}(A^*) \end{bmatrix} \to \begin{bmatrix} \mathcal{R}(B) \\ \mathcal{N}(B^*) \end{bmatrix}, \tag{3.9}$$

where X_1 and X_3 satisfy the following equality

$$A_1 X_1 + A_2 X_3 = I_{\mathcal{R}(A)}, \tag{3.10}$$

and X_2, X_4 are arbitrary operators from appropriate spaces.

(ii) *an arbitrary $\{1\}$-inverse of B is given by:*

$$B^{(1)} = \begin{bmatrix} B_1^{-1} & Y_2 \\ Y_3 & Y_4 \end{bmatrix} : \begin{bmatrix} \mathcal{R}(B) \\ \mathcal{N}(B^*) \end{bmatrix} \to \begin{bmatrix} \mathcal{R}(B^*) \\ \mathcal{N}(B) \end{bmatrix}, \tag{3.11}$$

where Y_2, Y_3 and Y_4 are arbitrary operators from appropriate spaces.

(iii) *if AB is regular, then so is $A_1 B_1$ and an arbitrary $\{1\}$-inverse of AB is given by:*

$$(AB)^{(1)} = \begin{bmatrix} (A_1B_1)^{(1)} & Z_2 \\ Z_3 & Z_4 \end{bmatrix} : \begin{bmatrix} \mathscr{R}(A) \\ \mathscr{N}(A^*) \end{bmatrix} \rightarrow \begin{bmatrix} \mathscr{R}(B^*) \\ \mathscr{N}(B) \end{bmatrix}, \quad (3.12)$$

where $(A_1B_1)^{(1)} \in (A_1B_1)\{1\}$ and Z_i, $i = \overline{2,4}$ are arbitrary operators from appropriate spaces.

Proof (i) Suppose a $\{1\}$-inverse of A is given by:

$$A^{(1)} = \begin{bmatrix} X_1 & X_2 \\ X_3 & X_4 \end{bmatrix} : \begin{bmatrix} \mathscr{R}(A) \\ \mathscr{N}(A^*) \end{bmatrix} \rightarrow \begin{bmatrix} \mathscr{R}(B) \\ \mathscr{N}(B^*) \end{bmatrix}.$$

From $AXA = A$ we get that $X \in A\{1\}$ if and only if X_1 and X_3 satisfy the following equations

$$(A_1X_1 + A_2X_3)A_1 = A_1,$$
$$(A_1X_1 + A_2X_3)A_2 = A_2. \quad (3.13)$$

Since $S = \begin{bmatrix} A_1 & A_2 \end{bmatrix} : \begin{bmatrix} \mathscr{R}(B^*) \\ \mathscr{N}(B) \end{bmatrix} \rightarrow \mathscr{R}(A)$ is a right invertible operator, there exists $S_r^{-1} : \mathscr{R}(A) \rightarrow \begin{bmatrix} \mathscr{R}(B^*) \\ \mathscr{N}(B) \end{bmatrix}$ such that $\begin{bmatrix} A_1 & A_2 \end{bmatrix} S_r^{-1} = I_{\mathscr{R}(A)}$. Notice that (3.13) is equivalent to

$$\begin{bmatrix} A_1 & A_2 \end{bmatrix} \begin{bmatrix} X_1 \\ X_3 \end{bmatrix} \begin{bmatrix} A_1 & A_2 \end{bmatrix} = \begin{bmatrix} A_1 & A_2 \end{bmatrix}. \quad (3.14)$$

Multiplying (3.14) by S_r^{-1} from the right, we get that (3.14) is equivalent with $\begin{bmatrix} A_1 & A_2 \end{bmatrix} \begin{bmatrix} X_1 \\ X_3 \end{bmatrix} = I_{\mathscr{R}(A)}$, i.e.,

$$A_1X_1 + A_2X_3 = I_{\mathscr{R}(A)}. \quad (3.15)$$

Note, that for X_1 and X_3 which satisfy (3.15), (3.13) also holds.
(ii) Suppose that a $\{1\}$-inverse of B is given by:

$$B^{(1)} = \begin{bmatrix} Y_1 & Y_2 \\ Y_3 & Y_4 \end{bmatrix} : \begin{bmatrix} \mathscr{R}(B) \\ \mathscr{N}(B^*) \end{bmatrix} \rightarrow \begin{bmatrix} \mathscr{R}(B^*) \\ \mathscr{N}(B) \end{bmatrix}.$$

From $BB^{(1)}B = B$ it follows that $B_1Y_1B_1 = B_1$ and since B_1 is invertible, $Y_1 = B_1^{-1}$.
(iii) Suppose that a $\{1\}$-inverse of AB is given by:

$$(AB)^{(1)} = \begin{bmatrix} Z_1 & Z_2 \\ Z_3 & Z_4 \end{bmatrix} : \begin{bmatrix} \mathscr{R}(A) \\ \mathscr{N}(A^*) \end{bmatrix} \rightarrow \begin{bmatrix} \mathscr{R}(B^*) \\ \mathscr{N}(B) \end{bmatrix}.$$

From $AB(AB)^{(1)}AB = AB$, we get

$$A_1 B_1 Z_1 A_1 B_1 = A_1 B_1, \tag{3.16}$$

and we also see that the operators Z_2, Z_3 and Z_4 can be arbitrary. Now, from (3.16) we see that $Z_1 \in (A_1 B_1)\{1\}$. $\square$

Lemma 3.6 *Let* $K_1 \in \mathcal{B}(\mathcal{H}_1, \mathcal{H}_3)$ *be left invertible and* $K_2 \in \mathcal{B}(\mathcal{H}_2, \mathcal{H}_3)$ *be arbitrary. If* $(I - K_1 K_1^{(1)}) K_2$ *is left invertible for some inner inverse* $K_1^{(1)}$ *of* K_1, *then*
$$\begin{bmatrix} K_1 & K_2 \end{bmatrix} : \begin{bmatrix} \mathcal{H}_1 \\ \mathcal{H}_2 \end{bmatrix} \to \mathcal{H}_3 \text{ is left invertible.}$$

Proof By our assumptions there are $X \in \mathcal{B}(\mathcal{H}_3, \mathcal{H}_1)$, an inner inverse $K_1^{(1)}$ of K_1 and $Y_0 \in \mathcal{B}(\mathcal{H}_3, \mathcal{H}_2)$ such that $X K_1 = I$ and $Y_0(I - K_1 K_1^{(1)}) K_2 = I$. It is easily verified that $D \begin{bmatrix} K_1 & K_2 \end{bmatrix} = I$, where

$$D = \begin{bmatrix} X - X K_2 Y \\ Y \end{bmatrix} : \mathcal{H}_3 \to \begin{bmatrix} \mathcal{H}_1 \\ \mathcal{H}_2 \end{bmatrix}$$

for $Y = Y_0(I - K_1 K_1^{(1)})$. $\square$

To enhance readability of the proof of our main result, we will first prove it under the assumption that $\dim \mathcal{N}(A^*) \le \dim \mathcal{N}(B)$, then directly derive from that the version in the remaining case $\dim \mathcal{N}(B) \le \dim \mathcal{N}(A^*)$, and finally simply combine the two results in Theorem 3.10 in which no assumptions are made.

The following auxiliary theorem will play a key role in the proof of our main result.

Theorem 3.7 *Let regular operators* $A \in \mathcal{B}(\mathcal{H}, \mathcal{K})$ *and* $B \in \mathcal{B}(\mathcal{L}, \mathcal{H})$ *be given by (3.7). If* $\dim \mathcal{N}(A^*) \le \dim \mathcal{N}(B)$ *and* AB *is regular, then the following conditions are equivalent:*

(i) $(AB)\{1\} \subseteq B\{1\}A\{1\}$,

(ii) *For any* $(A_1 B_1)^{(1)} \in (A_1 B_1)\{1\}$ *and* $Z_2 \in \mathcal{B}(\mathcal{N}(A^*), \mathcal{R}(B^*))$, *there exist operators* $W_3 \in \mathcal{B}(\mathcal{R}(A), \mathcal{N}(B^*))$ *with* $\mathcal{R}(I - A_2 W_3) \subseteq \mathcal{R}(A_1)$ *and* $W_4 \in \mathcal{B}(\mathcal{N}(A^*), \mathcal{N}(B^*))$ *such that*

$$X' = \begin{bmatrix} B_1(A_1 B_1)^{(1)} & B_1 Z_2 \\ W_3 & W_4 \end{bmatrix} : \begin{bmatrix} \mathcal{R}(A) \\ \mathcal{N}(A^*) \end{bmatrix} \to \begin{bmatrix} \mathcal{R}(B) \\ \mathcal{N}(B^*) \end{bmatrix} \tag{3.17}$$

is left invertible,

(iii) *For any* $(A_1 B_1)^{(1)} \in (A_1 B_1)\{1\}$ *and* $Z_2 \in \mathcal{B}(\mathcal{N}(A^*), \mathcal{R}(B^*))$, *there exists* $W_3 \in \mathcal{B}(\mathcal{R}(A), \mathcal{N}(B^*))$ *with* $\mathcal{R}(I - A_2 W_3) \subseteq \mathcal{R}(A_1)$ *such that at least one of the following two conditions is satisfied*

(1) $\mathcal{N}(W_3^* \mid (B_1(A_1 B_1)^{(1)})^*; \mathcal{N}(A^*))$ *contains a non-compact operator*

(2) $X_0 = \begin{bmatrix} B_1(A_1B_1)^{(1)} & B_1Z_2 \\ W_3 & 0 \end{bmatrix}$ is a left-Fredholm operator and

$$n(X_0) \leq d(W_3) + \dim\left(\mathscr{R}(W_3^*) \cap \mathscr{R}\left(((B_1(A_1B_1)^{(1)})^*|_{\mathscr{N}((B_1Z_2)^*)})\right)\right).$$

Proof Condition (i) states that for any $(AB)^{(1)} \in (AB)\{1\}$ there exist $A^{(1)} \in A\{1\}$ and $B^{(1)} \in B\{1\}$ such that
$$(AB)^{(1)} = B^{(1)}A^{(1)}$$

which is using Lemma 3.5, equivalent with the fact that for any $(A_1B_1)^{(1)} \in (A_1B_1)\{1\}$, $Z_2 \in \mathscr{B}(\mathscr{N}(A^*), \mathscr{R}(B^*))$, $Z_3 \in \mathscr{B}(\mathscr{R}(A), \mathscr{N}(B))$ and $Z_4 \in \mathscr{B}(\mathscr{N}(A^*), \mathscr{N}(B))$, there exist $Y_2 \in \mathscr{B}(\mathscr{N}(B^*), \mathscr{R}(B^*))$, $Y_3 \in \mathscr{B}(\mathscr{R}(B), \mathscr{N}(B))$, $Y_4 \in \mathscr{B}(\mathscr{N}(B^*), \mathscr{N}(B))$ and $X = \begin{bmatrix} X_1 & X_2 \\ X_3 & X_4 \end{bmatrix} : \begin{bmatrix} \mathscr{R}(A) \\ \mathscr{N}(A^*) \end{bmatrix} \rightarrow \begin{bmatrix} \mathscr{R}(B) \\ \mathscr{N}(B^*) \end{bmatrix}$ satisfying (3.10) such that

$$\begin{bmatrix} (A_1B_1)^{(1)} & Z_2 \\ Z_3 & Z_4 \end{bmatrix} = \begin{bmatrix} B_1^{-1} & Y_2 \\ Y_3 & Y_4 \end{bmatrix}\begin{bmatrix} X_1 & X_2 \\ X_3 & X_4 \end{bmatrix},$$

i.e.,

$$\begin{bmatrix} (A_1B_1)^{(1)} & Z_2 \end{bmatrix} = \begin{bmatrix} B_1^{-1} & Y_2 \end{bmatrix} X \tag{3.18}$$

$$\begin{bmatrix} Z_3 & Z_4 \end{bmatrix} = \begin{bmatrix} Y_3 & Y_4 \end{bmatrix} X. \tag{3.19}$$

In general for arbitrary but fixed Y_2 the Eq. (3.18) is solvable for X and the set of the solutions is given by

$$S = \left\{ \begin{bmatrix} B_1 \\ 0 \end{bmatrix}\begin{bmatrix} (A_1B_1)^{(1)} & Z_2 \end{bmatrix} + \left(I - \begin{bmatrix} B_1 \\ 0 \end{bmatrix}\begin{bmatrix} B_1^{-1} & Y_2 \end{bmatrix}\right) W : \right.$$
$$W \in \mathscr{B}(\mathscr{K}, \mathscr{H}) \}$$
$$= \left\{ \begin{bmatrix} B_1(A_1B_1)^{(1)} - B_1Y_2W_3 & B_1Z_2 - B_1Y_2W_4 \\ W_3 & W_4 \end{bmatrix} : \right. \tag{3.20}$$
$$\left. \begin{bmatrix} W_1 & W_2 \\ W_3 & W_4 \end{bmatrix} : \begin{bmatrix} \mathscr{R}(A) \\ \mathscr{N}(A^*) \end{bmatrix} \rightarrow \begin{bmatrix} \mathscr{R}(B) \\ \mathscr{N}(B^*) \end{bmatrix} \right\}.$$

Thus (i) is equivalent with the existence of at least one $X \in S \cap A\{1\}$ for which the Eq. (3.19) is solvable for $\begin{bmatrix} Y_3 & Y_4 \end{bmatrix}$. That is (i) holds if and only if for any $(A_1B_1)^{(1)} \in (A_1B_1)\{1\}$, $Z_2 \in \mathscr{B}(\mathscr{N}(A^*), \mathscr{R}(B^*))$, $Z_3 \in \mathscr{B}(\mathscr{R}(A), \mathscr{N}(B))$ and $Z_4 \in \mathscr{B}(\mathscr{N}(A^*), \mathscr{N}(B))$ there exist operators $W_3 \in \mathscr{B}(\mathscr{R}(A), \mathscr{N}(B^*))$, $W_4 \in \mathscr{B}(\mathscr{N}(A^*), \mathscr{N}(B^*))$ and $Y_2 \in \mathscr{B}(\mathscr{N}(B^*), \mathscr{R}(B^*))$ such that $K_1 = \begin{bmatrix} B_1(A_1B_1)^{(1)} - B_1Y_2W_3 \\ W_3 \end{bmatrix}$ is a right inverse of $\begin{bmatrix} A_1 & A_2 \end{bmatrix}$ and the following system

$$Z_3 = \begin{bmatrix} Y_3 & Y_4 \end{bmatrix} K_1 \tag{3.21}$$

$$Z_4 = \begin{bmatrix} Y_3 & Y_4 \end{bmatrix} K_2, \tag{3.22}$$

is solvable for $\begin{bmatrix} Y_3 & Y_4 \end{bmatrix}$, where $K_2 = \begin{bmatrix} B_1 Z_2 - B_1 Y_2 W_4 \\ W_4 \end{bmatrix}$.

This is the reformulation of the condition (i) that we will use in proving the implication (i) ⇒ (ii).

(i) ⇒ (ii): Let $(A_1 B_1)^{(1)} \in (A_1 B_1)\{1\}$ and $Z_2 \in \mathcal{B}(\mathcal{N}(A^*), \mathcal{R}(B^*))$. Taking $Z_3 = 0$ and a left invertible $Z_4 \in \mathcal{B}(\mathcal{N}(A^*), \mathcal{N}(B))$ (such Z_4 exists since $\dim \mathcal{N}(A^*) \leq \dim \mathcal{N}(B)$), the condition (i) yields an operator $\begin{bmatrix} K_1 & K_2 \end{bmatrix}$ as described above. Since the Eqs. (3.21) and (3.22) have a common solution and K_1 is regular, we get that

$$Z_4 = W(I - K_1 K_1^{(1)}) K_2,$$

for some $W \in \mathcal{B}(\mathcal{H}, \mathcal{N}(B))$ and some (any) $K_1^{(1)}$. Left invertibility of Z_4 implies left invertibility of $T = (I - K_1 K_1^{(1)}) K_2$ which, given that K_1 is left invertible, implies that $X = \begin{bmatrix} K_1 & K_2 \end{bmatrix}$ is a left invertible operator by Lemma 3.6. It can easily be checked that X is left invertible if and only if

$$X' = \begin{bmatrix} B_1 (A_1 B_1)^{(1)} & B_1 Z_2 \\ W_3 & W_4 \end{bmatrix}$$

is left invertible.

Finally, $\begin{bmatrix} A_1 & A_2 \end{bmatrix} K_1 = I$ means just that

$$A_1 B_1 Y_2 W_3 = A_2 W_3 - \left(I - (A_1 B_1)(A_1 B_1)^{(1)} \right)$$

which upon multiplication from the left by $I - (A_1 B_1)(A_1 B_1)^{(1)}$ gives

$$\left(I - (A_1 B_1)(A_1 B_1)^{(1)} \right) A_2 W_3 = I - (A_1 B_1)(A_1 B_1)^{(1)},$$

i.e., $\mathcal{R}(I - A_2 W_3) \subseteq \mathcal{R}(A_1 B_1) = \mathcal{R}(A_1)$.

(ii) ⇒ (i): Let $(A_1 B_1)^{(1)} \in (A_1 B_1)\{1\}$, $Z_2 \in \mathcal{B}(\mathcal{N}(A^*), \mathcal{R}(B^*))$, $Z_3 \in \mathcal{B}(\mathcal{R}(A), \mathcal{N}(B))$ and $Z_4 \in \mathcal{B}(\mathcal{N}(A^*), \mathcal{N}(B))$ be arbitrary. By our assumption, there are operators W_3 and W_4 acting between appropriate spaces such that X' given by (3.17) is left invertible and $\mathcal{R}(I - A_2 W_3) \subseteq \mathcal{R}(A_1 B_1)$. The latter condition implies that for $Y_2 = (A_1 B_1)^{(1)} A_2$ the operator

$$X = \begin{bmatrix} B_1 (A_1 B_1)^{(1)} - B_1 Y_2 W_3 & B_1 Z_2 - B_1 Y_2 W_4 \\ W_3 & W_4 \end{bmatrix}, \tag{3.23}$$

is an inner inverse of A. Also $X \in S$ so (3.18) is satisfied. As before, left invertibility of X' implies left invertibility of X, so the Eq. (3.19) is solvable for $\begin{bmatrix} Y_3 & Y_4 \end{bmatrix}$. Thus
$$\begin{bmatrix} (A_1 B_1)^{(1)} & Z_2 \\ L_3 & Z_4 \end{bmatrix} \in B\{1\}A\{1\}.$$

(ii) $\Leftrightarrow$ (iii): This follows from Corollary 3.1. $\qquad\square$

The following lemma of technical character will be needed later.

Lemma 3.7 *Let*
$$D = \begin{bmatrix} A_1 & A_2 \end{bmatrix} : \begin{bmatrix} \mathcal{H}_1 \\ \mathcal{H}_2 \end{bmatrix} \to \mathcal{H}_3$$

be a right invertible operator such that A_1 has closed range. Suppose $\mathcal{H}_2 = \mathcal{M} \oplus \mathcal{N}(A_2)$ and $\mathcal{H}_3 = \mathcal{R}(A_1) \oplus \mathcal{N}$, with $\mathcal{M}$ and $\mathcal{N}$ closed, and let

$$A_2 = \begin{bmatrix} A'_2 & 0 \\ A''_2 & 0 \end{bmatrix} : \begin{bmatrix} \mathcal{M} \\ \mathcal{N}(A_2) \end{bmatrix} \to \begin{bmatrix} \mathcal{R}(A_1) \\ \mathcal{N} \end{bmatrix}$$

(i) *An operator $W : \mathcal{H}_3 \to \mathcal{H}_2$ satisfies $\mathcal{R}(I - A_2 W) \subseteq \mathcal{R}(A_1)$ if and only if it has a representation*

$$W = \begin{bmatrix} W_1 & W_2 \\ W_3 & W_4 \end{bmatrix} : \begin{bmatrix} \mathcal{R}(A_1) \\ \mathcal{N} \end{bmatrix} \to \begin{bmatrix} \mathcal{M} \\ \mathcal{N}(A_2) \end{bmatrix} \qquad (3.24)$$

where $A''_2 W_1 = 0$ and $A''_2 W_2 = I$. There is at least one such operator.

(ii) $\dim \mathcal{N} \le \dim \mathcal{M}$.

Proof (i) Suppose W is given by (3.24). From

$$I - A_2 W = \begin{bmatrix} I - A'_2 W_1 & -A'_2 W_2 \\ -A''_2 W_1 & I - A''_2 W_2 \end{bmatrix} : \begin{bmatrix} \mathcal{R}(A_1) \\ \mathcal{N} \end{bmatrix} \to \begin{bmatrix} \mathcal{R}(A_1) \\ \mathcal{N} \end{bmatrix}$$

we see that $\mathcal{R}(I - A_2 W) \subseteq \mathcal{R}(A_1)$ holds if and only if $A''_2 W_1 = 0$ and $A''_2 W_2 = I$. One such operator is obtained by taking $W = X_2$ where

$$\begin{bmatrix} X_1 \\ H_2 \end{bmatrix} : \mathcal{H}_3 \to \begin{bmatrix} \mathcal{H}_1 \\ \mathcal{H}_2 \end{bmatrix}$$

is any right inverse of D.

(ii) The inequality follows from the fact that the existence of an operator W as in (i) implies that $A''_2 : \mathcal{M} \to \mathcal{N}$ is right invertible. It can also be trivially seen to hold true directly, without any recourse to (i). $\qquad\square$

The following theorem gives necessary and sufficient conditions for the inclusion $(AB)\{1\} \subseteq B\{1\}A\{1\}$ to hold under the additional assumption that $\dim \mathcal{N}(A^*) \le \dim \mathcal{N}(B)$. As we will explain later, the main result is practically a direct consequence of it.

Theorem 3.8 *Let regular operators $A \in \mathcal{B}(\mathcal{H}, \mathcal{K})$ and $B \in \mathcal{B}(\mathcal{L}, \mathcal{H})$ be given by (3.7). If* $\dim \mathcal{N}(A^*) \le \dim \mathcal{N}(B)$ *and AB is regular, then the following conditions are equivalent:*

(i) $(AB)\{1\} \subseteq B\{1\}A\{1\}$,

(ii) *One of the following conditions is satisfied:*

    (a) $\dim \mathcal{N}(B^*) < \infty$ *and* $\dim \mathcal{N}(A_1^*) + \dim \mathcal{N}(A^*) \le \dim \mathcal{N}(B^*)$

    (b) $\dim \mathcal{N}(B^*) = \infty$ *and* $\dim \mathcal{N}(A^*) \le \dim \mathcal{N}(A_2'') + \dim \mathcal{N}(A_2)$,

where $A_2'' = P_{\mathcal{N}(A_1^*)} A_2 |_{\mathcal{R}(A_2^*)}$.

Proof (i) $\Rightarrow$ (ii): We distinguish two cases:

Case 1. $\dim \mathcal{N}(B^*) < \infty$. Using Theorems 3.7 and 3.4 we see that

$$\dim \mathcal{N}\left(\begin{bmatrix} B_1(A_1 B_1)^\dagger & B_1 Z_2 \end{bmatrix} \right) \le \dim \mathcal{N}(B^*),$$

for any operator Z_2 which belongs to $\mathcal{B}(\mathcal{N}(A^*), \mathcal{R}(B^*))$, since by our assumption there always are $W_3 \in \mathcal{B}(\mathcal{R}(A), \mathcal{N}(B^*))$ and $W_4 \in \mathcal{B}(\mathcal{N}(A^*), \mathcal{N}(B^*))$ such that X' is left invertible. In particular, for $Z_2 = 0$ we have that $\mathcal{N}\left(\begin{bmatrix} B_1(A_1 B_1)^{(1)} & B_1 Z_2 \end{bmatrix} \right) = \mathcal{N}(A_1^*) \oplus \mathcal{N}(A^*)$, hence $\dim \mathcal{N}(A_1^*) + \dim \mathcal{N}(A^*) \le \dim \mathcal{N}(B^*)$. Thus (a) holds.

Case 2. $\dim \mathcal{N}(B^*) = \infty$. Taking $Z_2 = 0$ and $(A_1 B_1)^{(1)} = (A_1 B_1)^\dagger$ we obtain an operator W_3 such that $\mathcal{R}(I - A_2 W_3) \subseteq \mathcal{R}(A_1)$ for which one of the conditions (1) and (2) from (iii) of Theorem 3.7 is satisfied. From Lemma 3.7, we know that

$$W_3 = \begin{bmatrix} L & J \\ K & T \end{bmatrix} : \begin{bmatrix} \mathcal{R}(A_1) \\ \mathcal{N}(A_1^*) \end{bmatrix} \rightarrow \begin{bmatrix} \mathcal{R}(A_2^*) \\ \mathcal{N}(A_2) \end{bmatrix}, \qquad (3.25)$$

where $A_2'' L = 0$, $A_2'' J = I$ and

$$A_2 = \begin{bmatrix} A_2' & 0 \\ A_2'' & 0 \end{bmatrix} : \begin{bmatrix} \mathcal{R}(A_2^*) \\ \mathcal{N}(A_2) \end{bmatrix} \rightarrow \begin{bmatrix} \mathcal{R}(A_1) \\ \mathcal{N}(A_1^*) \end{bmatrix}.$$

If $\mathcal{N}(W_3^* \mid (B_1(A_1 B_1)^\dagger)^*)$ contains a non-compact operator, then there is a (closed) infinite dimensional subspace $\mathcal{M}$ of $\mathcal{N}(B^*)$ such that

$$\mathcal{R}\left(W_3^* \mid_{\mathcal{M}} \right) \subseteq \mathcal{R}\left((B_1(A_1 B_1)^\dagger)^* \right) = \mathcal{R}(A_1). \qquad (3.26)$$

From (3.26) it follows that $\mathcal{M} \subseteq \mathcal{N}\left(\begin{bmatrix} J^* & T^* \end{bmatrix} \right)$. Now

$$\dim \mathcal{N}\left(\begin{bmatrix} J^* & T^* \end{bmatrix} \right) \le \dim \mathcal{N}(J^*) + \dim \mathcal{N}(A_2) = \dim \mathcal{N}(A_2'') + \dim \mathcal{N}(A_2),$$

since $\dim \mathcal{N}(J^*) = \dim \mathcal{N}(A_2'')$, given that J is a right inverse of A_2'', so $\dim \mathcal{N}(A_2'') + \dim \mathcal{N}(A_2) = \infty$. Thus (b) holds.

Suppose the condition (2) from (iii) of Theorem 3.7 holds. We have

$$\mathscr{R}\big(((B_1(A_1B_1)^\dagger)^*|_{\mathscr{N}((B_1Z_2)^*)}\big) = \mathscr{R}(A_1)$$

and also

$$d(W_3) = n\left(\begin{bmatrix} L^* & K^* \end{bmatrix}|_{\mathscr{N}(\begin{bmatrix} J^* & T^* \end{bmatrix})}\right)$$

and

$$\mathscr{R}(W_3^*) \cap \mathscr{R}\big(((B_1(A_1B_1)^\dagger)^*|_{\mathscr{N}((B_1Z_2)^*)}\big) = \mathscr{R}\left(\begin{bmatrix} L^* & K^* \end{bmatrix}|_{\mathscr{N}(\begin{bmatrix} J^* & T^* \end{bmatrix})}\right).$$

The inclusion $\mathscr{R}(I - A_2W_3) \subseteq \mathscr{R}(A_1)$ implies that for $Y_2 = (A_1B_1)^{(1)}A_2$ the first column of the operator X given by (3.23) is left invertible. Thus the first column of the operator X_0 is also left invertible so $\mathscr{N}(X_0) = \mathscr{N}(A^*)$. Hence $n(A^*) \leq n\left(\begin{bmatrix} J^* & T^* \end{bmatrix}\right)$. Now using $A_2''J = I$ we get

$$n\left(\begin{bmatrix} J^* & T^* \end{bmatrix}\right) = n(J^*) + n(T^*) + \dim(\mathscr{R}(J^*) \cap \mathscr{R}(T^*))$$
$$= \dim\mathscr{N}(A_2'') + \dim\mathscr{N}(A_2).$$

Again, (b) holds.

We now turn to establishing the implication (ii) $\Rightarrow$ (i).

(a) $\Rightarrow$ (i): We will show that condition (ii) from Theorem 3.7 is satisfied. Let $(A_1B_1)^{(1)} \in (A_1B_1)\{1\}$ and $Z_2 \in \mathscr{B}(\mathscr{N}(A^*), \mathscr{R}(B^*))$ be given.

By Lemma 3.7, we can fix a right inverse $J : \mathscr{N}(A_1^*) \to \mathscr{R}(A_2^*)$ of $A_2'' = P_{\mathscr{N}(A_1^*)}A_2|_{\mathscr{R}(A_2^*)}$. Consider

$$W_3 = \begin{bmatrix} J & 0 \\ 0 & 0 \end{bmatrix} : \begin{bmatrix} \mathscr{N}(A_1^*) \\ \mathscr{R}(A_1) \end{bmatrix} \to \begin{bmatrix} \mathscr{R}(A_2^*) \\ \mathscr{N}(A_2) \end{bmatrix}.$$

Using Lemma 3.7, we have that $\mathscr{R}(I - A_2W_3) \subseteq \mathscr{R}(A_1)$. Put $\mathscr{M} = \mathscr{R}(W_3) = \mathscr{R}(J)$. Since J is left invertible, $\dim\mathscr{M} = \dim\mathscr{N}(A_1^*)$. Since $\dim\mathscr{N}(A_1^*) + \dim\mathscr{N}(A^*) \leq \dim\mathscr{N}(B^*) < \infty$ it follows that $\dim\mathscr{N}(A^*) \leq \dim\mathscr{M}^\perp$, where $\mathscr{M}^\perp$ is the orthogonal complement of $\mathscr{M}$ in $\mathscr{N}(B^*)$. Hence there is a left invertible $W_4 \in \mathscr{B}(\mathscr{N}(A^*), \mathscr{N}(B^*))$ such that $\mathscr{R}(W_4) \subseteq \mathscr{M}^\perp$. We will show that X' given by (3.17) is left invertible.

To see that X' has closed range suppose $x_n \in \mathscr{R}(A)$ and $y_n \in \mathscr{N}(A^*)$ for $n \in \mathbb{N}$ are such that

$$B_1(A_1B_1)^{(1)}x_n + B_1Z_2y_n \to u, \quad W_3x_n + W_4y_n \to v.$$

Since $\mathcal{R}(W_4) \subseteq \mathcal{R}(W_3)^{\perp}$ and W_4 is left invertible, it follows that $y_n \to y$ for some $y \in \mathcal{N}(A^*)$. Using left invertibility of $\begin{bmatrix} B_1(A_1B_1)^{(1)} \\ W_3 \end{bmatrix}$, we get that $x_n \to x$ for some $x \in \mathcal{R}(A)$. Hence $\begin{bmatrix} u \\ v \end{bmatrix} \in \mathcal{R}(X')$.

We now show that X' is injective. If $\begin{bmatrix} x \\ y \end{bmatrix} \in \mathcal{N}(X')$, then

$$B_1(A_1B_1)^{(1)}x + B_1 Z_2 y = 0$$
$$W_3 x + W_4 y = 0.$$

Since $\mathcal{M} = \mathcal{R}(W_3)$ and $\mathcal{R}(W_4) \subseteq \mathcal{M}^{\perp}$ it follows that $W_3 x = W_4 y = 0$. The injectivity of W_4 now gives $y = 0$ and so $B_1(A_1B_1)^{(1)}x = 0$. The inclusion $\mathcal{R}(I - A_2 W_3) \subseteq \mathcal{R}(A_1) = \mathcal{R}(A_1B_1)$ implies

$$A_1 B_1(A_1B_1)^{(1)}(I - A_2 W_3) = I - A_2 W_3$$

yielding $x = 0$.

(b) $\Rightarrow$ (i): Let $(A_1B_1)^{(1)} \in (A_1B_1)\{1\}$ and $Z_2 \in \mathcal{B}(\mathcal{N}(A^*), \mathcal{R}(B^*))$ be given. By Lemma 3.7 we have $\mathcal{R}(I - A_2 W_3) \subseteq \mathcal{R}(A_1)$ for the operator W_3 defined by (3.25) where $L = 0$, $K = 0$ and $T = 0$ and $J : \mathcal{N}(A_1^*) \to \mathcal{R}(A_2^*)$ is any right inverse of $A_2'' = P_{\mathcal{N}(A_1^*)} A_2|_{\mathcal{R}(A_2^*)}$ (Lemma 3.7 guaranties that there is one).

Since J is a right inverse of A_2'' we have that $\mathcal{R}(J) \oplus \mathcal{N}(A_2'') = \mathcal{R}(A_2^*)$, so $\dim \mathcal{R}(J)^{\perp} = \dim \mathcal{N}(A_2'') + \dim \mathcal{N}(A_2)$, where $\mathcal{R}(J)^{\perp}$ is the orthogonal comple-ment of $\mathcal{R}(J)$ in $\mathcal{N}(B^*)$. Hence there is a left invertible $W_4 : \mathcal{N}(A^*) \to \mathcal{N}(B^*)$ such that $\mathcal{R}(W_4) \subseteq \mathcal{R}(J)^{\perp}$. That the operator X' given by (3.17) is left invertible can now be proved exactly as in (a) $\Rightarrow$ (i).

We have thus shown that (ii) of Theorem 3.7 holds. $\square$

A standard argument allows us to easily turn the previous theorem into one dealing with the remaining case $\dim \mathcal{N}(B) \leq \dim \mathcal{N}(A^*)$.

Theorem 3.9 *Let regular operators $A \in \mathcal{B}(\mathcal{H}, \mathcal{K})$ and $B \in \mathcal{B}(\mathcal{L}, \mathcal{H})$ be given by (3.7). If $\dim \mathcal{N}(B) \leq \dim \mathcal{N}(A^*)$ and AB is regular, then the following conditions are equivalent:*
(i) *$(AB)\{1\} \subseteq B\{1\}A\{1\}$,*
(ii) *One of the following conditions is satisfied:*

(a) *$\dim \mathcal{N}(A) < \infty$ and $\dim \mathcal{N}(B_1^*) + \dim \mathcal{N}(B) \leq \dim \mathcal{N}(A)$*
(b) *$\dim \mathcal{N}(A) = \infty$ and $\dim \mathcal{N}(B) \leq \dim \mathcal{N}(B_2'') + \dim \mathcal{N}(B_2)$,*

where $B_1 = P_{\mathcal{R}(B^*)} B^*|_{\mathcal{R}(A^*)}$, $B_2 = P_{\mathcal{R}(B^*)} B^*|_{\mathcal{N}(A)}$ and $B_2'' = P_{\mathcal{N}(B_1^*)} B_2|_{\mathcal{R}(B_2^*)}$.

Proof Since (i) is equivalent with

$$(B^*A^*)\{1\} \subseteq A^*\{1\}B^*\{1\}, \tag{3.27}$$

we can apply Theorem 3.8 to the operators B^* and A^* instead of A and B, respectively. □

Combining Theorems 3.8 and 3.9 we are finally in the position to state the main result of this section.

Theorem 3.10 *Let regular operators $A \in \mathcal{B}(\mathcal{H}, \mathcal{K})$ and $B \in \mathcal{B}(\mathcal{L}, \mathcal{H})$ be given by (3.7) and let AB be regular. Then the following conditions are equivalent:*
(i) $(AB)\{1\} \subseteq B\{1\}A\{1\}$,
(ii) *One of the following conditions is satisfied:*

(a) $\dim \mathcal{N}(A^*) \le \dim \mathcal{N}(B)$, $\dim \mathcal{N}(A_1^*) + \dim \mathcal{N}(A^*) \le \dim \mathcal{N}(B^*)$ *and* $\dim \mathcal{N}(B^*) < \infty$,
(b) $\dim \mathcal{N}(A^*) \le \dim \mathcal{N}(B)$, $\dim \mathcal{N}(A^*) \le \dim \mathcal{N}(A_2'') + \dim \mathcal{N}(A_2)$ *and* $\dim \mathcal{N}(B^*) = \infty$,
(c) $\dim \mathcal{N}(B) \le \dim \mathcal{N}(A^*)$, $\dim \mathcal{N}(B_1^*) + \dim \mathcal{N}(B) \le \dim \mathcal{N}(A)$ *and* $\dim \mathcal{N}(A) < \infty$,
(d) $\dim \mathcal{N}(B) \le \dim \mathcal{N}(A^*)$, $\dim \mathcal{N}(B) \le \dim \mathcal{N}(B_2'') + \dim \mathcal{N}(B_2)$ *and* $\dim \mathcal{N}(A) = \infty$,

where $A_2'' = P_{\mathcal{N}(A_1^*)} A_2|_{\mathcal{R}(A_2^*)}$, $B_1 = P_{\mathcal{R}(B^*)} B^*|_{\mathcal{R}(A^*)}$, $B_2 = P_{\mathcal{R}(B^*)} B^*|_{\mathcal{N}(A)}$ *and* $B_2'' = P_{\mathcal{N}(B_1^*)} B_2|_{\mathcal{R}(B_2^*)}$.

As a corollary of the previous theorem, in the case of matrices we have the following already known result:

Corollary 3.2 *Let $A \in \mathbb{C}^{m \times n}$ and $B \in \mathbb{C}^{n \times p}$. The following conditions are equivalent:*
(i) $(AB)\{1\} \subseteq B\{1\}A\{1\}$,
(ii) $r(A) + r(B) - n \le r(AB) - \min\{m - r(A), p - r(B)\}$.

3.3 Applications of Completions of Operator Matrices to Invertibility of Linear Combination of Operators

In this section for given operators $A, B \in \mathcal{B}(\mathcal{H})$, we consider the problem of invertibility of the linear combination $\alpha A + \beta B$, $\alpha, \beta \in \mathbb{C} \setminus \{0\}$ using the results concerning the invertibility of an upper triangular operator matrix of the form M_C. The motivation behind this section was the paper of G. Hai et al. [8] where the invertibility of the linear combination $\alpha A + \beta B$, was considered in the case when $A, B \in \mathcal{B}(\mathcal{H})$ are regular operators and $\alpha, \beta \in \mathbb{C} \setminus \{0\}$ but also some recently published papers (see [16–20]) which considered the independence of the invertibility of the linear combination $\alpha A + \beta B$ in the cases, when $A, B \in \mathcal{B}(\mathcal{H})$ are projectors or orthogonal projectors.

Here, we will consider the general case, without the assumptions that $A, B \in \mathcal{B}(\mathcal{H})$ are closed range operators or that they belong to any particular classes of operators. As corollaries of our main result, we obtain results for certain special classes of operators. Hence, we completely solve the problem of invertibility of the linear combination $\alpha A + \beta B$ in each of the following cases:

- if $A, B \in \mathscr{B}(\mathscr{H})$ are regular operators,
- if $A, B \in \mathscr{B}(\mathscr{H})$ are projectors or orthogonal projectors,
- if $\mathscr{R}(A) \cap \mathscr{R}(B) = \{0\}$.
- if $\overline{\mathscr{R}\left(A P_{\mathscr{N}(B)}\right)} = \overline{\mathscr{R}(A)}$
- if either one of $A, B \in \mathscr{B}(\mathscr{H})$ is injective.

The following well-known lemma will be used throughout this section.

Lemma 3.8 *Let $\mathscr{M}$ and $\mathscr{N}$ be subspaces of a Hilbert space $\mathscr{H}$. Then*

$$(\mathscr{M} + \mathscr{N})^{\perp} = \mathscr{M}^{\perp} \cap \mathscr{N}^{\perp}.$$

In the following theorem we will reduce the problem of invertibility of the linear combination $\alpha A + \beta B$ to an equivalent one which concerns the invertibility of a certain upper triangular operator matrix. Of course, instead of the linear combination one could have simply considered the sum $A + B$ throughout the sequel.

Theorem 3.11 *Let $A, B \in \mathscr{B}(\mathscr{H})$ be given operators and $\alpha, \beta \in \mathbb{C} \setminus \{0\}$. Then $\alpha A + \beta B$ is invertible if and only if the following conditions hold:*

(i) $\mathscr{N}(A) \cap \mathscr{N}(B) = \{0\}$, $\mathscr{R}(A) + \mathscr{R}(B) = \mathscr{H}$,

(ii) $A|_{\mathscr{N}(B)}$ *has a closed range,*

(iii) $P_{\mathscr{S}, \mathscr{R}(A|_{\mathscr{N}(B)})} (\alpha A + \beta B)|_{\mathscr{T}}$ *is an injective operator with range $\mathscr{S}$,*

where $\overline{\mathscr{R}(A)} = \mathscr{R}(A|_{\mathscr{N}(B)}) \oplus \mathscr{S}$, $\mathscr{T} = B^{-1}\left(\overline{\mathscr{R}(A)}\right) \cap \mathscr{P}$ and $\mathscr{H} = \mathscr{N}(B) \oplus \mathscr{P}$.

Proof Let $\mathscr{H} = \mathscr{N}(B) \oplus \mathscr{P} = \overline{\mathscr{R}(A)} \oplus \mathscr{Q}$ be decompositions of the space $\mathscr{H}$. With respect to these decompositions the given operators $A, B \in \mathscr{B}(\mathscr{H})$ have the following representations:

$$A = \begin{bmatrix} A_1 & A_2 \\ 0 & 0 \end{bmatrix} : \begin{bmatrix} \mathscr{N}(B) \\ \mathscr{P} \end{bmatrix} \rightarrow \begin{bmatrix} \overline{\mathscr{R}(A)} \\ \mathscr{Q} \end{bmatrix}, \tag{3.28}$$

$$B = \begin{bmatrix} 0 & B_1 \\ 0 & B_2 \end{bmatrix} : \begin{bmatrix} \mathscr{N}(B) \\ \mathscr{P} \end{bmatrix} \rightarrow \begin{bmatrix} \overline{\mathscr{R}(A)} \\ \mathscr{Q} \end{bmatrix}. \tag{3.29}$$

Take arbitrary $\alpha, \beta \in \mathbb{C} \setminus \{0\}$. Using the above decompositions of $A, B \in \mathscr{B}(\mathscr{H})$, it follows that the linear combination $\alpha A + \beta B$ is invertible if and only if the operator matrix

$$\begin{bmatrix} \alpha A_1 & \alpha A_2 + \beta B_1 \\ 0 & \beta B_2 \end{bmatrix} : \begin{bmatrix} \mathscr{N}(B) \\ \mathscr{P} \end{bmatrix} \rightarrow \begin{bmatrix} \overline{\mathscr{R}(A)} \\ \mathscr{Q} \end{bmatrix} \tag{3.30}$$

is invertible. Using Theorem 3.2 we have that this holds if and only if the following three conditions are satisfied:

(i) αA_1 is left invertible
(ii) βB_2 is right invertible
(iii) $P_{\mathscr{S},\mathscr{R}(A_1)}(\alpha A_2 + \beta B_1)|_{\mathscr{N}(B_2)}$ is an injective operator with range $\mathscr{S}$, where $\overline{\mathscr{R}(A)} = \mathscr{R}(A_1) \oplus \mathscr{S}$.

Evidently, (i) holds if and only if $\mathscr{N}(A) \cap \mathscr{N}(B) = \{0\}$ and $\mathscr{R}(A|_{\mathscr{N}(B)})$ is closed. Also, (ii) is satisfied if and only if $\mathscr{R}(P_{\mathscr{Q},\overline{\mathscr{R}(A)}}B) = \mathscr{Q}$. Since

$$\mathscr{R}(P_{\mathscr{Q},\overline{\mathscr{R}(A)}}B) = \mathscr{Q} \Leftrightarrow \mathscr{Q} \subseteq \overline{\mathscr{R}(A)} + \mathscr{R}(B) \Leftrightarrow \overline{\mathscr{R}(A)} + \mathscr{R}(B) = \mathscr{H}$$

we have that (ii) is equivalent with $\overline{\mathscr{R}(A)} + \mathscr{R}(B) = \mathscr{H}$.

To discuss the third condition notice that

$$\mathscr{N}(B_2) = \mathscr{N}\left(P_{\mathscr{Q},\overline{\mathscr{R}(A)}}B\right) \cap \mathscr{P} = B^{-1}\left(\overline{\mathscr{R}(A)}\right) \cap \mathscr{P}$$

and let $\mathscr{T} = B^{-1}\left(\overline{\mathscr{R}(A)}\right) \cap \mathscr{P}$. Evidently,

$$\mathscr{N}\left(P_{\mathscr{S},\mathscr{R}(A_1)}(\alpha A_2 + \beta B_1)|_{\mathscr{T}}\right) = \mathscr{N}\left(P_{\mathscr{S},\mathscr{R}(A_1)}(\alpha A + \beta B)|_{\mathscr{T}}\right)$$

and

$$\mathscr{R}\left(P_{\mathscr{S},\mathscr{R}(A_1)}(\alpha A_2 + \beta B_1)|_{\mathscr{T}}\right) = \mathscr{R}\left(P_{\mathscr{S},\mathscr{R}(A_1)}(\alpha A + \beta B)|_{\mathscr{T}}\right).$$

Hence, we can conclude that $\alpha A + \beta B$ is invertible if and only if the following conditions hold:

(i) $\mathscr{N}(A) \cap \mathscr{N}(B) = \{0\}$, $\overline{\mathscr{R}(A)} + \mathscr{R}(B) = \mathscr{H}$,
(ii) $AP_{\mathscr{N}(B)}$ has closed range,
(iii) $P_{\mathscr{S},\mathscr{R}(A_1)}(\alpha A + \beta B)|_{\mathscr{T}}$ is an injective operator with range $\mathscr{S}$.

Notice that the second condition in (i), $\overline{\mathscr{R}(A)} + \mathscr{R}(B) = \mathscr{H}$, can be replaced by $\mathscr{R}(A) + \mathscr{R}(B) = \mathscr{H}$: Suppose that (i) $-$ (iii) are satisfied. Since

$$\mathscr{S} = \mathscr{R}(P_{\mathscr{S},\mathscr{R}(A_1)}(\alpha A + \beta B)|_{\mathscr{T}}),$$

we have that

$$\mathscr{S} \subseteq \mathscr{R}((\alpha A + \beta B)|_{\mathscr{T}}) + \mathscr{R}(A_1) \subseteq \mathscr{R}(A) + \mathscr{R}(B).$$

Now, $\overline{\mathscr{R}(A)} = \mathscr{R}(A_1) \oplus \mathscr{S}$ implies that $\overline{\mathscr{R}(A)} \subseteq \mathscr{R}(A) + \mathscr{R}(B)$. Hence, $\mathscr{R}(A) + \mathscr{R}(B) = \mathscr{H}$. (Also, directly from the invertibility of $\alpha A + \beta B$, we can conclude that $\mathscr{R}(A) + \mathscr{R}(B) = \mathscr{H}$). $\qquad\square$

In the special case, when $\mathscr{S}$ is the orthogonal complement of $\mathscr{R}(A|_{\mathscr{N}(B)}) = \mathscr{R}(AP_{\mathscr{N}(B)})$ in $\overline{\mathscr{R}(A)}$ and $\mathscr{P} = \mathscr{N}(B)^{\perp}$, applying Theorem 3.11 we get the following result:

Theorem 3.12 *Let $A, B \in \mathscr{B}(\mathscr{H})$ be given operators and $\alpha, \beta \in \mathbb{C} \setminus \{0\}$. Then $\alpha A + \beta B$ is invertible if and only if the following conditions hold:*

(i) $\mathscr{N}(A) \cap \mathscr{N}(B) = \{0\}$, $\mathscr{R}(A) + \mathscr{R}(B) = \mathscr{H}$,

(ii) $A|_{\mathscr{N}(B)}$ *has a closed range,*

(iii) $P_{\mathscr{S}}(\alpha A + \beta B)|_{\mathscr{T}}$ *is an injective operator with range $\mathscr{S}$,*

where $\mathscr{S} = \mathscr{R}\left(AP_{\mathscr{N}(B)}\right)^{\perp} \cap \overline{\mathscr{R}(A)}$ and $\mathscr{T} = \mathscr{N}\left(P_{\mathscr{R}(A)^{\perp}} B\right) \cap \mathscr{N}(B)^{\perp}$.

Evidently from the theorem given above, we can conclude that the invertibility of the linear combination $\alpha A + \beta B$ is possible for some constants $\alpha, \beta \in \mathbb{C} \setminus \{0\}$ only if

$$\dim \mathscr{N}\left(P_{\mathscr{R}(A)^{\perp}} B\right) \cap \mathscr{N}(B)^{\perp} = \dim \mathscr{R}\left(AP_{\mathscr{N}(B)}\right)^{\perp} \cap \overline{\mathscr{R}(A)},$$

so we get the following result:

Corollary 3.3 *Let $A, B \in \mathscr{B}(\mathscr{H})$ be given operators. If*

$$\dim \mathscr{N}\left(P_{\mathscr{R}(A)^{\perp}} B\right) \cap \mathscr{N}(B)^{\perp} \neq \dim \mathscr{R}\left(AP_{\mathscr{N}(B)}\right)^{\perp} \cap \overline{\mathscr{R}(A)},$$

then the linear combination $\alpha A + \beta B$ is not invertible for any $\alpha, \beta \in \mathbb{C} \setminus \{0\}$.

Now we will reconsider the condition (iii) from Theorem 3.12, which says that $\mathscr{R}(P_{\mathscr{S}}(\alpha A + \beta B)|_{\mathscr{T}}) = \mathscr{S}$ and $\mathscr{N}(P_{\mathscr{S}}(\alpha A + \beta B)|_{\mathscr{T}}) = \{0\}$. Suppose that $A, B \in \mathscr{B}(\mathscr{H})$ are given by (3.28) and (3.29), respectively, where $\mathscr{S}$ is the orthogonal complement of $\mathscr{R}(A|_{\mathscr{N}(B)}) = \mathscr{R}(AP_{\mathscr{N}(B)})$ in $\overline{\mathscr{R}(A)}$, $\mathscr{T} = \mathscr{N}\left(P_{\mathscr{R}(A)^{\perp}} B\right) \cap \mathscr{N}(B)^{\perp}$ and $\mathscr{P} = \mathscr{N}(B)^{\perp}$. The first condition is equivalent with

$$\overline{\mathscr{R}(A)} = \overline{\mathscr{R}(AP_{\mathscr{N}(B)})} + \overline{\mathscr{R}(A)} \cap \mathscr{R}\left((\alpha A + \beta B) P_{\mathscr{N}(B)^{\perp}}\right), \qquad (3.31)$$

since $\mathscr{R}((\alpha A + \beta B)|_{\mathscr{T}}) = \overline{\mathscr{R}(A)} \cap \mathscr{R}\left((\alpha A + \beta B) P_{\mathscr{N}(B)^{\perp}}\right)$. The second condition from (iii), $\mathscr{N}(P_{\mathscr{S}}(\alpha A + \beta B)|_{\mathscr{T}}) = \{0\}$ is equivalent with

$$\mathscr{N}(\alpha A_2 + \beta B_1) \cap \mathscr{N}(B_2) = \{0\},$$
$$\mathscr{R}((\alpha A_2 + \beta B_1)|_{\mathscr{T}}) \cap \overline{\mathscr{R}(AP_{\mathscr{N}(B)})} = \{0\}. \qquad (3.32)$$

Evidently the first condition from (3.32) is equivalent with

$$\mathscr{N}(\alpha A + \beta B) \cap \mathscr{N}(B)^{\perp} = \{0\}$$

while the second one is equivalent with

$$\mathscr{R}\left((\alpha A + \beta B) P_{\mathscr{N}(B)^{\perp}}\right) \cap \overline{\mathscr{R}(AP_{\mathscr{N}(B)})} = \{0\}.$$

Now, in view of the previous two conditions and (3.31), we can conclude that the condition (iii) from Theorem 3.12 is equivalent with

$$\overline{\mathscr{R}(A)} = \overline{\mathscr{R}(A P_{\mathscr{N}(B)})} \oplus \overline{\mathscr{R}(A)} \cap \mathscr{R}\left((\alpha A + \beta B) P_{\mathscr{N}(B)^{\perp}}\right)$$

and

$$\mathscr{N}\left(\alpha A + \beta B\right) \cap \mathscr{N}(B)^{\perp} = \{0\}$$

and we can formulate the following result:

Theorem 3.13 *Let* $A, B \in \mathscr{B}(\mathscr{H})$ *be given operators and* $\alpha, \beta \in \mathbb{C} \setminus \{0\}$*. Then the operator* $\alpha A + \beta B$ *is invertible if and only if the following conditions hold:*

(i) $\mathscr{N}(A) \cap \mathscr{N}(B) = \{0\}$, $\mathscr{R}(A) + \mathscr{R}(B) = \mathscr{H}$,
(ii) $A P_{\mathscr{N}(B)}$ *has closed range,*
(iii) $\overline{\mathscr{R}(A)} = \mathscr{R}(A P_{\mathscr{N}(B)}) \oplus \overline{\mathscr{R}(A)} \cap \mathscr{R}\left((\alpha A + \beta B) P_{\mathscr{N}(B)^{\perp}}\right)$, $\mathscr{N}\left(\alpha A + \beta B\right) \cap \mathscr{N}(B)^{\perp} = \{0\}$.

In Theorems 3.11 and 3.12, the problem of invertibility of a linear combination of two given operators is reduced to one in which yet another linear combination is required to be injective and to have a prescribed range, which at first glance might not strike the reader as much of an achievement. However, the conditions we have obtained (those given in Theorems 3.11 and 3.12) lend themselves for applications in further analysis of the initial problem for many special classes of operators where they will lead to its complete solution.

Since the condition that $\alpha A + \beta B$ be nonsingular is symmetrical in A and B, we can obtain new variants of the necessary and sufficient conditions in Theorems 3.11, 3.12 and 3.13 by interchanging the operators A and B in them.

Now, will be the focus of our attention on invertibility of linear combinations for some special classes of operators using the above mentioned results:
(**1**) The problem of invertibility of $\alpha A + \beta B$, in the case when $A, B \in \mathscr{B}(\mathscr{H})$ are regular operators and $\alpha, \beta \in \mathbb{C} \setminus \{0\}$ was considered in [9].

Theorem 3.14 ([9]) *Let* $A, B \in \mathscr{B}(\mathscr{H})$ *be given operators with closed ranges and* $\alpha, \beta \in \mathbb{C} \setminus \{0\}$*. The operator* $\alpha A + \beta B$ *is invertible if and only if the following conditions hold:*

(i') $\mathscr{N}(A) \cap \mathscr{N}(B) = \{0\}$, $\mathscr{R}(A)^{\perp} \cap \mathscr{R}(B)^{\perp} = \{0\}$,
(ii') *Both* $A^{\dagger}A(I - B^{\dagger}B)$ *and* $(I - AA^{\dagger})BB^{\dagger}$ *are closed range operators,*
(iii') $P'_{\mathscr{L}}\left(\alpha AB^{\dagger}B + \beta AA^{\dagger}B\right)|_{\mathscr{M}}$ *is an invertible,*

where $\mathscr{L} = (A^{*})^{\dagger}\left(\mathscr{R}(A^{*}) \cap \mathscr{R}(B^{*})\right)$, $\mathscr{M} = B^{\dagger}\left(\mathscr{R}(A) \cap \mathscr{R}(B)\right)$ *and* $P'_{\mathscr{L}} \in \mathscr{B}(\mathscr{H}, \mathscr{L})$ *is defined by* $P'_{\mathscr{L}}x = P_{\mathscr{L}}x$, $x \in \mathscr{H}$.

As a corollary of Theorem 3.12 we get some different conditions for the invertibility of $\alpha A + \beta B$ than the ones given in [9]. First give the following lemma.

Lemma 3.9 *Let $A, B \in \mathscr{B}(\mathscr{H})$ be given operators. If A and B have closed ranges then*

(i) $\mathscr{N}\left(P_{\mathscr{R}(A)^{\perp}}B\right) \cap \mathscr{N}(B)^{\perp} = B^{-1}\left(\mathscr{R}(A)\right) \cap \mathscr{N}(B)^{\perp} = B^{\dagger}\left(\mathscr{R}(A) \cap \mathscr{R}(B)\right)$

(ii) $\mathscr{R}(A) \cap \mathscr{R}\left(AP_{\mathscr{N}(B)}\right)^{\perp} = (A^*)^{\dagger}\left(\mathscr{R}(A^*) \cap \mathscr{R}(B^*)\right)$

Proof (i) The first equality is evident. Let $x \in B^{-1}\left(\mathscr{R}(A)\right) \cap \mathscr{N}(B)^{\perp}$. Then $Bx \in \mathscr{R}(A)$ and $x = B^{\dagger}Bx$. So, $x \in B^{\dagger}\left(\mathscr{R}(A) \cap \mathscr{R}(B)\right)$. Now, suppose that $x \in B^{\dagger}\left(\mathscr{R}(A) \cap \mathscr{R}(B)\right)$. Then for some $s, t \in \mathscr{H}$ we have that $x = B^{\dagger}Bt = B^{\dagger}As$ and $Bt = As$. Evidently, $x \in \mathscr{R}(B^*) = \mathscr{N}(B)^{\perp}$ and $Bx = Bt = As \in \mathscr{R}(A)$.

(ii) Let $y \in \mathscr{R}(A) \cap \mathscr{R}\left(AP_{\mathscr{N}(B)}\right)^{\perp}$. Then $y = AA^{\dagger}y$ and $A^*y = B^{\dagger}BA^*y$. Hence, $A^*y = B^{\dagger}BA^*y \in \mathscr{R}(A^*) \cap \mathscr{R}(B^*)$. Now

$$y = (A^{\dagger})^* A^* y = (A^{\dagger})^* B^{\dagger} B A^* y.$$

Now, suppose that $x \in (A^*)^{\dagger}\left(\mathscr{R}(A^*) \cap \mathscr{R}(B^*)\right)$. Then for some $s, t \in \mathscr{H}$ we have that $y = (A^{\dagger})^* A^* t = (A^{\dagger})^* B^* s$ and $A^* t = B^* s$. Evidently, $y \in \mathscr{R}(A)$ which implies $y = AA^{\dagger}y$. Now, we will prove that $y \in \mathscr{R}\left(AP_{\mathscr{N}(B)}\right)^{\perp} = \mathscr{N}(P_{\mathscr{N}(B)}A^*)$:

$$\begin{aligned}
B^{\dagger}BA^*y &= B^{\dagger}BA^*(A^{\dagger})^* B^* s = B^{\dagger}BA^{\dagger}AB^* s \\
&= B^{\dagger}BA^{\dagger}AA^* t = B^{\dagger}BB^* s = B^* s \\
&= A^* t = A^* y.
\end{aligned}$$

$\square$

Now, in the case when $A, B \in \mathscr{B}(\mathscr{H})$ are closed range operators, from Theorem 3.12 we get the following:

Theorem 3.15 *Let $A, B \in \mathscr{B}(\mathscr{H})$ be given closed range operators and $\alpha, \beta \in \mathbb{C} \setminus \{0\}$. Then $\alpha A + \beta B$ is invertible if and only if the following conditions hold:*

(i) $\mathscr{N}(A) \cap \mathscr{N}(B) = \{0\}$, $\mathscr{R}(A) + \mathscr{R}(B) = \mathscr{H}$,

(ii) $A|_{\mathscr{N}(B)}$ *has a closed range,*

(iii) $P_{\mathscr{S}, \mathscr{R}(A|_{\mathscr{N}(B)})} (\alpha A + \beta B)|_{\mathscr{T}}$ *is an injective operator with range $\mathscr{S}$,*

where $\mathscr{S} = (A^*)^{\dagger}\left(\mathscr{R}(A^*) \cap \mathscr{R}(B^*)\right)$ and $\mathscr{T} = B^{\dagger}\left(\mathscr{R}(A) \cap \mathscr{R}(B)\right)$.

(2) The problem of invertibility of projections (idempotents) has been considered in several papers. Coming from that line of research we can single out the result that the invertibility of any linear combination $\alpha P + \beta Q$, where $\alpha, \beta \in \mathbb{C} \setminus \{0\}, \alpha + \beta \neq 0$, is in fact equivalent to the invertibility of $P + Q$ which means that it is independent of the choice of the scalars α and β. For the first time, this was realized by J.K. Baksalary et al. [16] for the finite-dimesional case who proved that

$$\alpha P + \beta Q \text{ is nonsing.} \Leftrightarrow \mathscr{R}(P(I - Q)) \cap \mathscr{R}(Q(I - P)) = \mathscr{N}(P) \cap \mathscr{N}(Q) = \{0\}$$

and later generalized by Du et al. [21] to the case of idempotent operators on a Hilbert space and finally by Koliha et al. [19] to the Banach algebra case, without giving any necessary and sufficient conditions for the invertibility of $\alpha P + \beta Q$. The necessary and sufficient conditions for the invertibility of a linear combination of projections P and Q on a Hilbert space are given later in another paper by Koliha et al. [17] (as well as for the elements of a unital ring):

Theorem 3.16 ([17]) *Let* $P, Q \in \mathcal{B}(\mathcal{H})$ *be projections on a Hilbert space* $\mathcal{H}$. *Then the following conditions are equivalent:*

(i) $P + Q$ *is invertible.*
(ii) *The range of* $P + Q$ *is closed and*

$$\mathcal{R}(P) \cap \mathcal{R}(Q(I - P)) = \mathcal{N}(P) \cap \mathcal{N}(Q) = \{0\},$$

$$\mathcal{R}(P^*) \cap \mathcal{R}(Q^*(I - P^*)) = \mathcal{N}(P^*) \cap \mathcal{N}(Q^*) = \{0\}.$$

In the case when $P, Q \in \mathcal{B}(\mathcal{H})$ are projections, applying Theorem 3.11 to the decompositions $\mathcal{H} = \mathcal{N}(Q) \oplus \mathcal{R}(Q) = \mathcal{R}(P) \oplus \mathcal{N}(P)$ we get the main result from [21], which says that the invertibility of the linear combination $\alpha P + \beta Q$ is independent of the choice of the scalars $\alpha, \beta \in \mathbb{C}$, but additionally we also obtain necessary and sufficient conditions for the invertibility of the linear combination $\alpha P + \beta Q$ which are different from those given in Theorem 3.16.

Theorem 3.17 *Let* $P, Q \in \mathcal{B}(\mathcal{H})$ *be given projections and* $\alpha, \beta \in \mathbb{C} \backslash \{0\}, \alpha + \beta \neq 0$. *Then* $\alpha P + \beta Q$ *is an invertible operator if and only if the following conditions hold:*

(i) $\mathcal{N}(P) \cap \mathcal{N}(Q) = \{0\}, \mathcal{R}(P) + \mathcal{R}(Q) = \mathcal{H}$,
(ii) $\mathcal{R}(P) = \mathcal{R}(P) \cap \mathcal{R}(Q) \oplus \mathcal{R}(P|_{\mathcal{N}(Q)})$.

Proof Indeed, in this case the subspace $\mathcal{T}$ defined in Theorem 3.11 by $\mathcal{T} = Q^{-1}(\mathcal{R}(P)) \cap \mathcal{R}(Q)$ is equal to $\mathcal{T} = \mathcal{R}(P) \cap \mathcal{R}(Q)$. Hence, for any $x \in \mathcal{T}$, we have that $(\alpha P + \beta Q)x = (\alpha + \beta)x$ which implies that the injectivity of operator $P_{\mathcal{S}, \mathcal{R}(P|_{\mathcal{N}(Q)})}(\alpha P + \beta Q)|_{\mathcal{T}}$ is equivalent with $\mathcal{R}(P|_{\mathcal{N}(Q)}) \cap \mathcal{T} = \{0\}$. i.e.,

$$\mathcal{R}(P|_{\mathcal{N}(Q)}) \cap \mathcal{R}(Q) = \{0\}. \tag{3.33}$$

Also, operator $P_{\mathcal{S}, \mathcal{R}(P|_{\mathcal{N}(Q)})}(\alpha P + \beta Q)|_{\mathcal{T}}$ has range $\mathcal{S}$ if and only if $\mathcal{S} \subseteq \mathcal{T} + \mathcal{R}(P|_{\mathcal{N}(Q)})$, which is equivalent with $\mathcal{R}(P) = \mathcal{R}(P) \cap \mathcal{R}(Q) + \mathcal{R}(P|_{\mathcal{N}(Q)})$. Now, by (3.33), we have that

$$\mathcal{R}(P) = \mathcal{R}(P) \cap \mathcal{R}(Q) \oplus \mathcal{R}(P|_{\mathcal{N}(Q)}). \tag{3.34}$$

Using (3.34), the fact that the intersection of two operator ranges is an operator range and Theorem 2.3 [22], we conclude that $\mathcal{R}(P|_{\mathcal{N}(Q)})$ is closed. Now, the proof follows by Theorem 3.11. $\qquad \square$

Obviously, from Theorem 3.17 we get the following corollary:

Corollary 3.4 *Let $P, Q \in \mathcal{B}(\mathcal{H})$ be given projections and $\alpha, \beta \in \mathbb{C} \setminus \{0\}, \alpha + \beta \neq 0$. Then the invertibility of the linear combination $\alpha P + \beta Q$ is independent of the choice of the scalars $\alpha, \beta \in \mathbb{C} \setminus \{0\}, \alpha + \beta \neq 0$.*

(3) The problem of invertibility of the linear combination $\alpha P + \beta Q$ when P and Q are orthogonal projections has also received a lot of attention. In [20] Buckholtz considers the special case when $\alpha + \beta = 1$ and gives conditions under which the difference of projections on a Hilbert space is invertible, as well as an explicit formula for its inverse. In the paper of Koliha et al. [17], the invertibility of the sum of two orthogonal projections was considered which is, as we already know, equivalent with the invertibility of the linear combination $\alpha P + \beta Q$:

Theorem 3.18 ([17]) *Let $P, Q \in \mathcal{B}(\mathcal{H})$ be orthogonal projections on a Hilbert space $\mathcal{H}$. Then the following conditions are equivalent:*

(i) *$P + Q$ is invertible,*
(ii) *The range of $P + Q$ is closed and*

$$\mathcal{R}(P) \cap \mathcal{R}(Q(I - P)) = \mathcal{N}(P) \cap \mathcal{N}(Q) = \{0\}$$

Here, using Theorem 3.11 we obtain the following result:

Theorem 3.19 *Let $P, Q \in \mathcal{B}(\mathcal{H})$ be given orthogonal projections and $\alpha, \beta \in \mathbb{C} \setminus \{0\}, \alpha + \beta \neq 0$. Then $\alpha P + \beta Q$ is an invertible operator if and only if $\mathcal{R}(P) + \mathcal{R}(Q) = \mathcal{H}$.*

Proof Notice that in the case when $P, Q \in \mathcal{B}(\mathcal{H})$ are orthogonal projections, the subspaces $\mathcal{S}$ and $\mathcal{T}$ defined in Theorem 3.12 by $\mathcal{S} = \mathcal{R}\left(P P_{\mathcal{N}(Q)}\right)^{\perp} \cap \mathcal{R}(P)$ and $\mathcal{T} = \mathcal{N}\left(P_{\mathcal{R}(P)^{\perp}} Q\right) \cap \mathcal{N}(Q)^{\perp}$ coincide and $\mathcal{S} = \mathcal{T} = \mathcal{R}(P) \cap \mathcal{R}(Q)$. Indeed, if $P, Q \in \mathcal{B}(\mathcal{H})$ are orthogonal projections, then

$$\mathcal{S} = \mathcal{R}\left(P P_{\mathcal{N}(Q)}\right)^{\perp} \cap \mathcal{R}(P) = \mathcal{R}(P(I - Q))^{\perp} \cap \mathcal{R}(P)$$
$$= \mathcal{N}((I - Q)P) \cap \mathcal{R}(P) = \mathcal{R}(P) \cap \mathcal{R}(Q)$$

and

$$\mathcal{T} = \mathcal{N}\left(P_{\mathcal{R}(P)^{\perp}} Q\right) \cap \mathcal{N}(Q)^{\perp} = \mathcal{N}((I - P)Q) \cap \mathcal{R}(Q)$$
$$= \mathcal{R}(P) \cap \mathcal{R}(Q).$$

Hence, for any $x \in \mathcal{T}$, we have that $(\alpha P + \beta Q)x = (\alpha + \beta)x$ and $P_{\mathcal{S}}(\alpha P + \beta Q)x = (\alpha + \beta)x$. So, the operator $P_{\mathcal{S}}(\alpha P + \beta Q)|_{\mathcal{T}}$ from item (iii) of Theorem 3.12 is an injective operator with range $\mathcal{S}$ if and only if $\alpha + \beta \neq 0$. Also, the condition $\mathcal{R}(P) + \mathcal{R}(Q) = \mathcal{H}$ implies $\mathcal{N}(P) \cap \mathcal{N}(Q) = \{0\}$. Now, from Theorem 3.12 we can conclude that in the case when $P, Q \in \mathcal{B}(\mathcal{H})$ are orthogonal projections, $\alpha P + \beta Q$ is an invertible operator if and only if the following conditions hold:

(i) $\mathcal{R}(P) + \mathcal{R}(Q) = \mathcal{H}$,

(ii) $P|_{\mathcal{N}(Q)}$ has closed range.

Notice that the condition (ii) that $P|_{\mathcal{N}(Q)}$ has closed range can be replaced by the condition that $\mathcal{R}(P(I - Q))$ is closed. By Proposition 2.4 [23], we have that $\mathcal{R}(P(I - Q))$ is closed if and only if $\mathcal{R}(P + Q)$ is closed, which is by Corollary 3 [22] equivalent with the fact that $\mathcal{R}(P) + \mathcal{R}(Q)$ is closed. Since the condition (i) guarantees closedness of $\mathcal{R}(P) + \mathcal{R}(Q)$, we conclude that condition (i) is necessary and sufficient for the invertibility of $\alpha P + \beta Q$. $\square$.

If we compare Theorem 3.18 from [17] and our Theorem 3.19, it is evident that the condition $\mathcal{R}(P) \cap \mathcal{R}(Q(I - P)) = \{0\}$ is superfluous. In the following lemma we will give an explanation for that:

Lemma 3.10 *Let $P, Q \in \mathcal{B}(\mathcal{H})$ be orthogonal projections on a Hilbert space $\mathcal{H}$. Then*

$$\mathcal{R}(P) \cap \mathcal{R}(Q(I - P)) = \{0\}.$$

Proof First, let us observe that

$$\mathcal{R}(P) \cap \mathcal{R}(Q(I - P)) = \mathcal{R}(P) \cap \mathcal{R}(Q(I - P)) \cap \mathcal{R}(Q).$$

So, it is sufficient to prove that $\mathcal{R}(P) \cap \mathcal{R}(Q(I - P)) \cap \mathcal{R}(Q) = \{0\}$. It can be easy checked that

$$\mathcal{R}(Q(I - P))^{\perp} \cap \mathcal{R}(Q) = \mathcal{N}((I - P)Q) \cap \mathcal{R}(Q) = \mathcal{R}(P) \cap \mathcal{R}(Q),$$

implying that $\mathcal{R}(P) \cap \mathcal{R}(Q) \subseteq \mathcal{R}(Q(I - P))^{\perp}$, i.e., $\mathcal{R}(P) \cap \mathcal{R}(Q) \cap \mathcal{R}(Q(I - P)) = \{0\}$. $\square$

(4) Now we will consider the invertibility of the linear combination $\alpha A + \beta B$ for given operators $A, B \in \mathcal{B}(\mathcal{H})$ in two special cases: when $\mathcal{R}(A) \cap \mathcal{R}(B) = \{0\}$ and when $\overline{\mathcal{R}\left(AP_{\mathcal{N}(B)}\right)} = \overline{\mathcal{R}(A)}$. In both of these two cases, beside giving necessary and sufficient conditions for the the invertibility of $\alpha A + \beta B$, we will conclude that the invertibility of the linear combination $\alpha A + \beta B$ is independent of the choice of the scalars $\alpha, \beta \in \mathbb{C} \setminus \{0\}$.

In the special case when $A, B \in \mathcal{B}(\mathcal{H})$ are such that $\mathcal{R}(A) \cap \mathcal{R}(B) = \{0\}$ using Theorem 3.12 we get the following:

Theorem 3.20 *Let $A, B \in \mathcal{B}(\mathcal{H})$ be given operators and $\alpha, \beta \in \mathbb{C} \setminus \{0\}$. If $\mathcal{R}(A) \cap \mathcal{R}(B) = \{0\}$, then the operator $\alpha A + \beta B$ is invertible if and only if*

$$\mathcal{R}(A) \oplus \mathcal{R}(B) = \mathcal{H}, \quad \mathcal{N}(A) \oplus \mathcal{N}(B) = \mathcal{H}. \tag{3.35}$$

Proof Suppose that $\alpha A + \beta B$ is invertible. By Theorem 3.12, we have that $\mathcal{R}(A) \oplus \mathcal{R}(B) = \mathcal{H}$ which by Theorem 2.3 [22] gives that $\mathcal{R}(A)$ and $\mathcal{R}(B)$ are closed.

Now $\mathscr{R}(A) \cap \mathscr{R}(B) = \{0\}$ together with the fact that $\mathscr{R}(A)$ is closed implies that $\mathscr{T} = \mathscr{N}\left(P_{\mathscr{R}(A)^\perp} B\right) \cap \mathscr{N}(B)^\perp = \{0\}$ which by the condition (iii) from Theorem 3.12 gives that $\mathscr{S} = \mathscr{R}\left(AP_{\mathscr{N}(B)}\right)^\perp \cap \overline{\mathscr{R}(A)} = \{0\}$. Hence, $\mathscr{R}(A) = \mathscr{R}(AP_{\mathscr{N}(B)})$ which implies that $\mathscr{N}(A)^\perp \subseteq \mathscr{N}(B) + \mathscr{N}(A)$. So, $\mathscr{H} = \mathscr{N}(B) + \mathscr{N}(A)$. By the condition (i) of Theorem 3.12, we have that $\mathscr{N}(A) \cap \mathscr{N}(B) = \{0\}$, so $\mathscr{H} = \mathscr{N}(B) \oplus \mathscr{N}(A)$. On the other hand suppose that (3.35) holds. Evidently, $\mathscr{R}(A)$ and $\mathscr{R}(B)$ are closed and the first condition from Theorem 3.12 is satisfied. Also, $\mathscr{R}(A) = A(\mathscr{H}) = A(\mathscr{N}(A) \oplus \mathscr{N}(B)) = A(\mathscr{N}(B)) = \mathscr{R}(AP_{\mathscr{N}(B)})$, so (ii) of Theorem 3.12 is satisfied. To conclude that (iii) of Theorem 3.12 is true, simply notice that $\mathscr{T} = \mathscr{S} = \{0\}$.   □

Similarly, we get the following:

Theorem 3.21 Let $A, B \in \mathscr{B}(\mathscr{H})$ be given operators and $\alpha, \beta \in \mathbb{C} \setminus \{0\}$. If $\overline{\mathscr{R}\left(AP_{\mathscr{N}(B)}\right)} = \overline{\mathscr{R}(A)}$, then the operator $\alpha A + \beta B$ is invertible if and only if the following conditions hold:

(i) $\mathscr{N}(A) \cap \mathscr{N}(B) = \{0\}$, $\mathscr{R}(A) \oplus \mathscr{R}(B) = \mathscr{H}$,
(ii) $AP_{\mathscr{N}(B)}$ has closed range.

Proof If $\overline{\mathscr{R}\left(AP_{\mathscr{N}(B)}\right)} = \overline{\mathscr{R}(A)}$, then for $\mathscr{S}$ defined in Theorem 3.11 we have that $\mathscr{S} = \{0\}$. So the condition (iii) from Theorem 3.11 is satisfied if and only if $\mathscr{T} = \{0\}$, i.e., $\overline{\mathscr{R}(A)} \cap \mathscr{R}(B) = \{0\}$. Now, the proof follows directly from Theorem 3.11.   □

Corollary 3.5 Let $A, B \in \mathscr{B}(\mathscr{H})$ be given operators. If one of the conditions $\mathscr{R}(A) \cap \mathscr{R}(B) = \{0\}$ and $\overline{\mathscr{R}\left(AP_{\mathscr{N}(B)}\right)} = \overline{\mathscr{R}(A)}$ holds, then the invertibility of the linear combination $\alpha A + \beta B$ is independent of the choice of the scalars $\alpha, \beta \in \mathbb{C}\setminus\{0\}$.

(5) Now we will consider the case when either one of the operators $A, B \in \mathscr{B}(\mathscr{H})$ is injective.

Since the condition $\alpha A + \beta B$ is nonsingular is symmetrical in A and B, let us suppose that $B \in \mathscr{B}(\mathscr{H})$ is injective:

Theorem 3.22 Let $A, B \in \mathscr{B}(\mathscr{H})$ be given operators such that B is injective and $\alpha, \beta \in \mathbb{C} \setminus \{0\}$. Then $\alpha A + \beta B$ is invertible if and only if the following conditions hold:

(i) $\mathscr{R}(A) + \mathscr{R}(B) = \mathscr{H}$,
(ii) $(\alpha A + \beta B)|_{B^{-1}(\overline{\mathscr{R}(A)})}$ is an injective operator with range $\overline{\mathscr{R}(A)}$.

Considering some special classes of operators we have seen that the invertibility of the linear combination $\alpha A + \beta B$ is independent of the choice of the scalars $\alpha, \beta \in \mathbb{C} \setminus \{0\}$. Another instance of this phenomenon is provided by the following result.

Theorem 3.23 Let $A, B \in \mathscr{B}(\mathscr{H})$ be given operators and $\alpha, \beta \in \mathbb{C} \setminus \{0\}$. If there exists a closed subspace $\mathscr{P}$ such that $\mathscr{H} = \mathscr{N}(B) \oplus \mathscr{P}$ and $A|_{\mathscr{P}} = 0$ or $P_{\overline{\mathscr{R}(A)}} B = 0$, then the invertibility of the linear combination $\alpha A + \beta B$ is independent of the choice of the scalars.

Proof Using the representations (3.28) and (3.29) of the operators A and B, and the representation (3.30) of $\alpha A + \beta B$, from Theorem 3.11 the desired conclusion is immediately reached. $\qquad\square$

3.4 Drazin Invertible Completion of an Upper Triangular Operator Matrix

In this section we will consider the existence of a Drazin invertible completion of an upper triangular operator matrix of the form

$$\begin{bmatrix} A & ? \\ 0 & B \end{bmatrix} : \begin{bmatrix} \mathscr{H} \\ \mathscr{K} \end{bmatrix} \to \begin{bmatrix} \mathscr{H} \\ \mathscr{K} \end{bmatrix},$$

where $A \in \mathscr{B}(\mathscr{H})$ and $B \in \mathscr{B}(\mathscr{K})$ are given operators.

Throughout the section $\mathscr{H}$, $\mathscr{K}$ are infinite dimensional separable complex Hilbert spaces. For a given operator $A \in \mathscr{B}(\mathscr{H}, \mathscr{K})$, we set $n(A) = \dim \mathscr{N}(A)$ and $d(A) = \dim \mathscr{R}(A)^{\perp}$.

Let us recall that for $A \in \mathscr{B}(\mathscr{H})$, the smallest nonnegative integer k such that $\mathscr{N}(A^{k+1}) = \mathscr{N}(A^k)$ (resp. $\mathscr{R}(A^{k+1}) = \mathscr{R}(A^k)$), if one exists, is called the *ascent* (resp. *descent*) of the operator A and is denoted by $\mathrm{asc}(A)$ (resp. $\mathrm{dsc}(A)$); if there is no such integer k, the operator A is said to be of infinite ascent (resp. infinite descent), which is abbreviated by $\mathrm{asc}(A) = \infty$ (resp. $\mathrm{dsc}(A) = \infty$). Also $K(0, \delta) = \{\lambda \in \mathbb{C} : |\lambda| < \delta\}$ stands for the open disc with center 0 and radius δ.

An operator $A \in \mathscr{B}(\mathscr{H})$ is left Drazin invertible if $\mathrm{asc}(A) < \infty$ and $\mathscr{R}(A^{\mathrm{asc}(A)+1})$ is closed while $A \in \mathscr{B}(\mathscr{H})$ is right Drazin invertible if $\mathrm{dsc}(A) < \infty$ and $\mathscr{R}(A^{\mathrm{dsc}(A)})$ is closed.

The question of existence of Drazin invertible completions of the upper-triangular operator matrix

$$M_C = \begin{bmatrix} A & C \\ 0 & B \end{bmatrix} : \begin{bmatrix} \mathscr{H} \\ \mathscr{K} \end{bmatrix} \to \begin{bmatrix} \mathscr{H} \\ \mathscr{K} \end{bmatrix},$$

was addressed in [24] where some sufficient conditions were given but the proof of the result presented there is not correct as it is explained in [25].

Theorem 3.24 ([23]) *Let $\mathscr{H}$ and $\mathscr{K}$ be separable Hilbert spaces and $A \in \mathscr{B}(\mathscr{H})$ and $B \in \mathscr{B}(\mathscr{K})$ be given operators such that*

(i) *A is left Drazin invertible,*
(ii) *B is right Drazin invertible,*
(iii) *There exists a constant $\delta > 0$ such that $d(A - \lambda) = n(B - \lambda)$, for every $\lambda \in K(0, \delta) \setminus \{0\}$.*

Then there exists an operator $C \in \mathscr{B}(\mathscr{K}, \mathscr{H})$ such that M_C is Drazin invertible.

In order to give a correct proof of Theorem 3.24, we will first list some auxiliaries results:

Two completely different proofs of the following lemma that will be extensively used throughout the paper can be found in [26, 27]:

Lemma 3.11 For a Banach space $\mathscr{X}$, a given nonnegative integer m and $A \in \mathscr{B}(\mathscr{X})$, the following conditions are equivalent:

(i) $\mathrm{dsc}(A) \leq m < \infty$,
(ii) $\mathscr{N}(A^m) + \mathscr{R}(A^n) = \mathscr{X}$, for every $n \in \mathbb{N}$,
(iii) $\mathscr{N}(A^m) + \mathscr{R}(A^n) = \mathscr{X}$, for some $n \in \mathbb{N}$.

We will also need the following result which is proved in [27, 28].

Lemma 3.12 Let $A \in \mathscr{B}(\mathscr{X})$. We have the following

(1) If $\mathrm{dsc}(A) = m < \infty$, then there exists a constant $\delta > 0$ such that for every $\lambda \in K(0, \delta) \setminus \{0\}$:
   (i) $\mathrm{dsc}(A - \lambda) = d(A - \lambda) = 0$,
   (ii) $n(A - \lambda) = \dim \mathscr{N}(A) \cap \mathscr{R}(A^m)$.
(2) If $\mathrm{asc}(A) = m < \infty$ and $\mathscr{R}(A^{m+k})$ is closed for some $k \geq 1$, then there exists a constant $\delta > 0$ such that for every $\lambda \in K(0, \delta) \setminus \{0\}$:
   (i) $\mathrm{asc}(A - \lambda) = n(A - \lambda) = 0$,
   (ii) $d(A - \lambda) = \dim \left(\mathscr{R}(A^m)/\mathscr{R}(A^{m+1}) \right) = \dim \left(\mathscr{X}/(\mathscr{R}(A) + \mathscr{N}(A^m)) \right)$.

The following technical lemma will be used multiple times throughout this section.

Lemma 3.13 Suppose $B \in \mathscr{B}(\mathscr{K})$ and p is a positive integer such that $\mathscr{R}(B^p)$ is closed. If B is represented by

$$B = \begin{bmatrix} 0 & B_1 \\ 0 & B_2 \end{bmatrix} : \begin{bmatrix} \mathscr{N}(B) \cap \mathscr{R}(B^p) \\ (\mathscr{N}(B) \cap \mathscr{R}(B^p))^\perp \end{bmatrix} \to \begin{bmatrix} \mathscr{N}(B) \cap \mathscr{R}(B^p) \\ (\mathscr{N}(B) \cap \mathscr{R}(B^p))^\perp \end{bmatrix}, \quad (3.36)$$

then B_1 and B_2 must satisfy the following two conditions:

(i) The restriction of $B_1 B_2^{p-1}$ on $\mathscr{N}(B_2^p)$ is onto (equivalently: the restriction of B_1 to the subspace $\mathscr{R}(B_2^{p-1}) \cap \mathscr{N}(B_2)$ is onto)
(ii) $\mathscr{R}(B_2^p) \subseteq \mathscr{R}(B^p)$,
(iii) $\mathscr{R}(B_2^p) \cap \mathscr{N}(B_1) \cap \mathscr{N}(B_2) = \{0\}$ (equivalently: the restriction of B_1 to the subspace $\mathscr{R}(B_2^p) \cap \mathscr{N}(B_2)$ is injective).

Proof Put $\mathscr{S} := \mathscr{N}(B) \cap \mathscr{R}(B^p)$. To see that (i) is true, notice that if $y \in S$ then $\begin{bmatrix} y \\ 0 \end{bmatrix} = \begin{bmatrix} B_1 B_2^{p-1} x \\ B_2^p x \end{bmatrix}$ for some $x \in \mathscr{S}^\perp$. To see that (ii) is true, notice that for any $x \in S^\perp$ we have $\begin{bmatrix} 0 \\ B_2^p x \end{bmatrix} = \begin{bmatrix} B_1 B_2^{p-1} x \\ B_2^p x \end{bmatrix} - \begin{bmatrix} B_1 B_2^{p-1} x \\ 0 \end{bmatrix}$, and that by (i) we know that $\begin{bmatrix} B_1 B_2^{p-1} x \\ 0 \end{bmatrix} \in \mathscr{R}(B^p)$. Finally to show (iii), notice that if $y \in \mathscr{R}(B_2^p) \cap \mathscr{N}(B_1) \cap \mathscr{N}(B_2)$ then $y \in \mathscr{S}$ by (ii), and also $y \in \mathscr{S}^\perp$, so $y = 0$. $\qquad \square$

The following is a key lemma in the proof of our Theorem 3.24. Suppose that $A \in \mathscr{B}(\mathscr{H})$ is a left Drazin invertible operator, $B \in \mathscr{B}(\mathscr{K})$ is an operator with finite descent and suppose in addition that there exists a constant $\delta > 0$ such that $d(A - \lambda) = n(B - \lambda)$, for every $\lambda \in K(0, \delta) \setminus \{0\}$. Note that if p is any integer with $p \geq \max\{\mathrm{asc}(A), \mathrm{dsc}(B)\}$, then $\mathscr{R}(A) + \mathscr{N}(A^p) = A^{-\mathrm{asc}(A)}[\mathscr{R}(A^{\mathrm{asc}(A)+1})]$ is a closed subspace of codimension equal to the dimension of the subspace $\mathscr{N}(B) \cap \mathscr{R}(B^p)$, by Lemma 3.12. Thus we can fix an invertible operator $J \in \mathscr{B}(\overline{\mathscr{N}(B) \cap \mathscr{R}(B^p)}, (\mathscr{R}(A) + \mathscr{N}(A^p))^{\perp})$. Indeed, if $\mathscr{N}(B) \cap \mathscr{R}(B^p)$ is closed then this is clear. If it is not, then it must be infinite dimensional and so must be the closed subspace $(\mathscr{R}(A) + \mathscr{N}(A^p))^{\perp}$. But then $\overline{\mathscr{N}(B) \cap \mathscr{R}(B^p)}$ and $(\mathscr{R}(A) + \mathscr{N}(A^p))^{\perp})$ are both infinite dimensional separable Hilbert spaces and as such are isomorphic to one another.

Lemma 3.14 *Let $A \in \mathscr{B}(\mathscr{H})$, $B \in \mathscr{B}(\mathscr{K})$ be given operators such that*

(i) *A is left Drazin invertible,*
(ii) *$\mathrm{dsc}(B) < \infty$,*
(iii) *There exists a constant $\delta > 0$ such that $d(A - \lambda) = n(B - \lambda)$, for every $\lambda \in K(0, \delta) \setminus \{0\}$.*

Let $C \in \mathscr{B}(\mathscr{K}, \mathscr{H})$ be given by

$$C = \begin{bmatrix} J & 0 \\ 0 & 0 \end{bmatrix} : \begin{bmatrix} \overline{\mathscr{N}(B) \cap \mathscr{R}(B^p)} \\ (\mathscr{N}(B) \cap \mathscr{R}(B^p))^{\perp} \end{bmatrix} \to \begin{bmatrix} (\mathscr{R}(A) + \mathscr{N}(A^p))^{\perp} \\ \mathscr{R}(A) + \mathscr{N}(A^p) \end{bmatrix}, \quad (3.37)$$

where $p \in \mathbb{N}$ is such that $p \geq \max\{\mathrm{asc}(A), \mathrm{dsc}(B)\}$ and $J \in \mathscr{B}(\overline{\mathscr{N}(B) \cap \mathscr{R}(B^p)}, (\mathscr{R}(A) + \mathscr{N}(A^p))^{\perp})$ is any invertible operator. The following are equivalent:

(i) *$\mathrm{dsc}(M_C) \leq p$,*
(ii) *for any $x \in \mathscr{H}$ and $y \in \mathscr{K}$, there exist $x' \in \mathscr{H}$ and $y' \in \mathscr{K}$ such that*

$$A^p x = A^{p+1} x' + A^p C y', \quad (3.38)$$

and

$$y - By' \in \mathscr{N}(C) \cap \mathscr{N}(CB) \cap \ldots \cap \mathscr{N}(CB^{p-1}) \cap \mathscr{N}(B^p). \quad (3.39)$$

(iii) *$\mathscr{K} = \mathscr{R}(B) + \mathscr{N}(C) \cap \mathscr{N}(CB) \cap \mathscr{N}(CB^2) \cap \cdots \cap \mathscr{N}(CB^{p-1}) \cap \mathscr{N}(B^p)$.*

Proof (i) $\Leftrightarrow$ (ii) Since for any $k \in \mathbb{N}$

$$M_C^k = \begin{bmatrix} A^k & A^{k-1}C + A^{k-2}CB + \ldots + ACB^{k-2} + CB^{k-1} \\ 0 & B^k \end{bmatrix} : \begin{bmatrix} \mathscr{H} \\ \mathscr{K} \end{bmatrix} \to \begin{bmatrix} \mathscr{H} \\ \mathscr{K} \end{bmatrix},$$

it follows that $\mathrm{dsc}(M_C) \leq p$ if and only if for any $x \in \mathscr{H}$ and $y \in \mathscr{K}$, there exist $x' \in \mathscr{H}$ and $y' \in \mathscr{K}$ such that

$$A^p x + A^{p-1} Cy + A^{p-2} CBy + \ldots + ACB^{p-2} y + CB^{p-1} y =$$
$$A^{p+1} x' + A^p Cy' + A^{p-1} CBy' + \ldots + ACB^{p-1} y' + CB^p y' \qquad (3.40)$$
$$\text{and } B^p y = B^{p+1} y'.$$

The case $p = 1$ is evident, so suppose that $p > 1$. If we suppose that $\mathrm{dsc}(M_C) \leq p$, by the second equality in (3.32) we get that $y - By' \in \mathcal{N}(B^p)$. Since $\mathcal{R}(C) \subseteq \mathcal{R}(A)^\perp$, by the first equality in (3.32) we get that $y - By' \in \mathcal{N}(CB^{p-1})$ and

$$A^p x + A^{p-1} Cy + A^{p-2} CBy + \ldots + ACB^{p-2} y =$$
$$A^{p+1} x' + A^p Cy' + A^{p-1} CBy' + \ldots + ACB^{p-1} y'. \qquad (3.41)$$

By (3.41), we have that

$$A^{p-1} x + A^{p-2} Cy + \ldots + CB^{p-2} y - (A^p x' + A^{p-1} Cy' + \ldots + CB^{p-1} y') \in \mathcal{N}(A) \subseteq \mathcal{N}(A^p)$$

which implies that $CB^{p-2} y - CB^{p-1} y' \in \mathcal{N}(A^p) + \mathcal{R}(A)$, i.e., $y - By' \in \mathcal{N}(CB^{p-2})$. Continuing in the same manner, we get that (3.39) holds. Now, by (3.32) it follows that (3.31) is also satisfied.

If (ii) holds, then evidently (3.32) is satisfied, i.e., $\mathrm{dsc}(M_C) \leq p$.

(ii) $\Rightarrow$ (iii) Evidently (3.39) implies (iii).

(iii) $\Rightarrow$ (ii) Let $x \in \mathcal{H}$ and $y \in \mathcal{K}$ be arbitrary. Then there exists $y_0 \in \mathcal{K}$ such that

$$y - By_0 \in \mathcal{N}(C) \cap \mathcal{N}(CB) \cap \ldots \cap \mathcal{N}(CB^{p-1}) \cap \mathcal{N}(B^p).$$

Let $\mathcal{S} = \mathcal{R}(A) + \mathcal{N}(A^p)$. By the definition of the operator C, for given x there exists $y_{00} \in \mathcal{N}(B) \cap \mathcal{R}(B^p)$ such that $(I - P_{\mathcal{S}})x = Jy_{00} = Cy_{00}$. Since $\mathcal{N}(B)$ is closed we have $By_{00} = 0$. Define $y' = P_{\mathcal{N}(B)^\perp} y_0 + y_{00}$. Then $By' = By_0$ and $Cy' = Cy_{00}$ which implies that

$$y - By' \in \mathcal{N}(C) \cap \mathcal{N}(CB) \cap \ldots \cap \mathcal{N}(CB^{p-1}) \cap \mathcal{N}(B^p)$$

and that

$$(I - P_{\mathcal{S}})x = Cy'. \qquad (3.42)$$

Now, $A^p x = A^p Cy' + A^p P_{\mathcal{S}} x$. Since $P_{\mathcal{S}} x \in \mathcal{R}(A) + \mathcal{N}(A^p)$ it follows that $A^p P_{\mathcal{S}} x \in \mathcal{R}(A^{p+1})$ so there exists $x' \in \mathcal{H}$ such that $A^p P_{\mathcal{S}} x = A^{p+1} x'$. Now,

$$A^p x = A^{p+1} x' + A^p Cy'.$$

$\square$

Now, we are ready to make clear which conditions on the operators A and B are necessary for the existence of some $C \in \mathcal{B}(\mathcal{K}, \mathcal{H})$ such that the operator M_C is

Drazin invertible. Combining Lemma 2.6 from [29], Lemma 3.12 and Theorem 3.24, we obtain the following result:

Theorem 3.25 *Let $A \in \mathcal{B}(\mathcal{H})$ and $B \in \mathcal{B}(\mathcal{K})$ be given operators. If there exists an operator $C \in \mathcal{B}(\mathcal{K}, \mathcal{H})$ such that M_C is Drazin invertible, then the following hold:*

(i) $\mathrm{asc}(A) < \infty$,

(ii) $\mathrm{dsc}(B) < \infty$,

(iii) *There exists a constant $\delta > 0$ such that $A - \lambda$ is left invertible, $B - \lambda$ is right invertible and*

$$d(A - \lambda) = n(B - \lambda) = \dim \mathcal{N}(B) \cap \mathcal{R}(B^{\mathrm{dsc}(B)}),$$

for every $\lambda \in K(0, \delta) \setminus \{0\}$.

We will show that the three conditions above together with the assumption that both subspaces $\mathcal{R}(A^{\mathrm{asc}(A)+1})$ and $\mathcal{R}(B^{\mathrm{dsc}(B)})$ are closed (thus meaning that A is left Drazin invertible and B is right Drazin invertible) are actually sufficient for the existence of a Drazin completion of the operator matrix in question.

In [16], the authors correctly showed that $\mathrm{asc}(M_C) < \infty$, for $C \in \mathcal{B}(\mathcal{K}, \mathcal{H})$ given by the following:

$$C = \begin{bmatrix} J & 0 \\ 0 & 0 \end{bmatrix} : \begin{bmatrix} \mathcal{N}(B) \cap \mathcal{R}(B^p) \\ (\mathcal{N}(B) \cap \mathcal{R}(B^p))^{\perp} \end{bmatrix} \rightarrow \begin{bmatrix} (\mathcal{R}(A) + \mathcal{N}(A^p))^{\perp} \\ \mathcal{R}(A) + \mathcal{N}(A^p) \end{bmatrix}, \quad (3.43)$$

where $p \geq \max\{\mathrm{asc}(A), \mathrm{dsc}(B)\}$ and J is an invertible operator. However we will show that the operator C as defined in (3.37) by the authors indeed does the trick. To properly show that, we first give an equivalent description of when exactly the operator M_C is Drazin invertible for this particular choice of C.

Theorem 3.26 *Let $A \in \mathcal{B}(\mathcal{H})$, $B \in \mathcal{B}(\mathcal{K})$ be given operators such that*

(i) *A is left Drazin invertible,*

(ii) $\mathrm{dsc}(B) < \infty$,

(iii) *There exists a constant $\delta > 0$ such that $d(A - \lambda) = n(B - \lambda)$, for every $\lambda \in K(0, \delta) \setminus \{0\}$.*

Then M_C is Drazin invertible for $C \in \mathcal{B}(\mathcal{K}, \mathcal{H})$ given by (3.37) if and only if

$$\mathcal{K} = \mathcal{R}(B) + \mathcal{N}(C) \cap \mathcal{N}(CB) \cap \mathcal{N}(CB^2) \cap \cdots \cap \mathcal{N}(CB^{p-1}) \cap \mathcal{N}(B^p).$$

Proof In [23] it is proved that $\mathrm{asc}(M_C) \leq p$. Thus we can conclude that M_C is Drazin invertible if and only if $\mathrm{dsc}(M_C) \leq p$. Now the assertion follows by Lemma 3.14.   □

Remark If $B \in \mathcal{B}(\mathcal{K})$ is right Drazin invertible and is given by (3.36), where $p = \mathrm{dsc}(B)$, and if $C \in \mathcal{B}(\mathcal{K}, \mathcal{H})$ is given by (3.37), then

$$\mathcal{N}(C) \cap \mathcal{N}(CB) \cap \mathcal{N}(CB^2) \cap \cdots \cap \mathcal{N}(CB^{p-1}) \cap \mathcal{N}(B^p) =$$
$$\mathcal{N}(B_1) \cap \mathcal{N}(B_1 B_2) \cap \cdots \cap \mathcal{N}(B_1 B_2^{p-1}) \cap \mathcal{N}(B_2^p). \tag{3.44}$$

Indeed, this is a consequence of the following equalities:

$$\mathcal{N}(C) = [\mathcal{N}(B) \cap \mathcal{R}(B^p)]^{\perp}, \quad \mathcal{N}(CB^k) = [\mathcal{N}(B) \cap \mathcal{R}(B^p)] \oplus \mathcal{N}(B_1 B_2^{k-1}),$$

the latter of which follows from the representation

$$CB^k = \begin{bmatrix} 0 & JB_1 B_2^{k-1} \\ 0 & 0 \end{bmatrix} : \begin{bmatrix} \mathcal{N}(B) \cap \mathcal{R}(B^p) \\ (\mathcal{N}(B) \cap \mathcal{R}(B^p))^{\perp} \end{bmatrix} \rightarrow \begin{bmatrix} (\mathcal{R}(A) + \mathcal{N}(A^p))^{\perp} \\ \mathcal{R}(A) + \mathcal{N}(A^p) \end{bmatrix}.$$

Since we make use of Lemma 3.13 in the following theorem, in contrast to the previous auxiliary results here we must assume that B is right Drazin invertible.

Theorem 3.27 *Let $A \in \mathcal{B}(\mathcal{H})$, $B \in \mathcal{B}(\mathcal{K})$ be given operators such that*

(i) *A is left Drazin invertible,*
(ii) *B is right Drazin invertible,*
(iii) *There exists a constant $\delta > 0$ such that $d(A - \lambda) = n(B - \lambda)$, for every $\lambda \in K(0, \delta) \setminus \{0\}$.*

Then M_C is Drazin invertible for $C \in \mathcal{B}(\mathcal{K}, \mathcal{H})$ given by (3.37).

Proof Let B be given by (3.36). Suppose first that $p = 1$. By Theorem 3.26, to prove that M_C is Drazin invertible for $C \in \mathcal{B}(\mathcal{K}, \mathcal{H})$ given by (3.37) it is sufficient to prove that $\mathcal{K} = \mathcal{R}(B) + \mathcal{N}(C) \cap \mathcal{N}(B)$. Since $dsc(B) = 1$, by Lemma 3.11 it follows that $\mathcal{K} = \mathcal{R}(B) + \mathcal{N}(B)$. Put $\mathcal{S} = \mathcal{N}(B) \cap \mathcal{R}(B)$. As $\mathcal{N}(B) = \mathcal{S} \oplus \mathcal{N}(B_1) \cap \mathcal{N}(B_2)$, and $\mathcal{S} \subseteq \mathcal{R}(B)$, it follows that $\mathcal{K} = \mathcal{R}(B) + \mathcal{N}(B_1) \cap \mathcal{N}(B_2)$. Since $\mathcal{N}(C) \cap \mathcal{N}(B) = \mathcal{N}(B_1) \cap \mathcal{N}(B_2)$, we have $\mathcal{K} = \mathcal{R}(B) + \mathcal{N}(C) \cap \mathcal{N}(B)$.

Now, consider the case when $p > 1$. By Theorem 3.26, we have to prove that

$$\mathcal{K} = \mathcal{R}(B) + \mathcal{N}(C) \cap \mathcal{N}(CB) \cap \mathcal{N}(CB^2) \cap \cdots \cap \mathcal{N}(CB^{p-1}) \cap \mathcal{N}(B^p)$$

which is by (3.44) from the preceding remark equivalent with

$$\mathcal{K} = \mathcal{R}(B) + \mathcal{N}(B_1) \cap \mathcal{N}(B_1 B_2) \cap \cdots \cap \mathcal{N}(B_1 B_2^{p-1}) \cap \mathcal{N}(B_2^p).$$

Since $\mathcal{K} = \mathcal{R}(B) + \mathcal{N}(B^p)$, which is equivalent with

$$\mathcal{K} = \mathcal{R}(B) + \mathcal{N}(B_1 B_2^{p-1}) \cap \mathcal{N}(B_2^p),$$

it is sufficient to prove that

$$\mathcal{N}(B_1 B_2^{p-1}) \cap \mathcal{N}(B_2^p) \subseteq$$
$$\mathcal{R}(B) + \mathcal{N}(B_1) \cap \mathcal{N}(B_1 B_2) \cap \cdots \cap \mathcal{N}(B_1 B_2^{p-1}) \cap \mathcal{N}(B_2^p). \tag{3.45}$$

Take arbitrary $x \in \mathcal{N}(B_1 B_2^{p-1}) \cap \mathcal{N}(B_2^p)$. Now $B_1 B_2^{p-2} \in \mathcal{B}((\mathcal{N}(B) \cap \mathcal{R}(B^p))^{\perp}$, $\mathcal{N}(B) \cap \mathcal{R}(B^p))$ so $B_1 B_2^{p-2} x \in \mathcal{N}(B) \cap \mathcal{R}(B^p)$. Lemma 6.13 says that the operator $B_1 B_2^{p-1} \in \mathcal{B}((\mathcal{N}(B) \cap \mathcal{R}(B^p))^{\perp}, \mathcal{N}(B) \cap \mathcal{R}(B^p))$ maps the subspace $\mathcal{N}(B_2^p)$ onto $\mathcal{N}(B) \cap \mathcal{R}(B^p)$. Hence there exists $y \in \mathcal{N}(B_2^p)$ such that $B_1 B_2^{p-2} x = B_1 B_2^{p-1} y$. Now, $x - B_2 y \in \mathcal{N}(B_1 B_2^{p-2}) \cap \mathcal{N}(B_1 B_2^{p-1}) \cap \mathcal{N}(B_2^p)$ which together with (ii) of Lemma 6.13 gives that $x \in \mathcal{R}(B) + \mathcal{N}(B_1 B_2^{p-2}) \cap \mathcal{N}(B_1 B_2^{p-1}) \cap \mathcal{N}(B_2^p)$. We have thus shown that $\mathcal{N}(B_1 B_2^{p-1}) \cap \mathcal{N}(B_2^p) \subseteq \mathcal{R}(B) + \mathcal{N}(B_1 B_2^{p-2}) \cap \mathcal{N}(B_1 B_2^{p-1}) \cap \mathcal{N}(B_2^p)$.

Continuing in the same manner we further obtain consecutively

$$\mathcal{N}(B_1 B_2^{p-2}) \cap \mathcal{N}(B_1 B_2^{p-1}) \cap \mathcal{N}(B_2^p) \subseteq$$
$$\mathcal{R}(B) + \mathcal{N}(B_1 B_2^{p-3}) \cap \mathcal{N}(B_1 B_2^{p-2}) \cap \mathcal{N}(B_1 B_2^{p-1}) \cap \mathcal{N}(B_2^p),$$

..., and finally

$$\mathcal{N}(B_1 B_2) \cap \cdots \cap \mathcal{N}(B_1 B_2^{p-1}) \cap \mathcal{N}(B_2^p) \subseteq$$
$$\mathcal{R}(B) + \mathcal{N}(B_1) \cap \mathcal{N}(B_1 B_2) \cap \cdots \cap \mathcal{N}(B_1 B_2^{p-1}) \cap \mathcal{N}(B_2^p).$$

Taking into account all these inclusions, we immediately get (3.45).   □

Open question: We wonder if at least one of the conditions (if not both) (i) and (ii) in Theorem 3.27 could be relaxed to the requirement that simply $asc(A) < \infty$ and $dsc(B) < \infty$, respectively?

References

1. Pavlović, V., Cvetković-Ilić, D.S.: Applications of completions of operator matrices to reverse order law for {1}-inverses of operators on Hilbert spaces. Linear Algebra Appl. **484**, 219–236 (2015)
2. Du, H.K., Pan, J.: Perturbation of spectrums of 2×2 operator matrices. Proc. Am. Math. Soc. **121**, 761–776 (1994)
3. Han, J.K., Lee, H.Y., Lee, W.Y.: Invertible completions of 2×2 upper triangular operator matrices. Proc. Am. Math. Soc. **128**, 119–123 (1999)
4. Cvetković-Ilić, D.S.: Completion of operator matrices to the invertibility and regularity. Electron. J. Linear Algebra **30**, 530–549 (2015)
5. Chen, A., Hai, G.: Perturbations of the right and left spectra for operator matrices. J. Oper. Theory **67**(1), 207–214 (2012)
6. Takahashi, K.: Invertible completions of operator matrices. Integral Equ. Oper. Theory **21**, 355–361 (1995)
7. Chen, A., Hou, G.L., Hai, G.J.: Perturbation of spectra for a class of 2×2 operator matrices. Acta Math. Appl. Sin. **28**(4), 711–720 (2012)
8. Hai, G., Bao, C., Chen, A.: Invertibility for linear combinations of bounded linear operators with closed range, Operators and Matrices, **11**(3), 715–723 (2017)
9. Hai, G., Chen, A.: On the right(left) invertible completions for operator matrices. Integral Equ. Oper. Theory **67**, 79–93 (2010)

10. Hai, G., Chen, A.: The semi-fredholmness of 2×2 operator matrices. J. Math. Anal. Appl. **352**, 733–738 (2009)
11. Li, Y., Du, H.: The intersection of essential approximate point spectra of operator matrices. J. Math. Anal. Appl. **323**, 1171–1183 (2006)
12. Li, Y., Sun, X., Du, H.: The intersection of left (right) spectra of 2×2 upper triangular operator matrices. Linear Algebra Appl. **418**, 112–121 (2006)
13. Yao, X.: A spectral completions of 2×2 operator matrices. Int. Math. Forum **2**(55), 2737–2745 (2007)
14. Yuan, L., Du, H.K.: On essential spectra of 2×2 operator matrices. J. Math. Res. Expo. **28**(2), 359–365 (2008)
15. Douglas, R.G.: Banach Algebra Techniques in Operator Theory, 2nd edn. Springer, New York (1998)
16. Baksalary, J.K., Baksalary, O.M.: Nonsingularity of linear combinations of idempotent matrices. Linear Algebra Appl. **388**, 25–29 (2004)
17. Koliha, J.J., Rakočević, V.: Invertibility of the sum of idempotents. Linear Multilinear Algebra **50**, 285–292 (2002)
18. Du, H.K., Yao, X.Y., Deng, C.Y.: Invertibility of linear combinations of two idempotents. Proc. Am. Math. Soc. **134**, 1451–1457 (2005)
19. Koliha, J.J., Rakočević, V.: Stability theorems for linear combinations of idempotents. Integral Equ. Oper. Theory **58**, 597–601 (2007)
20. Buckholtz, D.: Inverting the difference of Hilbert space projections. Am. Math. Mon. **104**, 60–61 (1997)
21. Du, H., Deng, C.: The representation and characterization of Drazin inverses of operators on a Hilbert space. Linear Algebra Appl. **407**, 117–124 (2005)
22. Fillmore, P.A., Williams, J.P.: On operator ranges. Adv. Math. **7**, 254–281 (1971)
23. Izumino, S.: Convergence of generalized inverses and spline projectors. J. Approx. Theory **38**(3), 269–278 (1983)
24. Boumazgour, M.: Drazin invertibility of upper triangular operator matrices. Linear Multilinear Algebra **61**(5), 627–634 (2013)
25. Cvetković-Ilić, D.S., Pavlović, V.: Drazin invertibility of upper triangular operator matrices, Linear Multilinear Algebra. (accepted)
26. Grabiner, S., Zemanek, J.: Ascent, descent and ergodic properties of linear operators. J. Oper. Theory **48**, 69–81 (2002)
27. Taylor, A.E., Lay, D.C.: Introduction to Functional Analysis, 2nd edn. Wiley, New York (1980)
28. Lay, D.C.: Spectral properties of generalized inverses of linear operators. SIAM J. Appl. Math. **29**, 103–109 (1975)
29. Zhang, S.F., Zhong, H.J., Jiang, Q.: Drazin spectrum of operator matrices on the Banach space. Linear Algebra Appl. **429**, 2067–2075 (2008)

Chapter 4
Generalized Inverses and Idempotents

In the recent years, a number of researchers have focused their attention to questions concerning ordinary invertibility of differences, sums and linear combinations of idempotents (see [1–7]). Consequently, this topic was considered in some papers, extending ordinary invertibility to Drazin invertibility, for instance [3, 8–13], and differences and sums to linear combinations of idempotents [8–10, 13].

In this chapter, we will present related results on Drazin invertibility of the product and difference of two idempotent operators, Drazin and generalized Drazin invertibility of linear combinations of idempotents, commutators and anticommutators in Banach algebras as well as on the Moore-Penrose inverse of a linear combination of commuting generalized and hypergeneralized projectors.

4.1 Drazin Invertibility of the Product and Difference of Two Idempotent Operators

The aim of this section is to present, using the spectral theory of linear operators, necessary and sufficient conditions for Drazin invertibility of the product and difference of two idempotents on an infinite dimensional Hilbert space.

Throughout this section, we assume that P, Q are idempotents in $\mathscr{B}(\mathscr{H})$. The matrix forms of P, Q with respect to the decomposition $\mathscr{H} = \mathscr{R}(P) \oplus \mathscr{N}(P)$ are given by

$$P = \begin{bmatrix} I & 0 \\ 0 & 0 \end{bmatrix}, \quad Q = \begin{bmatrix} Q_1 & Q_2 \\ Q_3 & Q_4 \end{bmatrix}. \tag{4.1}$$

It is interesting to remark that without loss of generality, we can assume that one of the operators P and Q is an orthogonal projection: If P and Q are idempotents, by Lemma 1.1 [14], there exists an invertible operator $S \in \mathscr{B}(\mathscr{H})$ such that SPS^{-1} is an orthogonal projection. Hence, we can consider $P_1 = SPS^{-1}$ and $Q_1 = SQS^{-1}$

© Springer Nature Singapore Pte Ltd. 2017   89
D. Cvetković Ilić and Y. Wei, *Algebraic Properties of Generalized Inverses*,
Developments in Mathematics 52, DOI 10.1007/978-981-10-6349-7_4

instead of P and Q. Obviously, Q_1 is an idempotent and Drazin invertibility of PQ, $P \pm Q$, $P + Q - PQ$ is equivalent to Drazin invertibility of $P_1 Q_1$, $P_1 \pm Q_1$, $P_1 + Q_1 - P_1 Q_1$, respectively. So, from now on in the proofs we always assume that P is an idempotent and Q is an orthogonal projection.

We shall begin with some lemmas, the first of which is proved in [15] (see also [16, 17] for the finite dimensional case and [18] for the elements in a Banach algebra).

Lemma 4.1 *If $A \in \mathscr{B}(\mathscr{X})$ and $B \in \mathscr{B}(\mathscr{Y})$ are Drazin invertible, $C \in \mathscr{B}(\mathscr{Y}, \mathscr{X})$ and $D \in \mathscr{B}(\mathscr{X}, \mathscr{Y})$, then*

$$M = \begin{bmatrix} A & C \\ 0 & B \end{bmatrix} \text{ and } N = \begin{bmatrix} A & 0 \\ D & B \end{bmatrix}$$

are also Drazin invertible and

$$M^d = \begin{bmatrix} A^d & S \\ 0 & B^d \end{bmatrix}, \quad N^d = \begin{bmatrix} B^d & 0 \\ S & A^d \end{bmatrix}, \tag{4.2}$$

where $S = \sum_{n=0}^{\infty} (A^d)^{n+2} C B^n B^\pi + \sum_{n=0}^{\infty} A^\pi A^n C (B^d)^{n+2} - A^d C B^d$.

Lemma 4.2 *Let $M \in \mathscr{B}(H \oplus K)$ have the operator matrix form*

$$M = \begin{pmatrix} A & B \\ 0 & C \end{pmatrix}. \tag{4.3}$$

If two of the elements M, A and C are Drazin invertible, then the third element is also Drazin invertible. In particular, if $B = 0$, then M is Drazin invertible if and only if A and C are Drazin invertible.

Proof The first part of the assertion is just a special case of Theorem 3.2 (ii) [18]. As for the second part we just have to note that if $B = 0$, then $\sigma(M) = \sigma(A) \cup \sigma(C)$. $\square$

The following lemma is proved in [19] for the finite dimensional case but the proof is similar for bounded linear operators.

Lemma 4.3 *Let $M \in \mathscr{B}(H \oplus K)$ have the operator matrix form*

$$M = \begin{pmatrix} 0 & A \\ B & 0 \end{pmatrix}. \tag{4.4}$$

Then M is Drazin invertible if and only if AB (or BA) is Drazin invertible. In this case,

$$M^d = \begin{pmatrix} 0 & (AB)^d A \\ B(AB)^d & 0 \end{pmatrix} = \begin{pmatrix} 0 & A(BA)^d \\ (BA)^d B & 0 \end{pmatrix}.$$

In the following theorem, we can find some equivalents of Drazin invertibility of PQ based on the facts that $\sigma(PQ) \cup \{0\} = \sigma(QP) \cup \{0\}$ and that P is Drazin invertible if and only if P^* is Drazin invertible.

Theorem 4.1 *Let P be an idempotent and Q be an orthogonal projection in $\mathscr{B}(\mathscr{H})$. The following statements are equivalent:*

(1) PQ *is Drazin invertible,*
(2) QP *is Drazin invertible,*
(3) PQP *is Drazin invertible,*
(4) QPQ *is Drazin invertible,*
(5) P^*Q *is Drazin invertible,*
(6) QP^* *is Drazin invertible,*
(7) P^*QP^* *is Drazin invertible,*
(8) QP^*Q *is Drazin invertible.*

If in Theorem 4.1 we replace P and Q by $I - P$ and $I - Q$, respectively, we get the following:

Corollary 4.1 *Let P be an idempotent and Q be an orthogonal projection in $\mathscr{B}(\mathscr{H})$. The following statements are equivalent:*

(1) $(I - P)(I - Q)$ *is Drazin invertible,*
(2) $(I - Q)(I - P)$ *is Drazin invertible,*
(3) $(I - P)(I - Q)(I - P)$ *is Drazin invertible,*
(4) $(I - Q)(I - P)(I - Q)$ *is Drazin invertible,*
(5) $(I - P)^*(I - Q)$ *is Drazin invertible,*
(6) $(I - Q)(I - P)^*$ *is Drazin invertible,*
(7) $(I - P)^*(I - Q)(I - P)^*$ *is Drazin invertible,*
(8) $(I - Q)(I - P)^*(I - Q)$ *is Drazin invertible.*

Lemma 4.4 *Let $A, B \in \mathscr{B}(\mathscr{H})$. Then $I - AB$ is Drazin invertible if and only if $I - BA$ is Drazin invertible.*

Proof Since, $\sigma(AB) \cup \{0\} = \sigma(BA) \cup \{0\}$ the assertion follows.   □

Let us remark that for ordinary invertibility the result analogous to that given by the next theorem is proved in Theorem 3.2 [6].

Theorem 4.2 *Let P, Q be idempotents in $\mathscr{B}(\mathscr{H})$. Then $P - Q$ is Drazin invertible if and only if $I - PQ$ and $P + Q - PQ$ are Drazin invertible.*

Proof The matrix forms of P, Q with respect to the decomposition $\mathscr{H} = \mathscr{R}(P) \oplus \mathscr{N}(P)$ are given by (4.1). If $I - PQ$ and $P + Q - PQ$ are Drazin invertible, by Lemma 4.2 we conclude that $I_{\mathscr{R}(P)} - Q_1$ is Drazin invertible in $\mathscr{R}(P)$ and Q_4 is Drazin invertible in $\mathscr{N}(P)$. Since,

$$(P - Q)^2 = \begin{bmatrix} I - Q_1 & 0 \\ 0 & Q_4 \end{bmatrix} : \mathscr{R}(P) \oplus \mathscr{N}(P) \to \mathscr{R}(P) \oplus \mathscr{N}(P)$$

and

$$\sigma((P - Q)^2) = \{\lambda^2 : \lambda \in \sigma(P - Q)\}, \qquad (4.5)$$

we have that $P - Q$ is Drazin invertible.

If $P - Q$ is Drazin invertible, by (4.5) we have that $(P - Q)^2$ is Drazin invertible. Now, we can check that

$$(I - PQP)^d = \left((P - Q)^d\right)^2 P + I - P$$

also, by Lemma 4.4, $I - PQ$ is Drazin invertible. Similarly, from the Drazin invertibility of $(I-P)-(I-Q) = -(P-Q)$, it follows that $I-(I-P)(I-Q) = P+Q-PQ$ is Drazin invertible. $\qquad \square$

Also, we have the following result:

Corollary 4.2 *Let P, Q be idempotents in $\mathcal{B}(\mathcal{H})$. The following statements are equivalent:*

(1) $I - PQ$ *is Drazin invertible,*
(2) $P - PQ$ *is Drazin invertible,*
(3) $I - PQP$ *is Drazin invertible,*
(4) $P - PQP$ *is Drazin invertible,*
(5) $I - QP$ *is Drazin invertible,*
(6) $Q - QP$ *is Drazin invertible,*
(7) $I - QPQ$ *is Drazin invertible,*
(8) $Q - QPQ$ *is Drazin invertible.*

Proof Using the matrix forms of P, Q given by (4.1), it is easy to see that $(1)-(4)$ are all equivalent to the fact that $I_{\mathcal{R}(P)} - Q_1$ is Drazin invertible in $\mathcal{R}(P)$. Analogously, $(5) - (8)$ are equivalent. Evidently, (1) is equivalent to (5). $\qquad \square$

As before, if in Corollary 3.4 we replace P and Q by $I - P$ and $I - Q$, respectively, we have the following:

Corollary 4.3 *Let P, Q be idempotents in $\mathcal{B}(\mathcal{H})$. The following statements are equivalent:*

(1) $P + Q - PQ$ *is Drazin invertible,*
(2) $Q - PQ$ *is Drazin invertible,*
(3) $P + (I - P)Q - (I - P)QP$ *is Drazin invertible,*
(4) $(I - P)Q(I - P)$ *is Drazin invertible,*
(5) $P + Q - QP$ *is Drazin invertible,*
(6) $P - QP$ *is Drazin invertible,*
(7) $Q + (I - Q)P - (I - Q)PQ$ *is Drazin invertible,*
(8) $(I - Q)P(I - Q)$ *is Drazin invertible.*

Corollary 4.4 (1) *([4]) Let P and Q be orthogonal projections in $\mathscr{B}(\mathscr{H})$. Then the conditions in Theorem 4.2, Corollary 4.2 and Corollary 4.3 are all equivalent to the fact that $P + Q$ is Drazin invertible.*
(2) *Let $P, Q \in \mathscr{B}(\mathscr{H})$ be idempotents. Then $P - Q$ is Drazin invertible if and only if one of the conditions from Corollary 4.2 and one of the conditions from Corollary 4.3 hold.*

4.2 Drazin and Generalized Drazin Invertibility of Combinations of Idempotents, Commutators and Anticommutators in Banach Algebra

The basic motivation for the results presented in this section was the result of Koliha and Rakočević concerning invertibility of the difference and the sum of idempotents in the setting of rings:

$$p - q \in \mathscr{R}^{inv} \iff 1 - pq \in \mathscr{R}^{inv} \text{ and } p + q \in \mathscr{R}^{inv}.$$

In this section, we will show that in the case of Drazin and generalized Drazin invertibility in a Banach algebra $\mathscr{A}$ we have the equivalence between the following three conditions

$$p - q \in \mathscr{A}^D, \quad p + q \in \mathscr{A}^D \text{ and } 1 - pq \in \mathscr{A}^D.$$

More generally we will look at linear combinations $\alpha p + \beta q$ and consider generalized Drazin and Drazin invertibility of them.

For the sake of brevity we will sometimes use the terms 'd-invertible' and 'D-invertible' for 'Drazin invertible' and 'generalized Drazin invertible', respectively.

First, we gather various known results we will rely on.

Lemma 4.5 *[20] Let $a \in \mathscr{A}$. Then $a \in \mathscr{A}^D$ if and only if 0 is not an accumulation point of $\sigma(a)$, and $a \in \mathscr{A}^d$ if and only if 0 is not an essential singularity of the resolvent $(\lambda - a)^{-1}$ of a.*

Lemma 4.6 *If $b, c \in \mathscr{A}$, then $\sigma(bc) \setminus \{0\} = \sigma(cb) \setminus \{0\}$. Further,*

$$bc \in \mathscr{A}^{qnil} \leftrightarrow cb \in \mathscr{A}^{qnil}, \qquad bc \in \mathscr{A}^D \leftrightarrow cb \in \mathscr{A}^D,$$
$$bc \in \mathscr{A}^{nil} \leftrightarrow cb \in \mathscr{A}^{nil}, \qquad bc \in \mathscr{A}^d \leftrightarrow cb \in \mathscr{A}^d.$$

Proof The spectral relation is well known; the rest follows from it on application of the preceding lemma and the equation

$$\lambda(\lambda - cb)^{-1} = 1 + c(\lambda - bc)^{-1}b, \quad \lambda \neq 0.$$

$\square$

Combining the results of [21] on isolated spectral points and Exercise VII.5.21 in [22] on poles of the resolvent (interpreted for elements of a Banach algebra in place of operators), we obtain the following result.

Lemma 4.7 *Let* $a \in \mathscr{A}$, *let* f *be a function holomorphic in an open neighborhood of* $\sigma(a)$, *and let* $f^{-1}(0) \cap \sigma(a) = \{\lambda_1, \ldots, \lambda_m\}$ *(a finite set). Then*

$$f(a) \in \mathscr{A}^{\mathsf{D}} \iff \lambda_i - a \in \mathscr{A}^{\mathsf{D}} \text{ for all } i = 1, \ldots, m,$$
$$f(a) \in \mathscr{A}^{\mathsf{d}} \iff \lambda_i - a \in \mathscr{A}^{\mathsf{d}} \text{ for all } i = 1, \ldots, m.$$

If $p \in \mathscr{A}^{\mathsf{idem}}$, then each element a of $\mathscr{A}$ has a matrix representation

$$a = \begin{bmatrix} a_{11} & a_{12} \\ a_{21} & a_{22} \end{bmatrix}_p, \quad \text{where } a_{ij} = p_i a p_j, \ p_1 = p, \ p_2 = 1 - p, \ i, j = \overline{1, 2}.$$

Lemma 4.8 *Let* $a \in \mathscr{A}$, $p \in \mathscr{A}^{\mathsf{idem}}$, $\delta \in \{\mathsf{D}, \mathsf{d}\}$, *and let*

$$a = \begin{bmatrix} b & d \\ 0 & c \end{bmatrix}_p.$$

If two of the elements a, b, c *are in* $\mathscr{A}^{\delta}$, *then so is the third. In each case,*

$$a^{\delta} = \begin{bmatrix} b^{\delta} & u \\ 0 & c^{\delta} \end{bmatrix}_p$$

with a uniquely determined $u \in p\mathscr{A}(1 - p)$.

Proof The case $\delta = \mathsf{D}$ follows from Theorem 2.3 of [18]. Let b and c be Drazin invertible. By the preceding part of the proof, $a \in \mathscr{A}^{\mathsf{D}}$. Let $k \in \mathbb{N}$. Then

$$(a(1 - aa^{\mathsf{D}}))^{2k} = \begin{bmatrix} (b(1 - bb^{\mathsf{d}}))^k & u_k \\ 0 & (c(1 - cc^{\mathsf{d}}))^k \end{bmatrix}_p^2$$

with a uniquely determined $u_k \in p\mathscr{A}(1 - p)$. If $k \geq \max\{\mathrm{ind}(b), \mathrm{ind}(c)\}$, then $(b(1 - bb^{\mathsf{d}}))^k = 0 = (c(1 - cc^{\mathsf{d}}))^k$, and $(a(1 - aa^{\mathsf{D}}))^{2k} = 0$ and $a^{\mathsf{D}} = a^{\mathsf{d}}$ with $\mathrm{ind}(a) \leq 2\max\{\mathrm{ind}(b), \mathrm{ind}(c)\}$.

Next assume that a and c are Drazin invertible. Again $b \in \mathscr{A}^{\mathsf{D}}$ by the first part of the proof. The matrix representation then ensures that $(a(1 - aa^{\mathsf{d}}))^k = 0 =$

$(c(1 - cc^d))^k$ implies $(b(1 - bb^D))^k = 0$, that is, $b^D = b^d$. We note that in this case $\text{ind}(b) \le \max\{\text{ind}(a), \text{ind}(c)\}$. The case $a, b \in \mathscr{A}^d$ is symmetric.

Remark 4.1 There is also a lower triangular version of the preceding lemma.

In the next lemma we extend Lemma 2.1 in [10] to the generalized Drazin inverse. In our proof we succeed in avoiding the use of Cline's formula $(ba)^D = b((ab)^D)^2 a$ originally employed in [10], which would require a proof for the generalized Drazin inverse.

Lemma 4.9 *Let $a \in \mathscr{A}$, $p \in \mathscr{A}^{\text{idem}}$, and let*

$$a = \begin{bmatrix} 0 & b \\ c & 0 \end{bmatrix}_p.$$

If $\delta \in \{D, d\}$, then $a \in \mathscr{A}^\delta$ if and only if $bc \in \mathscr{A}^\delta$, and in this case

$$a^\delta = \begin{bmatrix} 0 & (bc)^\delta b \\ c(bc)^\delta & 0 \end{bmatrix}_p. \tag{4.6}$$

Proof Let $\delta = D$. Let bc be D-invertible and let x be defined by the matrix on the right-hand side of (4.6). A direct calculation shows that $ax = xa$ and $ax^2 = x$. Write $w = a - a^2 x$. Then

$$w = \begin{bmatrix} 0 & u \\ v & 0 \end{bmatrix}_p, \qquad w^2 = \begin{bmatrix} uv & 0 \\ 0 & vu \end{bmatrix}_p,$$

where $u = b - bc(bc)^D b$, $v = c - cbc(bc)^D$ and $uv = bc - (bc)^2(bc)^D$. Since uv is *quasi-nilpotent* by the definition of $(bc)^D$, vu is *quasi-nilpotent* by Lemma 4.6. Hence w^2, and also w, is *quasi-nilpotent*, and x is the D-inverse of a.

Conversely assume that a is D-invertible. Then so is

$$a^2 = \begin{bmatrix} bc & 0 \\ 0 & cb \end{bmatrix}_p,$$

and $\sigma(a^2) \setminus \{0\} = \sigma(bc) \setminus \{0\} = \sigma(cb) \setminus \{0\}$. This implies that bc and cb are D-invertible.

Let $\delta = d$. Suppose that bc is Drazin invertible and put $x = \begin{bmatrix} 0 & (bc)^d b \\ c(bc)^d & 0 \end{bmatrix}_p$.

By computation, we get

$$ax = xa = \begin{bmatrix} bc(bc)^d & 0 \\ 0 & c(bc)^d b \end{bmatrix}_p.$$

and $xax = x$. Since $a^{2k+2} = \begin{bmatrix} (bc)^{k+1} & 0 \\ 0 & (cb)^{k+1} \end{bmatrix}_p$, for $k = \text{ind}(bc)$ it follows that $a^{2k+2}x = a^{2k+1}$. Hence, a is Drazin invertible and $a^d = x$. Moreover, $\text{ind}(a) \leq 2\text{ind}(bc) + 1$.

If a is Drazin invertible, then a^2 is Drazin invertible. Since $a^2 = \begin{bmatrix} bc & 0 \\ 0 & cb \end{bmatrix}_p$, we conclude that bc and cb are Drazin invertible. $\qquad\square$

Lemma 4.10 *Let $p, q \in \mathscr{A}^{\text{idem}}$ and let $\alpha, \beta \in \mathbb{C} \setminus \{0\}$. If $\lambda \in \mathbb{C} \setminus \{0, \alpha, \beta\}$, then*

$$\lambda \in \sigma(\alpha p + \beta q) \iff (1 - \alpha^{-1}\lambda)(1 - \beta^{-1}\lambda) \in \sigma(pq). \tag{4.7}$$

Proof Let $\lambda \in \mathbb{C} \setminus \{0, \alpha, \beta\}$. Observe that

$$(1 - \alpha^{-1}\lambda - p)(\lambda - (\alpha p + \beta q))(1 - \beta^{-1}\lambda - q) = \lambda((1 - \alpha^{-1}\lambda)(1 - \beta^{-1}\lambda) - pq). \tag{4.8}$$

This implies (4.7). $\qquad\square$

Theorem 4.3 *Let $p, q \in \mathscr{A}^{\text{idem}}$ and $\alpha, \beta \in \mathbb{C} \setminus \{0\}$. If $\delta \in \{D, d\}$, then*

$$1 - pq \in \mathscr{A}^\delta \implies \alpha p + \beta q \in \mathscr{A}^\delta. \tag{4.9}$$

Proof Suppose that $1 - pq$ is D-invertible, and for a proof by contradiction assume that 0 is an accumulation point of $\sigma(\alpha p + \beta q)$. There is a sequence (λ_n) in $\sigma(\alpha p + \beta q) \setminus \{0\}$ convergent to zero. According to the preceding lemma, $\mu_n = (1 - \alpha^{-1}\lambda_n)(1 - \beta^{-1}\lambda_n)$ is a sequence in $\sigma(pq) \setminus \{1\}$ convergent to 1; hence $(1 - \mu_n)$ is a sequence in $\sigma(1 - pq) \setminus \{0\}$ convergent to 0 contrary to the assumption about $1 - pq$. This proves $\alpha p + \beta q \in \mathscr{A}^D$.

Assume that $1 - pq \in \mathscr{A}^d$ and write

$$\mu = 1 - (1 - \alpha^{-1}\lambda)(1 - \beta^{-1}\lambda) = (\alpha\beta)^{-1}\lambda(\alpha + \beta - \lambda).$$

Taking into account (4.8) and the inclusion $\sigma(t) \subset \{0, 1\}$ valid for any idempotent t, we conclude that there is $\rho > 0$ such that

$$(\lambda - (\alpha p + \beta q))^{-1} = (1 - \beta^{-1}\lambda - q)\lambda^{-1}(1 - pq - \mu)^{-1}(1 - \alpha^{-1}\lambda - p) \tag{4.10}$$

for all λ satisfying $0 < |\lambda| < \rho$. The resolvent $(\mu - (1 - pq))^{-1}$ has a pole at $\mu = 0$. Expanding $(1 - pq - \mu)^{-1}$ in a Laurent series at $\mu = 0$ and substituting $\mu = (\alpha\beta)^{-1}\lambda(\alpha + \beta - \lambda)$, we obtain the right hand side of (4.10) as a Laurent series in λ with only a finite number of nonzero terms in negative powers of λ. Thus 0 is a pole of the resolvent of $\alpha p + \beta q$, and $\alpha p + \beta q$ is Drazin invertible. $\qquad\square$

We now come to the main result which contrasts the case of the ordinary inverse. In view of the following theorem, the generalized Drazin and Drazin invertibility of the linear combination $\alpha p + \beta q$ of idempotents are independent of the choice of scalars $\alpha, \beta \in \mathbb{C} \setminus \{0\}$, and the case $\alpha + \beta = 0$ is allowed.

Theorem 4.4 *Let $p, q \in \mathscr{A}^{\mathrm{idem}}$, $\alpha, \beta \in \mathbb{C} \setminus \{0\}$, and $\delta \in \{\mathrm{D}, \mathrm{d}\}$. Then*

$$\alpha p + \beta q \in \mathscr{A}^\delta \Longleftrightarrow 1 - pq \in \mathscr{A}^\delta.$$

In particular,

$$p - q \in \mathscr{A}^{\mathrm{D}} \Longleftrightarrow p + q \in \mathscr{A}^{\mathrm{D}} \Longleftrightarrow 1 - pq \in \mathscr{A}^{\mathrm{D}},$$
$$p - q \in \mathscr{A}^{\mathrm{d}} \Longleftrightarrow p + q \in \mathscr{A}^{\mathrm{d}} \Longleftrightarrow 1 - pq \in \mathscr{A}^{\mathrm{d}}.$$

Proof In view of the preceding theorem we only need to prove that $p - q \in \mathscr{A}^\delta$ implies $1 - pq \in \mathscr{A}^\delta$. For this we could turn to the proof of Lemma 3.1 in [5]. Instead, we offer an alternative proof based on a matrix representation which will then be useful in the proof of the next theorem. Relative to the idempotent p,

$$p = \begin{bmatrix} p & 0 \\ 0 & 0 \end{bmatrix}, \ q = \begin{bmatrix} q_1 & q_2 \\ q_3 & q_4 \end{bmatrix}, \ (p-q)^2 = \begin{bmatrix} p - q_1 & 0 \\ 0 & q_4 \end{bmatrix}, \ 1 - pq = \begin{bmatrix} p - q_1 & -q_2 \\ 0 & 1 - p \end{bmatrix}. \quad (4.11)$$

Clearly $p - q \in \mathscr{A}^\delta$ if and only if $(p-q)^2 \in \mathscr{A}^\delta$. The implication $p - q \in \mathscr{A}^\delta \implies 1 - pq \in \mathscr{A}^\delta$ then follows from Lemma 4.8. ∎

Theorem 4.5 *Let $p, q \in \mathscr{A}^{\mathrm{idem}}$ and let*

$$\mathscr{L}_1 = \{p - q, \ p + q, \ 1 - pq, \ p - pq, \ p - qp, \ 1 - pqp, \ p - pqp\},$$
$$\mathscr{L}_2 = \{q - qp, \ q - pq, \ 1 - qpq, \ q - qpq\},$$
$$\mathscr{L}_3 = \{p + q - pq, \ p + q - pq - qp + pqp, \ q - pq - qp + pqp\},$$
$$\mathscr{K} = \{p + q - 1, \ 1 + p - q, \ 1 - p + pq, \ pq, \ qp, \ 1 - p + pqp, \ pqp\}.$$

Let $\delta \in \{\mathrm{D}, \mathrm{d}\}$. If one of the elements of the set $\mathscr{L}_1 \cup \mathscr{L}_2 \cup \mathscr{L}_3$ (resp. $\mathscr{K}$) is in $\mathscr{A}^\delta$, then they all are.

Proof Theorem 4.4 accounts for the first three elements of $\mathscr{L}_1$. For the rest of $\mathscr{L}_1$ use the matrix representations as in (4.11),

$$p - pq = \begin{bmatrix} p - q_1 & -q_2 \\ 0 & 0 \end{bmatrix}, \qquad p - qp = \begin{bmatrix} p - q_1 & 0 \\ -q_3 & 0 \end{bmatrix}$$
$$1 - pqp = \begin{bmatrix} p - q_1 & 0 \\ 0 & 1 - p \end{bmatrix}, \qquad p - pqp = \begin{bmatrix} p - q_1 & 0 \\ 0 & 0 \end{bmatrix},$$

and Lemma 4.8. $\mathscr{L}_2$ is obtained from $\mathscr{L}_1$ by interchanging p and q, and $\mathscr{L}_3$ by replacing p by $1 - p$ and simultaneously q by $1 - q$ in $\mathscr{L}_1$ (note that $(1 - p) - (1 - q) = q - p$). $\mathscr{K}$ is obtained from $\mathscr{L}_1$ by replacing q by $1 - q$. ∎

Further equivalences are obtained when we interchange p and q in $\mathcal{K}$.

In the following result, for given idempotents p and q in $\mathcal{A}$, we consider the generalized Drazin and Drazin invertibility of their commutator $pq - qp$ and anti-commutator $pq + qp$.

Theorem 4.6 *Let* $p, q \in \mathcal{A}^{\text{idem}}$ *and* $\delta \in \{D, d\}$. *Then*

$$pq - qp \in \mathcal{A}^{\delta} \iff pq + qp \in \mathcal{A}^{\delta} \iff (p - q \in \mathcal{A}^{\delta} \text{ and } pq \in \mathcal{A}^{\delta}).$$

Proof (a) First we prove that

$$pq - qp \in \mathcal{A}^{\delta} \iff (p - q \in \mathcal{A}^{\delta} \text{ and } pq \in \mathcal{A}^{\delta}).$$

Represent p and q by matrices as in (4.11). Then

$$pq - qp = \begin{bmatrix} 0 & q_2 \\ -q_3 & 0 \end{bmatrix}.$$

By Lemma 4.9, $pq - qp \in \mathcal{A}^{\delta}$ if and only if $q_2 q_3 \in \mathcal{A}^{\delta}$. We have $q_2 q_3 = pq(1 - p)qp = pqp(1 - pqp) = f(pqp)$ with $f(\lambda) = \lambda(1 - \lambda)$. Thus $pq - qp \in \mathcal{A}^{\delta} \iff f(pqp) \in \mathcal{A}^{\delta}$. Since $f^{-1}(0) = \{0, 1\}$, Lemma 4.7 implies that $f(pqp) \in \mathcal{A}^{\delta}$ if and only if $pqp \in \mathcal{A}^{\delta}$ and $1 - pqp \in \mathcal{A}^{\delta}$. The conclusion follows by Theorem 4.5 as $pqp \in \mathcal{A}^{\delta} \iff pq \in \mathcal{A}^{\delta}$ and $1 - pqp \in \mathcal{A}^{\delta} \iff p - q \in \mathcal{A}^{\delta}$.

(b) Next we show that

$$pq + qp \in \mathcal{A}^{\delta} \iff (p \overset{\cdot}{+} q \in \mathcal{A}^{\delta} \text{ and } 1 - p - q \in \mathcal{A}^{\delta}).$$

A straightforward check shows that $pq + qp = f(p + q)$, where $f(\lambda) = \lambda(\lambda - 1)$. Since $f^{-1}(0) = \{0, 1\}$, Lemma 4.7 says that $f(p+q) \in \mathcal{A}^{\delta}$ if and only if $p+q \in \mathcal{A}^{\delta}$ and $1 - p - q \in \mathcal{A}^{\delta}$.

By Theorem 4.5, $p - q \in \mathcal{A}^{\delta} \iff p + q \in \mathcal{A}^{\delta}$, and $pq \in \mathcal{A}^{\delta} \iff 1 - p - q \in \mathcal{A}^{\delta}$. This completes the proof of the theorem. $\qquad\square$

4.3   Some Results for Idempotent Operators on Banach Spaces

Let $\mathcal{X}$ be a Banach space and $\mathcal{B}(\mathcal{X})$ the Banach algebra of all bounded linear operators on $\mathcal{X}$. We recall that an operator $A \in \mathcal{B}(\mathcal{X})$ is generalized Drazin invertible if and only if $A = A_1 \oplus A_2$, where A_1, A_2 are bounded linear operators acting on closed subspaces of $\mathcal{X}$, with A_1 invertible and A_2 *quasi-nilpotent*. If A_2 is nilpotent, A is Drazin invertible.

Let $P, Q \in \mathcal{B}(\mathcal{X})$ be idempotent. Relative to the space decomposition $\mathcal{X} = \mathcal{R}(P) \oplus \mathcal{N}(P)$, where $\mathcal{R}(P)$ and $\mathcal{N}(P)$ are the range and nullspace of P, we have

$$P = \begin{bmatrix} I_1 & 0 \\ 0 & 0 \end{bmatrix}, \qquad Q = \begin{bmatrix} Q_1 & Q_2 \\ Q_3 & Q_4 \end{bmatrix}.$$

Theorem 4.7 *If P, $Q \in \mathscr{B}(\mathscr{X})$ are idempotent and $\delta \in \{D, d\}$, then the following conditions are equivalent:*

(i)   *$I - PQ$ is δ-invertible.*
(ii) *$\alpha P + \beta Q$ is δ-invertible for any $\alpha, \beta \in \mathbb{C} \setminus \{0\}$.*
(iii) *$I_1 - Q_1$ is δ-invertible in $\mathscr{B}(\mathscr{R}(P))$.*
(iv) *Q_4 is δ-invertible in $\mathscr{B}(\mathscr{N}(P))$.*

Proof The equivalence of (i) and (ii) follows from Theorem 4.4. For the proof of (iii) and (iv), consider

$$I - PQ = \begin{bmatrix} I_1 - Q_1 & -Q_2 \\ 0 & I_2 \end{bmatrix}, \quad P + Q - PQ = \begin{bmatrix} I_1 & 0 \\ Q_3 & Q_4 \end{bmatrix}.$$

By Theorem 4.5, $I - PQ$ is δ-invertible if and only if $P + Q - PQ$ is δ-invertible. Lemma 4.8 modified for operator matrices then shows that the δ-invertibility of $I - PQ$ is equivalent to the δ-invertibility of either $I_1 - Q_1$ or Q_4 in the appropriate spaces. The Drazin inverse case follows similarly. □

Remark 4.2 If $Q \in \mathscr{B}(\mathscr{X})$ is an idempotent with the operator matrix

$$Q = \begin{bmatrix} Q_1 & Q_2 \\ Q_3 & Q_4 \end{bmatrix}$$

relative to the space decomposition $\mathscr{X} = \mathscr{X}_1 \oplus \mathscr{X}_2$, then the generalized Drazin or Drazin invertibility of $I_1 - Q_1$ and Q_4 are closely linked, in fact, one is generalized Drazin or Drazin invertible if and only if the other is. This follows from the preceding theorem when we define P as the projection in $\mathscr{B}(\mathscr{X})$ with $\mathscr{R}(P) = \mathscr{X}_1$ and $\mathscr{N}(P) = \mathscr{X}_2$.

Remark 4.3 (i) For projections P, Q in a Hilbert space, Böttcher and Spitkovsky [8] gave criteria for the Drazin invertibility of operators in the von Neumann algebra generated by two orthogonal projections along with explicit representations for the corresponding inverses.

(ii) Theorem 3.2 of [11] proves the equivalence of the Drazin invertibility of the difference $P - Q$ of two Hilbert space idempotent operators with the simultaneous Drazin invertibility of $P + Q$ and $I - PQ$. This is now strengthened to provide the equivalence of all three conditions.

For the commutator and anticommutator of P, Q we have the following result.

Theorem 4.8 *Let P, $Q \in \mathscr{B}(\mathscr{X})$ be idempotent and let $\delta \in \{D, d\}$. Then the following are equivalent:*

(i)   *$PQ - QP$ is δ-invertible.*

(ii) $PQ + QP$ is δ-invertible.

(iii)   Both $P - Q$ and PQ are δ-invertible.

4.4   Moore-Penrose Inverse of a Linear Combination of Commuting Generalized and Hypergeneralized Projectors

The concepts of generalized and hypergeneralized projectors were introduced by Groß and Trenkler [23] who presented interesting properties of these classes of projectors. Some related results concerning this subject can be found in the papers of Baksalary et al. [24], Baksalary [25], Baksalary et al. [26], Benítez [27] and Stewart [28].

In this section we give a form for the Moore-Penrose inverse, i.e., the group inverse of a linear combination $c_1 A + c_2 B$ of two commuting generalized or hypergeneralized projectors. Also, we study invertibility of $c_1 A + c_2 B$ and $c_1 A + c_2 B + c_3 C$, where A, B and C are commuting generalized or hypergeneralized projectors under various conditions.

We use the notations $\mathbb{C}_n^P$, $\mathbb{C}_n^{OP}$, $\mathbb{C}_n^{EP}$, $\mathbb{C}_n^{GP}$ and $\mathbb{C}_n^{HGP}$ for the subsets of $\mathbb{C}^{n\times n}$ consisting of projectors (idempotent matrices), orthogonal projectors (Hermitian idempotent matrices), EP (range-Hermitian) matrices, generalized and hypergeneralized projectors, respectively, i.e.,

$$
\begin{aligned}
\mathbb{C}_n^P &= \{A \in \mathbb{C}^{n\times n} : A^2 = A\}, \\
\mathbb{C}_n^{OP} &= \{A \in \mathbb{C}^{n\times n} : A^2 = A = A^*\}, \\
\mathbb{C}_n^{EP} &= \{A \in \mathbb{C}^{n\times n} : \mathscr{R}(A) = \mathscr{R}(A^*)\} = \{A \in \mathbb{C}^{n\times n} : AA^\dagger = A^\dagger A\}, \\
\mathbb{C}_n^{GP} &= \{A \in \mathbb{C}^{n\times n} : A^2 = A^*\}, \\
\mathbb{C}_n^{HGP} &= \{A \in \mathbb{C}^{n\times n} : A^2 = A^\dagger\}.
\end{aligned}
$$

Baksalary et al. ([24]), proved that any generalized projector $A \in \mathbb{C}_r^{n\times n}$ can be represented by

$$
A = U \begin{bmatrix} K & 0 \\ 0 & 0 \end{bmatrix} U^*, \tag{4.12}
$$

where $U \in \mathbb{C}^{n\times n}$ is unitary and $K \in \mathbb{C}^{r\times r}$ is such that $K^3 = I_r$ and $K^* = K^{-1}$. Any hypergeneralized projector $A \in \mathbb{C}_r^{n\times n}$ has the form

$$
A = U \begin{bmatrix} \Sigma K & 0 \\ 0 & 0 \end{bmatrix} U^*, \tag{4.13}
$$

where $U \in \mathbb{C}^{n \times n}$ is unitary, $\sum = \mathrm{diag}(\sigma_1 I_{r_1}, \ldots, \sigma_t I_{r_t})$ is a diagonal matrix of singular values of A, $\sigma_1 > \sigma_2 > \cdots > \sigma_t > 0$, $r_1 + r_2 + \cdots + r_t = r$ and $K \in \mathbb{C}^{r \times r}$ satisfies $(\sum K)^3 = I_r$ and $KK^* = I_r$.

There are also some other very useful representations for generalized and hyper-generalized projectors: Using the fact that any generalized projector $A \in \mathbb{C}_r^{n \times n}$ is a normal matrix, by the spectral theorem we have that $A = U\mathrm{diag}(\lambda_1, \lambda_2, \ldots, \lambda_n)U^*$, where U is a unitary matrix and λ_j, $j = \overline{1, n}$ are the eigenvalues of A. By Theorem 2.1 [27] we have that $\lambda_j \in \{0, 1, \omega, \overline{\omega}\}$, $j = \overline{1, n}$, where $\omega = \exp 2\pi i/3$. Hence,

$$A \in \mathbb{C}_n^{GP} \Leftrightarrow A = U\mathrm{diag}(\lambda_1, \lambda_2, \ldots, \lambda_n)U^*, \tag{4.14}$$

where $U^* = U^{-1}$ and $\lambda_j \in \{0, 1, \omega, \overline{\omega}\}$, $j = \overline{1, n}$, $\omega = \exp 2\pi i/3$.

Similarly, for $A \in \mathbb{C}_n^{HGP}$ using the fact that A is EP-matrix, by Theorem 4.3.1 [29] we can conclude that

$$A \in \mathbb{C}_n^{HGP} \Leftrightarrow A = U(K \oplus 0)U^*, \tag{4.15}$$

where $U \in \mathbb{C}^{n \times n}$ is a unitary matrix and $K \in \mathbb{C}^{r \times r}$ is invertible such that $K^3 = I_r$, where $r = \mathrm{r}(A)$.

From the above representations it is obvious that any generalized projector is a hypergeneralized projector.

The following fact will be used very often:

If $X, Y \in \mathbb{C}^{n \times n}$ and $c_1, c_2 \in \mathbb{C}$, then

$$X^3 = Y^3 = I_n, \ XY = YX \Rightarrow$$
$$(c_1 X + c_2 Y)(c_1^2 X^2 - c_1 c_2 XY + c_2^2 Y^2) = (c_1^3 + c_2^3)I_n. \tag{4.16}$$

In this section, we first present a form for the Moore-Penrose inverse, i.e., the group inverse of $c_1 A + c_2 B$, where A, B are two commuting generalized or hypergeneralized projectors and $c_1, c_2 \in \mathbb{C} \setminus \{0\}$ are such that $c_1^3 + c_2^3 \neq 0$.

Theorem 4.9 Let $A \in \mathbb{C}^{n \times n}$ and $B \in \mathbb{C}^{n \times n}$ be commuting hypergeneralized projectors, and let $c_1, c_2 \in \mathbb{C} \setminus \{0\}$ be such that $c_1^3 + c_2^3 \neq 0$. Then

$$(c_1 A + c_2 B)^\dagger = \frac{1}{c_1^3 + c_2^3}\left(c_1^2 A^2 B^3 - c_1 c_2 AB + c_2^2 A^3 B^2\right) + \frac{1}{c_1}A^2(I_n - B^3)$$

$$+ \frac{1}{c_2}B^2(I_n - A^3). \tag{4.17}$$

Furthermore, $c_1 A + c_2 B$ is invertible if and only if $n = \mathrm{r}(A) + \mathrm{r}(B) - \mathrm{r}(AB)$ and in this case $(c_1 A + c_2 B)^{-1}$ is given by (4.17).

Proof Since A and B are two commuting EP-matrices, by Corollary 3.9 from [30], we have that

$$A = U(A_1 \oplus A_2 \oplus 0_{t,t} \oplus 0)U^*, \quad B = U(B_1 \oplus 0_{s,s} \oplus B_2 \oplus 0)U^*,$$

where $A_1, B_1 \in \mathbb{C}^{r \times r}$, $A_2 \in \mathbb{C}^{s \times s}$, $B_2 \in \mathbb{C}^{t \times t}$ are invertible and $A_1 B_1 = B_1 A_1$. If in addition A and B are hypergeneralized projectors, then $A_1^3 = B_1^3 = I_r$, $A_2^3 = I_s$ and $B_2^3 = I_t$. Since

$$c_1 A + c_2 B = U\Big((c_1 A_1 + c_2 B_1) \oplus c_1 A_2 \oplus c_2 B_2 \oplus 0\Big)U^*, \tag{4.18}$$

we can use (4.16) to get the expression for $(c_1 A + c_2 B)^\dagger$. Thus, by (4.16) we get that $c_1 A_1 + c_2 B_1$ is invertible and that

$$(c_1 A_1 + c_2 B_1)^{-1} = \frac{1}{c_1^3 + c_2^3}\Big(c_1^2 A_1^2 - c_1 c_2 A_1 B_1 + c_2^2 B_1^2\Big).$$

Now, using that

$$I_n - A^3 = U(0 \oplus 0 \oplus I_t \oplus I_{n-(r+t+s)})U^*, \quad A^3 B^3 = U(I_r \oplus 0 \oplus 0 \oplus 0)U^*$$

and

$$I_n - B^3 = U(0 \oplus I_s \oplus 0 \oplus I_{n-(r+t+s)})U^*,$$

we have

$$(c_1 A + c_2 B)^\dagger = U((c_1 A_1 + c_2 B_1)^{-1} \oplus \frac{1}{c_1} A_2^2 \oplus \frac{1}{c_2} B_2^2 \oplus 0)U^*$$

$$= \frac{1}{c_1^3 + c_2^3}\Big(c_1^2 A^2 - c_1 c_2 AB + c_2^2 B^2\Big)A^3 B^3$$

$$+ \frac{1}{c_1} A^2 (I_n - B^3) + \frac{1}{c_2} B^2 (I_n - A^3).$$

Since $A^4 = A$ and $B^4 = B$, we get that (4.17) holds. Also, it is evident that $r(A) = r + s$, $r(B) = r + t$ and $r(AB) = r$. So, the last summand in the direct sum in (4.18) does not appear if and only if $n = r(A) + r(B) - r(AB)$ which is a necessary and sufficient condition for the invertibility of $c_1 A + c_2 B$. $\qquad \square$

As a corollary we get that in the case when A is a hypergeneralized projector and $c_1, c_2 \in \mathbb{C}$, $c_1 \neq 0$, $c_1^3 + c_2^3 \neq 0$, the linear combination $c_1 I_n + c_2 A$ is always invertible.

Theorem 4.10 *Let $A \in \mathbb{C}^{n \times n}$ be a hypergeneralized projector, $c_1, c_2 \in \mathbb{C}$, $c_1 \neq 0$, $c_1^3 + c_2^3 \neq 0$. Then $c_1 I_n + c_2 A$ is invertible and*

$$(c_1 I_n + c_2 A)^{-1} = \frac{1}{c_1^3 + c_2^3} \left(c_1^2 A^3 - c_1 c_2 A + c_2^2 A^2 \right) + \frac{1}{c_1} (I_n - A^3).$$

We say that a family $\mathcal{G} \subset \mathbb{C}_n$ of matrices is commuting if each pair of elements of the set $\mathcal{G}$ commutes under multiplication. If we consider a finite commuting family $\{A_i\}_{i=1}^m$ where all of the members are hypergeneralized projectors, then $\prod_{i=1}^m A_i^{k_i}$, where $k_1, \ldots, k_m \in \mathbb{N}$, is also a hypergeneralized projector. Hence, we have the following result:

Proposition 4.1 *Let $A_i \in \mathbb{C}^{n \times n}$, $i = \overline{1, m}$ be pairwise commuting generalized or hypergeneralized projectors, $c_1, c_2 \in \mathbb{C}$, $c_1 \neq 0$, $c_1^3 + c_2^3 \neq 0$ and $k_1, \ldots, k_m \in \mathbb{N}$. Then $c_1 I_n + c_2 \prod_{i=1}^m A_i^{k_i}$ is invertible.*

With the additional requirements of Theorem 4.9 it is possible to give a more precise form of the Moore-Penrose inverse, i.e., the group inverse.

Corollary 4.5 *Let $c_1, c_2 \in \mathbb{C} \setminus \{0\}$. If A, B are commuting hypergeneralized projectors such that $AB = 0$, then*

$$(c_1 A + c_2 B)^\dagger = \frac{1}{c_1} A^2 + \frac{1}{c_2} B^2. \tag{4.19}$$

In the next result, we present a form of Moore-Penrose inverse, i.e., the group inverse of $c_1 A^m + c_2 A^k$, where $m, k \in \mathbb{N}$ and A is a hypergeneralized projector. It is a corollary of Theorem 4.9.

Corollary 4.6 *Let $A \in \mathbb{C}_r^{n \times n}$ be a hypergeneralized projector and let $c_1, c_2 \in \mathbb{C}$, $c_1^3 + c_2^3 \neq 0$ and $m, k \in \mathbb{N}$. Then*

$$(c_1 A^m + c_2 A^k)^\dagger = \frac{1}{c_1^3 + c_2^3} \left(c_1^2 A^{2m} - c_1 c_2 A^{m+k} + c_2 A^{2k} \right),$$

where $A^t = \begin{cases} A^3, & t \equiv_3 0, \\ A, & t \equiv_3 1 \\ A^2, & t \equiv_3 2 \end{cases}$. *Furthermore, $c_1 A^m + c_2 A^k$ is invertible if and only if A is invertible and in this case the inverse of $c_1 A^m + c_2 A^k$ is given by*

$$(c_1 A^m + c_2 A^k)^{-1} = \frac{1}{c_1^3 + c_2^3} \left(c_1^2 A^p - c_1 c_2 A^q + c_2 A^r \right),$$

where $2m \equiv_3 p$, $m + k \equiv_3 q$ and $2k \equiv_3 r$.

Proof This follows from Theorem 4.9 and the fact that $r(A^p) = r(A)$, for any $p \in \mathbb{N}$.   □

As a corollary we get a result from [24]:

Corollary 4.7 ([24]) Let $A \in \mathbb{C}_r^{n \times n}$ be a generalized projector and let $c_1, c_2 \in \mathbb{C}$, $c_1^3 + c_2^3 \neq 0$. Then

$$(c_1 A + c_2 A^*)^\dagger = \frac{1}{c_1^3 + c_2^3}(c_1^2 A^2 - c_1 c_2 A^3 + c_2^2 A).$$

Let us recall that for matrices $A, B \in \mathbb{C}^{n \times m}$, the matrix A is less than or equal to the matrix B with respect to the star partial ordering, which is denoted by $A \overset{*}{\leq} B$ (see [31]), if

$$A^* A = A^* B \text{ and } AA^* = BA^*.$$

If $A \in \mathbb{C}_n^{EP}$, then for any $B \in \mathbb{C}^{n \times n}$,

$$A \overset{*}{\leq} B \Leftrightarrow AB = A^2 = BA.$$

In the next theorem, we present a form of Moore-Penrose inverse, i.e., the group inverse of $c_1 A^m + c_2 B^k$ under the condition that A, B are generalized projectors and $AB = BA = A^2$. Remark that the same result holds if we suppose that A, B are generalized projectors such that $B - A \in \mathbb{C}_n^{GP}$ or if $A \in \mathbb{C}_n^{EP}$, $B \in \mathbb{C}_n^{HGP}$ are such that $A \overset{*}{\leq} B$.

Theorem 4.11 Let $c_1, c_2 \in \mathbb{C}$, $c_2 \neq 0$, $c_1^3 + c_2^3 \neq 0$ and $m, k \in \mathbb{N}$. If $A \in \mathbb{C}^{n \times n}$ and $B \in \mathbb{C}^{n \times n}$ are hypergeneralized projectors such that $AB = BA = A^2$, then

$$(c_1 A^m + c_2 B^k)^\dagger = \frac{1}{c_1^3 + c_2^3}(c_1^2 A^{2m} - c_1 c_2 A^{m+k} + c_2^2 A^{2k}) + \frac{1}{c_2} B^{2k}(I_n - A^3) \quad (4.20)$$

where $A^t = \begin{cases} A^3, & t \equiv_3 0 \\ A, & t \equiv_3 1 \\ A^2, & t \equiv_3 2 \end{cases}$ and $B^s = \begin{cases} B^3, & s \equiv_3 0 \\ B, & s \equiv_3 1 \\ B^2, & s \equiv_3 2 \end{cases}$.

Proof By Corollary 3.9 from [30] and the fact that $AB = BA = A^2$, we have that

$$A = U(A_1 \oplus 0_{t,t} \oplus 0)U^*, \quad B = U(B_1 \oplus B_2 \oplus 0)U^*,$$

where $A_1, B_1 \in \mathbb{C}^{r \times r}$, $B_2 \in \mathbb{C}^{t \times t}$ are invertible and $A_1 B_1 = B_1 A_1 = A_1^2$. Evidently $A_1 = B_1$. If in addition A and B are hypergeneralized projectors, then $A_1^3 = I_r$ and $B_2^3 = I_t$. Hence,

$$c_1 A^m + c_2 B^k = U\Big((c_1 A_1^m + c_2 A_1^k) \oplus c_2 B_2^k \oplus 0\Big)U^*.$$

By (4.16), we get that $c_1 A_1^m + c_2 A_1^k$ is invertible and that

$$(c_1 A_1^m + c_2 A_1^k)^{-1} = \frac{1}{c_1^3 + c_2^3}\Big(c_1^2 A_1^{2m} - c_1 c_2 A_1^{m+k} + c_2^2 A_1^{2k}\Big).$$

Now, using the equalities

$$A^3 = U(I_r \oplus 0 \oplus 0)U^*, \quad B^3 - A^3 = U(0 \oplus I_t \oplus 0)U^*,$$

we have that (4.20) holds. $\qquad \square$

Remark: If $A \in \mathbb{C}_n^{EP}$, $B \in \mathbb{C}_n^{HGP}$ and $A \overset{*}{\le} B$, we can conclude that $B - A$ is a hypergeneralized projector (if A and B are hypergeneralized projectors, then $A \overset{*}{\le} B$ or $AB = A^2 = BA$ are sufficient for $B - A$ to be a hypergeneralized projector (see [23]).

Theorem 4.12 *Let $A \in \mathbb{C}_r^{n \times n}$ and $B \in \mathbb{C}^{n \times n}$ be commuting hypergeneralized projectors. Let $c_1, c_2 \in \mathbb{C} \setminus \{0\}$, $c_1^3 + c_2^3 \ne 0$ and $m, k, l \in \mathbb{N}$. Then*

$$\left[A^m (c_1 A^k + c_2 B^l) \right]^\dagger = \frac{1}{c_1^3 + c_2^3} \left(c_1^2 A^{2(m+k)} - c_1 c_2 A^{(m+k)} B^l + c_2^2 A^3 B^{2l} \right)$$
$$+ \frac{1}{c_1} A^{2(m+k)} (I_n - B^3),$$

where $A^t = \begin{cases} A^3, & t \equiv_3 0 \\ A, & t \equiv_3 1 \\ A^2, & t \equiv_3 2 \end{cases}$ *and* $B^s = \begin{cases} B^3, & s \equiv_3 0 \\ B, & s \equiv_3 1 \\ B^2, & s \equiv_3 2 \end{cases}$.

Proof The proof is similar to that of Theorem 4.11 and for that reason is left to the reader. $\qquad \square$

The following theorem presents necessary and sufficient conditions for the invertibility of $c_1 A + c_2 B + c_3 C$ in the case when A, B, C are pairwise commuting hypergeneralized projectors such that $BC = 0$ and $c_1, c_2, c_3 \in \mathbb{C} \setminus \{0\}$, $c_1^3 + c_2^3 \ne 0$, $c_1^3 + c_3^3 \ne 0$.

Theorem 4.13 *Let $c_1, c_2, c_3 \in \mathbb{C} \setminus \{0\}$, $c_1^3 + c_2^3 \ne 0$, $c_1^3 + c_3^3 \ne 0$. If $A, B, C \in \mathbb{C}^{n \times n}$ are pairwise commuting hypergeneralized projectors such that $BC = 0$, then the following conditions are equivalent:*
  (i) $c_1 A + c_2 B + c_3 C$ *is invertible,*
  (ii) $B^3 + C^3 + A(I_n - B^3 - C^3)$ *is invertible,*
  (iii) $r(A(I_n - B^3 - C^3)) = n - (r(B) + r(C))$.

Proof By [30, Corollary 3.9], we have that

$$B = U(B_1 \oplus 0_{s,s} \oplus 0)U^*, \quad C = U(0_{r,r} \oplus C_1 \oplus 0)U^*, \qquad (4.21)$$

where $B_1 \in \mathbb{C}^{r \times r}$, $C_1 \in \mathbb{C}^{s \times s}$ are invertible and U is unitary. Since $B^2 = B^\dagger$ and $C^2 = C^\dagger$, we get that $B_1^3 = I_r$ and $C_1^3 = I_s$. Also, since A commutes with B and C, it follows that $A = U(A_1 \oplus A_2 \oplus A_3)U^*$ where A_1, A_2, A_3 are hypergeneralized projectors and $A_1 B_1 = B_1 A_1$, $A_2 C_1 = C_1 A_2$.

Now,

$$c_1 A + c_2 B + c_3 C = U\Big((c_1 A_1 + c_2 B_1) \oplus (c_1 A_2 + c_3 C_1) \oplus c_1 A_3\Big)U^*, \quad (4.22)$$

so $c_1 A + c_2 B + c_3 C$ is invertible if and only if $c_1 A_1 + c_2 B_1$, $c_1 A_2 + c_3 C_1$ and A_3 are invertible. By Proposition 4.1 we get that $c_1 A_1 B_1^2 + c_2 I$ is invertible. Now, by $c_1 A_1 + c_2 B_1 = (c_1 A_1 B_1^2 + c_2 I)B_1$ it follows that $c_1 A_1 + c_2 B_1$ is invertible. Similarly, we get that $c_1 A_2 + c_3 C_1$ is invertible. Thus, $c_1 A + c_2 B + c_3 C$ is invertible if and only if A_3 is invertible i.e., $B^3 + C^3 + A(I_n - B^3 - C^3)$ is invertible. Hence, (i) $\Leftrightarrow$ (ii). Also, we have that A_3 is invertible if and only if $r(A_3) = n - (r + s)$ which is equivalent with the fact that $r(A(I_n - B^3 - C^3)) = n - (r + s) = n - (r(B) + r(C))$. So, (i) $\Leftrightarrow$ (iii). $\square$

Remark that from the proofs of Theorems 4.13 and 4.9, if one of the conditions (i) − (iii) holds, we get the following formula for the inverse of $c_1 A + c_2 B + c_3 C$:

$$
\begin{aligned}
(c_1 A + c_2 B + c_3 C)^{-1} ={} & \frac{1}{c_1^3 + c_2^3}\Big(c_1^2 A^2 B^3 - c_1 c_2 AB + c_2^2 A^3 B^2\Big) + \frac{1}{c_2} B^2 (I_n - A^3) \\
& + \frac{1}{c_1^3 + c_3^3}\Big(c_1^2 A^2 C^3 - c_1 c_3 AC + c_2^2 A^3 C^2\Big) + \frac{1}{c_3} C^2 (I_n - A^3) \\
& + \frac{1}{c_1}\Big(B^3 + C^3 + A(I_n - B^3 - C^3)\Big)^{-1}(I_n - B^3 - C^3),
\end{aligned}
$$
$$(4.23)$$

which will be useful later.

In the remainder of the section we will be concerned with the notion of star-orthogonality which was introduced by Hestenes [32]. Let us recall that matrices $A, B \in \mathbb{C}^{n \times m}$ are star-orthogonal, which is denoted by $A \overset{*}{\perp} B$, if $AB^* = 0$ and $A^* B = 0$. It is well-known that for $A, B \in \mathbb{C}_n^{EP}$,

$$A \overset{*}{\perp} B \Leftrightarrow AB = 0 \Leftrightarrow BA = 0.$$

If A, B are hypergeneralized projectors, then $A \overset{*}{\perp} B$ or $AB = BA = 0$ are sufficient for $A + B$ to be a hypergeneralized projector (see [23]).

Remark that the conclusion of Theorem 4.13 remains valid if we suppose that A, B, C are pairwise commuting generalized projectors such that $B + C \in \mathbb{C}_n^{GP}$ or if A, B, C are pairwise commuting hypergeneralized projectors such that $B \overset{*}{\perp} C$ with the same conditions for the scalars c_1, c_2, c_3.

In the following theorem, under the assumption that $c_1, c_2, c_3 \in \mathbb{C}$, $c_1 \neq 0$, $c_1^3 + c_2^3 \neq 0$, $c_1^3 + c_3^3 \neq 0$, we show that $c_1 I_n + c_2 A + c_3 B$ is invertible, in the case when A, B are commuting hypergeneralized projectors such that $AB = 0$. Remark that the theorem remains true if we suppose that A, B are generalized projectors such that $A + B \in \mathbb{C}_n^{GP}$ or when A, B are hypergeneralized projectors such that $A \overset{*}{\perp} B$.

Theorem 4.14 *Let $c_1, c_2, c_3 \in \mathbb{C}$, $c_1 \neq 0$, $c_1^3 + c_2^3 \neq 0$, $c_1^3 + c_3^3 \neq 0$. If $A, B \in \mathbb{C}^{n \times n}$ are commuting hypergeneralized projectors such that $AB = 0$, then $c_1 I_n + c_2 A + c_3 B$ is invertible and*

$$(c_1 I_n + c_2 A + c_3 B)^{-1} = \frac{1}{c_1^3 + c_2^3} \left(c_1^2 A^3 - c_1 c_2 A + c_2^2 A^2 \right)$$

$$+ \frac{1}{c_1^3 + c_3^3} \left(c_1^2 B^3 - c_1 c_3 B + c_3^2 B^2 \right) + \frac{1}{c_1} (I_n - A^3 - B^3).$$

Proof The proof follows by Theorem 4.13 and (4.23).   □

Corollary 4.8 *Let $c_1, c_2, c_3 \in \mathbb{C} \setminus \{0\}$, $c_1^3 + c_2^3 \neq 0$, $c_1^3 + c_3^3 \neq 0$. If $A, B, C \in \mathbb{C}^{n \times n}$ are pairwise commuting hypergeneralized projectors such that $BC = 0$, then the invertibility of $c_1 A + c_2 B + c_3 C$ is independent of the choice of the scalars c_1, c_2, c_3.*

Corollary 4.9 *Let $c_1, c_2, c_3 \in \mathbb{C} \setminus \{0\}$, $c_1^3 + c_2^3 \neq 0$, $c_1^3 + c_3^3 \neq 0$. If $A, B, C \in \mathbb{C}^{n \times n}$ are pairwise commuting generalized projectors such that $B + C \in \mathbb{C}_n^{GP}$ or $A, B, C \in \mathbb{C}^{n \times n}$ are pairwise commuting hypergeneralized projectors such that $B \perp\!\!\!\!^* C$, then the invertibility of $c_1 A + c_2 B + c_3 C$ is independent of the choice of the scalars c_1, c_2, c_3.*

References

1. Baksalary, J.K., Baksalary, O.M., Kik, P.: Generalizations of a property of orthogonal projectors. Linear Algebra Appl. **420**, 1–8 (2007)
2. Baksalary, O.M., Benítez, J.: Idempotency of linear combinations of three idempotent matrices, two of which are commuting. Linear Algebra Appl. **424**, 320–337 (2007)
3. Deng, C.Y.: The Drazin inverses of products and differences of orthogonal projections. J. Math. Anal. Appl. **335**, 64–71 (2007)
4. Du, H.K., Yao, X.Y., Deng, C.Y.: Invertibility of linear combinations of two idempotents. Proc. Am. Math. Soc. **134**, 1451–1457 (2005)
5. Koliha, J.J., Rakočević, V.: Invertibility of the sum of idempotents. Linear Multilinear Algebra **50**, 285–292 (2002)
6. Koliha, J.J., Rakočević, V.: Invertibility of the difference of idempotents. Linear Multilinear Algebra **51**, 97–110 (2003)
7. Koliha, J.J., Rakočević, V., Strakraba, I.: The difference and sum of projectors. Linear Algebra Appl. **388**, 279–288 (2004)
8. Böttcher, A., Spitkovsky, I.M.: Drazin inversion in the von Neumann algebra generated by two orthogonal projections. J. Math. Anal. Appl. **358**, 403–409 (2009)
9. Böttcher, A., Spitkovsky, I.M.: A gentle guide to the basics of two projection theory. Linear Algebra Appl. **432**, 1412–1459 (2010)
10. Cvetković-Ilić, D.S., Deng, C.Y.: Some results on the Drazin invertibility and idempotents. J. Math. Anal. Appl. **359**, 731–738 (2009)
11. Cvetković-Ilić, D.S., Deng, C.Y.: The Drazin invertibility of the difference and the sum of two idempotent operators. J. Comp. Appl. Math. **233**, 1717–1732 (2010)
12. Deng, C.Y.: Characterizations and representations of the group inverse involving idempotents. Linear Algebra Appl. **434**, 1067–1079 (2011)
13. Zhang, S., Wu, J.: The Drazin inverse of the linear combinations of two idempotents in the Banach algebra. Linear Algebra Appl. **436**, 3132–31381 (2012)

14. Deng, C.Y.: The Drazin inverses of sum and difference of idempotents. Linear Algebra Appl. **430**, 1282–1291 (2009)
15. Djordjević, D.S., Stanimirović, P.S.: On the generalized Drazin inverse and generalized resolvent. Czechoslovak Math. J. **51**(126), 617–634 (2001)
16. Hartwig, R.E., Shoaf, J.M.: Group inverse and Drazin inverse of bidiagonal and triangular Toeplitz matrices. Austr. J. Math. **24A**, 10–34 (1977)
17. Meyer, C.D., Rose, N.J.: The index and the Drazin inverse of block triangular matrices. SIAM J. Appl. Math. **33**, 1–7 (1977)
18. Castro-González, N., Koliha, J.J.: New additive results for the g-Drazin inverse. Proc. R. Soc. Edinburgh **134A**, 1085–1097 (2004)
19. Catral, M., Olesky, D.D., Den, Van: Driessche P,: Block representations of the Drazin inverse of bipartite matrix. Electronic J. Linear Algebra **18**, 98–107 (2009)
20. Koliha, J.J.: A generalized Drazin inverse. Glasgow Math. J. **38**, 367–381 (1996)
21. Koliha, J.J.: Spectral sets. Studia Math. **123**, 97–107 (1997)
22. Dunford, N., Schwartz, J.: Linear Operators I. Wiley Interscience, New York (1957)
23. Groß, J., Trenkler, G.: Generalized and hypergeneralized projectors. Linear Algebra Appl. **264**, 463–474 (1997)
24. Baksalary, J.K., Baksalary, O.M., Liu, X., Trenkler, G.: Further results on generalized and hypergeneralized projectors. Linear Algebra Appl. **429**, 1038–1050 (2008)
25. Baksalary, O.M.: Revisitation of generalized and hypergeneralized projectors. In: Schipp, B., Krämer, W. (eds.) Statistical Inference, pp. 317–324. Econometric Analysis and Matrix Algebra-Festschrift in Honour of Götz Trenkler. Springer, Heidelberg (2008)
26. Baksalary, J.K., Baksalary, O.M., Groß, J.: On some linear combinations of hypergeneralized projectors. Linear Algebra Appl. **413**, 264–273 (2006)
27. Benítez, J., Thome, N.: Characterizations and linear combinations of k-generalized projectors. Linear Algebra Appl. **410**, 150–159 (2005)
28. Stewart, G.W.: A note on generalized and hypergeneralized projectors. Linear Algebra Appl. **412**, 408–411 (2006)
29. Campbell, S.L., Meyer, C.D.: Generalized Inverse of Linear Transformations, Pitman, London (1979). Dover, New York (1991)
30. Benítez, J.: Moore-Penrose inverses and commuting elements of C*-algebras. J. Math. Anal. Appl. **345**, 766–770 (2008)
31. Drazin, M.P.: Natural structures on semigroups with involution. Bull. Am. Math. Soc. **84**, 139–141 (1978)
32. Hestenes, M.R.: Relative hermitian matrices. Pacific J. Math. **11**, 225–245 (1961)

Chapter 5
Drazin Inverse of a 2 × 2 Block Matrix

The problem of finding representations of the Drazin inverse of a 2×2 block matrix is of great significance primarily due to its applications in solving systems of linear differential equations, linear difference equations [1–4] and perturbation theory of the Drazin inverse [5–7].

We begin by a brief exposition of an application of the Drazin inverse of a 2×2 block matrix: suppose that $E, F \in \mathbb{C}^{n \times n}$ and that E is singular. Assume that there exists a scalar μ such that $\mu E + F$ is invertible. Then the general solution of the singular system of differential equations

$$Ex'(t) + Fx(t) = 0, \quad t \geq t_0, \tag{5.1}$$

is given by

$$x(t) = e^{-\widehat{E}^D \widehat{F}(t-t_0)} \widehat{E}^D \widehat{E} q, \tag{5.2}$$

where $\widehat{E} = (\mu E + F)^{-1} E$, $\widehat{F} = (\mu E + F)^{-1} F$ and $q \in \mathbb{C}^n$ (for more details see [8]).

Now consider the following second-order system where $G \in \mathbb{C}^{n \times n}$ is invertible,

$$Ex''(t) + Fx'(t) + Gx(t) = 0. \tag{5.3}$$

Evidently there is a nonzero λ such that $\lambda^2 E + \lambda F + G$ is invertible. With $x(t) = e^{\lambda t} y(t)$, we have that (5.3) is equivalent to

$$(\lambda^2 E + \lambda F + G)^{-1} E y''(t) + (\lambda^2 E + \lambda F + G)^{-1}(F + 2\lambda E) y'(t) + y(t) = 0.$$

© Springer Nature Singapore Pte Ltd. 2017
D. Cvetković Ilić and Y. Wei, *Algebraic Properties of Generalized Inverses*,
Developments in Mathematics 52, DOI 10.1007/978-981-10-6349-7_5

Upon letting $w(t) = y'(t)$ the above system becomes equivalent to the following first-order system,

$$\begin{bmatrix} 0 & -I \\ \widetilde{E} & \widetilde{F} \end{bmatrix} \begin{bmatrix} w \\ y \end{bmatrix}' + \begin{bmatrix} I & 0 \\ 0 & I \end{bmatrix} \begin{bmatrix} w \\ y \end{bmatrix} = \begin{bmatrix} 0 \\ 0 \end{bmatrix}, \tag{5.4}$$

where

$$\widetilde{E} = (\lambda^2 E + \lambda F + G)^{-1} E, \qquad \widetilde{F} = (\lambda^2 E + \lambda F + G)^{-1}(F + 2\lambda E). \tag{5.5}$$

As obviously $\mu \begin{bmatrix} 0 & -I \\ \widetilde{E} & \widetilde{F} \end{bmatrix} + \begin{bmatrix} I & 0 \\ 0 & I \end{bmatrix}$ is invertible for sufficiently small μ, the previous remarks about the system (5.1) apply. In order to express the solutions of (5.4) explicitly in terms of $\widetilde{E}$ and $\widetilde{F}$, we need to find an explicit representation for the Drazin inverse of a 2 × 2 block matrix,

$$\widehat{E} = \begin{bmatrix} I & -\mu I \\ \mu\widetilde{E} & \mu\widetilde{F} + I \end{bmatrix}^{-1} \begin{bmatrix} 0 & -\mu I \\ \mu\widetilde{E} & \mu\widetilde{F} \end{bmatrix}. \tag{5.6}$$

Yet another differential equation example with the Drazin inverse of a 2 × 2 block matrix is given in [9].

5.1   General Representations for the Drazin Inverse of a 2 × 2 Block Matrix

In 1979 Campbell and Meyer [2] posed the problem of finding an explicit representation for the Drazin inverse of a 2 × 2 complex block-matrix

$$M = \begin{bmatrix} A & B \\ C & D \end{bmatrix}, \tag{5.7}$$

in terms of its blocks, where A and D are square matrices, not necessarily of the same size. To this day no formula has still been found for M^D without any side conditions for the blocks of the matrix M. However, many papers studied special cases of this open problem and offered a formula for M^D under some specific conditions for the blocks of M. Here we list some of them:

(i) $B = 0$ (or $C = 0$) (see [10, 11]);
(ii) $BC = 0$, $BD = 0$ and $DC = 0$ (see [3]);
(iii) $BC = 0$, $DC = 0$ (or $BD = 0$) and D is nilpotent (see [12]);
(iv) $BC = 0$ and $DC = 0$ (see [13]);
(v) $CB = 0$ and $AB = 0$ (or $CA = 0$) (see [13, 14]);
(vi) $BCA = 0$, $BCB = 0$, $DCA = 0$ and $DCB = 0$ (see [15]);

(vii) $ABC = 0$, $CBC = 0$, $ABD = 0$ and $CBD = 0$ (see [15]);
(viii) $BCA = 0$, $BCB = 0$, $ABD = 0$ and $CBD = 0$ (see [16]);
(ix) $BCA = 0$, $DCA = 0$, $CBC = 0$, and $CBD = 0$ (see [16]);
(x) $BCA = 0$, $BD = 0$ and $DC = 0$ (or BC is nilpotent) (see [17]);
(xi) $BCA = 0$, $DC = 0$ and D is nilpotent (see [17]);
(xii) $ABC = 0$, $DC = 0$ and $BD = 0$ (or BC is nilpotent, or D is nilpotent)
         (see [18]);
(xiii) $BCA = 0$ and $BD = 0$ (see [19]);
(xiv) $ABC = 0$ and $DC = 0$ (or $BD = 0$) (see [19, 20]).

In this section we will derive expressions for M^D under less restrictive assumptions than those listed above.

First, we derive an explicit representation for M^D under the conditions $BCA = 0$, $DCA = 0$ and $DCB = 0$. Therefore we can see that the condition $BCB = 0$ from [15] is superfluous.

Theorem 5.1 *Let M be a matrix of the form (5.7) such that $BCA = 0$, $DCA = 0$ and $DCB = 0$. Then*

$$M^D = \begin{bmatrix} A^D + \Sigma_0 C & B\Psi + A\Sigma_0 \\ \Psi C + CA\Sigma_1 C + C(A^D)^2 \\ -CA^D(B\Psi^2 D + AB\Psi^2)C & D^D + C\Sigma_0 \end{bmatrix},$$

where

$$\Sigma_k = \left(V_1\Psi^k + (A^D)^{2k}V_2\right)D + A\left(V_1\Psi^k + (A^D)^{2k}V_2\right), \; for\; k = 0, 1, \quad (5.8)$$

$$V_1 = \sum_{i=0}^{\nu_1-1} A^\pi A^{2i} B\Psi^{i+2}, \quad (5.9)$$

$$V_2 = \sum_{i=0}^{\mu_1-1} (A^D)^{2i+4} B(D^2 + CB)^i D^\pi - \sum_{i=0}^{\mu_1} (A^D)^{2i+2} B(CB)^i \Psi, \quad (5.10)$$

$$\Psi = (D^2 + CB)^D = \sum_{i=0}^{t_2-1} (CB)^\pi (CB)^i (D^D)^{2i+2} + \sum_{i=0}^{\nu_2-1} ((CB)^D)^{i+1} D^{2i} D^\pi, \quad (5.11)$$

$\nu_1 = \mathrm{Ind}(A^2)$, $\mu_1 = \mathrm{Ind}(D^2 + CB)$, $t_2 = \mathrm{Ind}(CB)$ *and* $\nu_2 = \mathrm{Ind}(D^2)$.

Proof Consider the splitting of matrix M

$$M = \begin{bmatrix} A & B \\ C & D \end{bmatrix} = \begin{bmatrix} 0 & B \\ C & D \end{bmatrix} + \begin{bmatrix} A & 0 \\ 0 & 0 \end{bmatrix} := P + Q.$$

Since $BCA = 0$ and $DCA = 0$ we get $P^2Q = 0$ and $QPQ = 0$. Hence matrices P and Q satisfy the conditions of Lemma 2.2 [21] and

$$(P+Q)^D = Y_1 + Y_2 + PQY_1(P^D)^2 + PQ^DY_2 - PQQ^D(P^D)^2 - PQ^DP^D, \quad (5.12)$$

where Y_1 and Y_2 are defined by

$$Y_1 = \sum_{i=0}^{s-1} Q^\pi Q^i (P^D)^{i+1}, \quad Y_2 = \sum_{i=0}^{r-1} (Q^D)^{i+1} P^i P^\pi. \quad (5.13)$$

By one of the assumptions of the theorem $DCB = 0$, so we have that the matrix P satisfies the conditions of Lemma 2.6 [21]. After applying Lemma 2.6 [21], we get

$$Y_1 = \begin{bmatrix} (V_1D + AV_1)C & A^\pi B\Psi + A(V_1D + AV_1) \\ \Psi C & \Psi D \end{bmatrix}, \quad (5.14)$$

$$Y_2 = \begin{bmatrix} A^D + (V_2D + AV_2)C & B\Psi - A^\pi B\Psi + A(V_2D + AV_2) \\ 0 & 0 \end{bmatrix}, \quad (5.15)$$

where V_1 and V_2 are defined by (5.9) and (5.10), respectively. After substituting (5.14) and (5.15) into (5.12) and computing all the terms in (5.12) we obtain the result. □

As a direct corollary of the previous theorem, we get the following result.

Corollary 5.1 *Let M be as in (5.7). If $DCB = 0$ and $CA = 0$, then*

$$M^D = \begin{bmatrix} A^D + \Sigma_0 C & B\Psi + A\Sigma_0 \\ \Psi C & \Psi D \end{bmatrix},$$

where Σ_0 and Ψ are defined by (5.8) and (5.11), respectively.

Notice that Corollary 5.1, therefore Theorem 5.1 also, is a generalization of the result about the representation for M^D under the conditions $CB = 0$ and $CA = 0$ which is given in [14].

The next result which is a corollary of Theorem 5.1, also follows using the splitting $M = \begin{bmatrix} 0 & 0 \\ 0 & D \end{bmatrix} + \begin{bmatrix} A & B \\ C & 0 \end{bmatrix} := P + Q$ and applying Lemmas 2.1 and 2.5 from [21].

Corollary 5.2 *Let M be a matrix of the form (5.7). If $BCA = 0$ and $DC = 0$, then*

$$M^D = \begin{bmatrix} A\Omega & \Omega B + RD \\ C\Omega & D^D + CR \end{bmatrix},$$

where

$$R = (R_1 + R_2)D + A(R_1 + R_2),$$

$$R_1 = \sum_{i=0}^{\mu_2-1} A^\pi (A^2 + BC)^i B(D^D)^{2i+4} - \sum_{i=0}^{\mu_2} \Omega(BC)^i B(D^D)^{2i+2},$$

$$R_2 = \sum_{i=0}^{\nu_2-1} \Omega^{i+2} B D^{2i} D^\pi,$$

$$\Omega = (A^2 + BC)^D = \sum_{i=0}^{t_1-1} (A^D)^{2i+2} (BC)^i (BC)^\pi + \sum_{i=0}^{\nu_1-1} A^\pi A^{2i} ((BC)^D)^{i+1},$$

$\nu_2 = \mathrm{Ind}(D^2)$, $\mu_2 = \mathrm{Ind}(A^2 + BC)$, $t_1 = \mathrm{Ind}(BC)$, $\nu_1 = \mathrm{Ind}(A^2)$.

We remark that Corollary 5.2, hence also Theorem 5.1, is an extension of the results from [17], where beside the conditions $BCA = 0$ and $DC = 0$ the additional condition $BD = 0$ (or that D is nilpotent) is required.

Castro–González et al. [17] gave an explicit representation for M^D under the conditions $BCA = 0$, $BD = 0$ and that BC is nilpotent (or $DC = 0$). This result was extended to the case when $BCA = 0$ and $BD = 0$ (see [19]). The following theorem is a common extension of both of these results.

Theorem 5.2 *Let M be a matrix of the form (5.7) such that $BCA = 0$, $ABD = 0$ and $CBD = 0$. Then*

$$M^D = \begin{bmatrix} A\Omega + B(F_1 + F_2) & \begin{array}{c} \Omega B + BD(F_1\Omega + (D^D)^2 F_2)B \\ + B(D^D)^2 - BD^D(CA + DC)\Omega^2 B \end{array} \\ C\Omega + D(F_1 + F_2) & D^D + (F_1 + F_2)B \end{bmatrix}, \quad (5.16)$$

where

$$F_1 = \sum_{i=0}^{\nu_2-1} D^\pi D^{2i} (CA + DC)\Omega^{i+2},$$

$$F_2 = \sum_{i=0}^{\mu_2-1} (D^D)^{2i+4}(CA + DC)(A^2 + BC)^i (BC)^\pi - \sum_{i=0}^{\mu_2} (D^D)^{2i+2}(CA + DC)A^{2i}\Omega,$$

$$\Omega = (A^2 + BC)^D = \sum_{i=0}^{t_1-1} (A^D)^{2i+2}(BC)^i (BC)^\pi + \sum_{i=0}^{\nu_1-1} A^\pi A^{2i} ((BC)^D)^{i+1},$$

$\nu_2 = \mathrm{Ind}(D^2)$, $\mu_2 = \mathrm{Ind}(A^2 + BC)$.

Proof If we split the matrix M as

$$M = \begin{bmatrix} A & B \\ C & 0 \end{bmatrix} + \begin{bmatrix} 0 & 0 \\ 0 & D \end{bmatrix} := P + Q.$$

we have that $QPQ = 0$ and $P^2 Q = 0$. Hence, the matrices P and Q satisfy the conditions of Lemma 2.2 [21]. Since $BCA = 0$, the matrix P satisfies the conditions of Lemma 2.5 [21]. Using a method similar to that used in the proof of Theorem 5.1, after applying Lemmas 2.2 and 2.5 from [21], we get that (5.16) holds.   □

In [16] a formula for M^D is given under the conditions $BCA = 0$, $DCA = 0$, $CBD = 0$ and $CBC = 0$. In the next theorem we offer a representation for M^D under the conditions $BCA = 0$, $DCA = 0$ and $CBD = 0$, without the additional condition $CBC = 0$.

Theorem 5.3 *Let M be given by (5.7). If $BCA = 0$, $DCA = 0$ and $CBD = 0$, then*

$$M^D = \begin{bmatrix} A^D + (G_1 + G_2)C & B\Gamma + A(G_1 + G_2) \\ \Gamma C + CA(G_1\Gamma + (A^D)^2 G_2)C & \\ + C(A^D)^2 - CA^D(AB + BD)\Gamma^2 C & D\Gamma + C(G_1 + G_2) \end{bmatrix},$$

where

$$G_1 = \sum_{i=0}^{\nu_1 - 1} A^\pi A^{2i}(AB + BD)\Gamma^{i+2}, \tag{5.17}$$

$$G_2 = \sum_{i=0}^{\mu_1 - 1}(A^D)^{2i+4}(AB+BD)(D^2+CB)^i(CB)^\pi - \sum_{i=0}^{\mu_1}(A^D)^{2i+2}(AB+BD)D^{2i}\Gamma, \tag{5.18}$$

$$\Gamma = \sum_{i=0}^{t_2 - 1}(D^D)^{2i+2}(CB)^i(CB)^\pi + \sum_{i=0}^{\nu_2 - 1} D^\pi D^{2i}((CB)^D)^{i+1}, \tag{5.19}$$

$\nu_1 = \text{Ind}(A^2)$, $\mu_1 = \text{Ind}(D^2 + CB)$, $t_2 = \text{Ind}(CB)$ *and* $\nu_2 = \text{Ind}(D^2)$.

Proof Using the splitting of matrix M

$$M = \begin{bmatrix} 0 & B \\ C & D \end{bmatrix} + \begin{bmatrix} A & 0 \\ 0 & 0 \end{bmatrix} := P + Q,$$

we get that conditions of Lemma 2.6 [21] are satisfied. Also, we have that the matrix P satisfies the conditions of Lemma 2.7 [21]. Using these lemmas, we get that the statement of the theorem is valid.   □

We can see that Theorem 5.3 is an extension of the result from [13] giving a representation for M^D under the conditions $CB = 0$ and $CA = 0$.

In [13] a representation for M^D is offered under the conditions $AB = 0$ and $CB = 0$. This result is extended in [15], where a formula for M^D is given under the conditions $ABC = 0$, $ABD = 0$, $CBD = 0$ and $CBC = 0$. In our following result we derive a representation for M^D under the conditions $ABC = 0$, $ABD = 0$ and $CBD = 0$, without the additional condition $CBC = 0$.

Theorem 5.4 *Let M be a matrix of the form (5.7). If* $ABC = 0$, $ABD = 0$ *and* $CBD = 0$, *then*

$$M^D = \begin{bmatrix} A^D + B\Theta_0 & B\Gamma + B\Theta_1 AB + (A^D)^2 B \\ & -B(\Gamma^2 CA + D\Gamma^2 C)A^D B \\ \Gamma C + \Theta_0 A & D^D + \Theta_0 B \end{bmatrix}, \qquad (5.20)$$

where

$$\Theta_k = \left(K_1(A^D)^{2k} + \Gamma^k K_2\right) A + D\left(K_1(A^D)^{2k} + \Gamma^k K_2\right), \ for \ k = 0, 1, \quad (5.21)$$

$$K_1 = \sum_{i=0}^{\mu_1 - 1} D^\pi (D^2 + CB)^i C(A^D)^{2i+4} - \sum_{i=0}^{\mu_1} \Gamma(CB)^i C(A^D)^{2i+2}, \qquad (5.22)$$

$$K_2 = \sum_{i=0}^{\nu_1 - 1} \Gamma^{i+2} CA^{2i} A^\pi, \qquad (5.23)$$

$\nu_1 = \mathrm{Ind}(A^2)$, $\mu_1 = \mathrm{Ind}(D^2 + CB)$, $t_1 = \mathrm{Ind}(BC)$.

Proof We can split the matrix M as $M = P + Q$, where

$$P = \begin{bmatrix} A & 0 \\ 0 & 0 \end{bmatrix}, \ Q = \begin{bmatrix} 0 & B \\ C & D \end{bmatrix}.$$

According to the assumptions of the theorem, we have that $PQP = 0$ and $PQ^2 = 0$. Hence we can apply Lemma 2.1 [21] and we have

$$(P+Q)^D = Y_1 + Y_2 + \left(Y_1(P^D)^2 + (Q^D)^2 Y_2 - Q^D(P^D)^2 - (Q^D)^2 P^D\right) PQ, \quad (5.24)$$

where Y_1 and Y_2 are defined by (5.13). Since $CBD = 0$, the matrix Q satisfies the conditions of Lemma 2.7 [21]. After applying Lemma 2.7 [21], we get

$$Y_1 = \begin{bmatrix} A^D + B(K_1 A + DK_1) & 0 \\ \Gamma C - \Gamma CA^\pi + (K_1 A + DK_1)A \ 0 \end{bmatrix}, \qquad (5.25)$$

$$Y_2 = \begin{bmatrix} B(K_2 A + DK_2) & B\Gamma \\ \Gamma CA^\pi + (K_2 A + DK_2)A & D\Gamma \end{bmatrix}, \qquad (5.26)$$

where K_1 and K_2 are given by (5.22) and (5.23), respectively. Now, substituting (5.26) and (5.25) into (5.24) we get that (5.20) holds.   □

Notice that Theorem 5.4 generalizes the result found in [21], where it is assumed that $ABC = 0$ and $BD = 0$.

As another extension of a result from [13], where a formula for M^D is given under the conditions $AB = 0$ and $CB = 0$, we offer the following theorem and its corollary.

Theorem 5.5 *Let M be a matrix of the form (5.7). If $ABC = 0$, $ABD = 0$ and $DCB = 0$, then*

$$
M^D = \begin{bmatrix} A^D + B(N_1 + N_2) & \begin{aligned} B\Psi + B(N_1(A^D)^2 + \Psi N_2)AB \\ +(A^D)^2B - B\Psi^2(CA + DC)A^DB \end{aligned} \\ \Psi C + (N_1 + N_2)A & \Psi D + (N_1 + N_2)B \end{bmatrix}, \quad (5.27)
$$

where

$$
N_1 = \sum_{i=0}^{\mu_1-1}(CB)^\pi(D^2+CB)^i(CA+DC)(A^D)^{2i+4} - \sum_{i=0}^{\mu_1}\Psi D^{2i}(CA+DC)(A^D)^{2i+2},
$$
$$
(5.28)
$$

$$
N_2 = \sum_{i=0}^{\nu_1-1}\Psi^{i+2}(CA + DC)A^{2i}A^\pi, \quad (5.29)
$$

$\nu_1 = \mathrm{Ind}(A^2)$, $\mu_1 = \mathrm{Ind}(D^2 + CB)$ *and* Ψ *is defined by (5.11).*

Proof Using the splitting

$$
M = \begin{bmatrix} A & 0 \\ 0 & 0 \end{bmatrix} + \begin{bmatrix} 0 & B \\ C & D \end{bmatrix} := P + Q,
$$

we get that the matrices P and Q satisfy the conditions of Lemma 2.1 [21]. Furthermore, the matrix Q satisfies the conditions of Lemma 2.6 from [21]. After applying these lemmas and computing, we get that (5.27) holds.   □

Cvetković and Milovanović [18] offered a representation for M^D under the conditions $ABC = 0$, $DC = 0$, along with a third condition $BD = 0$ (or that BC is nilpotent, or that D is nilpotent). In [19] a formula for M^D was given under the conditions $ABC = 0$ and $DC = 0$, without any additional conditions. In our next result we replace the second condition $DC = 0$ from [19] with two weaker conditions.

Theorem 5.6 *Let M be a matrix of the form (5.7), such that $ABC = 0$, $DCA = 0$ and $DCB = 0$. Then*

$$M^D = \begin{bmatrix} \Phi A + (U_1 + U_2)C & \Phi B + (U_1 + U_2)D \\ C\Phi + C(U_1(D^D)^2 + \Phi U_2)DC & D^D + C(U_1 + U_2) \\ +(D^D)^2 C - C\Phi^2(AB + BD)D^D C \end{bmatrix},$$

where

$$U_1 = \sum_{i=0}^{\mu_2-1} (BC)^\pi (A^2 + BC)^i (AB + BD)(D^D)^{2i+4} - \sum_{i=0}^{\mu_2} \Phi A^{2i}(AB + BD)(D^D)^{2i+2},$$

$$U_2 = \sum_{i=0}^{\nu_2-1} \Phi^{i+2}(AB + BD)D^{2i}D^\pi,$$

$$\Phi = (A^2 + BC)^D = \sum_{i=0}^{t_1-1} (BC)^\pi (BC)^i (A^D)^{2i+2} + \sum_{i=0}^{\nu_1-1} ((BC)^D)^{i+1} A^{2i} A^\pi,$$

$\nu_2 = \text{Ind}(D^2)$, $\mu_2 = \text{Ind}(A^2 + BC)$, $t_1 = \text{Ind}(BC)$ *and* $\nu_1 = \text{Ind}(A^2)$.

Proof If we split the matrix M as

$$M = \begin{bmatrix} 0 & 0 \\ 0 & D \end{bmatrix} + \begin{bmatrix} A & B \\ C & 0 \end{bmatrix} := P + Q,$$

we have $PQP = 0$ and $PQ^2 = 0$. Also, the matrix P satisfies conditions of Lemma 2.4 [21]. After applying Lemmas 2.1 and 2.4 from [21] and computing we get that the statement of the theorem is valid.   □

Example 5.1 [15] Consider the block matrix $M = \begin{bmatrix} A & B \\ C & D \end{bmatrix}$, where $A = \begin{bmatrix} 1 & 1 \\ 0 & 0 \end{bmatrix}$, $B = \begin{bmatrix} 1 & -1 \\ 0 & 0 \end{bmatrix}$, $C = \begin{bmatrix} 1 & 1 \\ 1 & -1 \end{bmatrix}$, $D = \begin{bmatrix} 0 & 0 \\ -1 & 1 \end{bmatrix}$.

We observe that $BCA = 0$, $BCB = 0$, $DCA = 0$ and $DCB = 0$. Hence, by applying Theorem 5.1, we obtain

$$M^D = \begin{bmatrix} 1 & -1 & -1 & 1 \\ 0 & -2 & 0 & 0 \\ 1 & -5 & -2 & 2 \\ 1 & -7 & -3 & 3 \end{bmatrix}.$$

5.2 Representations for the Drazin Inverse of a 2 × 2 Block Matrix Involving the Schur Complement

Let M be a 2×2 complex block matrix given by

$$M = \begin{bmatrix} A & C \\ B & D \end{bmatrix}, \tag{5.30}$$

where A and D are square matrices, not necessarily of the same size.

Definition 5.1 If M is given by (5.30), the generalized Schur complement of M, is defined by

$$Z = D - BA^{D}C.$$

In this section, we present some representations for the Drazin inverse of M, in terms of its blocks under conditions that involve its generalized Schur complement. These results will fall into two classes: those that assume that the Schur complement of M is invertible and those where it is assumed to be to zero.

Throughout this section, we adopt some notation as follows. Let

$$K = A^{D}C, \quad H = BA^{D}, \quad \Gamma = HK, \quad Z = D - BA^{D}C, \tag{5.31}$$

and

$$P = (I - AA^{D})C, \quad Q = B(I - A^{D}A). \tag{5.32}$$

First, we will present some representations for M^{D} in the case when the generalized Schur complement of M is invertible.

The Drazin inverse of a 2×2 block matrix in the case when the generalized Schur complement $Z = D - CA^{D}B$ is invertible has been studied by Wei [22] and Miao [23].

Theorem 5.7 ([22, 23]) *Let M be given by (5.30). If*

$$C(I - AA^{D}) = 0, \quad (I - AA^{D})B = 0, \tag{5.33}$$

and the generalized Schur complement $Z = D - CA^{D}B$ is invertible, then

$$M^{D} = \begin{bmatrix} A^{D} + A^{D}BZ^{-1}CA^{D} & -A^{D}BZ^{-1} \\ -Z^{-1}CA^{D} & Z^{-1} \end{bmatrix}. \tag{5.34}$$

Here, we will derive formulae for the Drazin inverse of a 2×2 block matrix under conditions weaker than those in the theorem above.

First we present an auxiliary additive result for the Drazin inverse:

Lemma 5.1 ([4]) *Let P and Q be square matrices of the same order. If $PQ = 0$ and $Q^2 = 0$, then*

$$(P + Q)^D = P^D + Q(P^D)^2.$$

We introduce another simple lemma which will be used later.

Lemma 5.2 *Let $M \in \mathbb{C}^{n \times n}$, $G \in \mathbb{C}^{m \times n}$ and $H \in \mathbb{C}^{n \times m}$ be such that $HG = I_n$ and $m \geq n$. Then*

$$(GMH)^D = GM^D H,$$

and

$$\operatorname{Ind}(GMH) = 0, \quad \text{if } M \text{ is invertible and } m = n;$$
$$\operatorname{Ind}(GMH) = 1, \quad \text{if } M \text{ is invertible and } m > n;$$
$$\operatorname{Ind}(GMH) = \operatorname{Ind}(M), \quad \text{if } M \text{ is singular.}$$

Proof Notice that

$$r((GMH)^p) = r(GM^p H) = r(M^p),$$

for any positive integer p. The case when M is invertible and $m = n$ is trivial. Let M be invertible and $m > n$. It is easy to verify that

$$r(GMH) = r(M) = n < m,$$

but

$$r((GMH)^2) = r(M^2) = r(M) = r(GMH).$$

Thus, $\operatorname{Ind}(GMH) = 1$ and we can check that $GM^{-1}H$ is the group inverse of GMH. Now we consider the case where M is singular (or $\operatorname{Ind}(M) = k > 0$). We have

$$r(GMH) = r(M) < n < m,$$

which implies $\operatorname{Ind}(GMH) > 0$. Thus k is the smallest positive integer such that

$$r((GMH)^{k+1}) = r(M^{k+1}) = r(M^k) = r((GMH)^k).$$

In other words, $\operatorname{Ind}(GMH) = \operatorname{Ind}(M) = k$. It can be verified that

$$(GMH)^k(GM^D H)(GMH) = (GMH)^k;$$
$$(GM^D H)(GMH)(GM^D H) = GM^D H;$$
$$(GMH)(GM^D H) = (GM^D H)(GMH).$$

Therefore, $(GMH)^D = GM^D H$.   □

Suppose that A has a Jordan canonical form,

$$A = X \begin{bmatrix} \Sigma & 0 \\ 0 & N \end{bmatrix} X^{-1}, \tag{5.35}$$

where X and Σ are invertible matrices, and N is nilpotent (for more details see [2, 24]). Then

$$A^D = X \begin{bmatrix} \Sigma^{-1} & 0 \\ 0 & 0 \end{bmatrix} X^{-1}, \qquad \mathrm{Ind}(N) = \mathrm{Ind}(A) = k. \tag{5.36}$$

Let X and X^{-1} in (5.35) be partitioned compatibly as

$$X = \begin{bmatrix} U_1 & U_2 \end{bmatrix}, \qquad X^{-1} = \begin{bmatrix} V_1 \\ V_2 \end{bmatrix}, \tag{5.37}$$

where the column dimensions of U_1 and V_1^* are the same as that of Σ. Then the following holds:

$$A = U_1 \Sigma V_1 + U_2 N V_2, \qquad A^D = U_1 \Sigma^{-1} V_1,$$
$$A A^D = U_1 V_1, \qquad I - A A^D = U_2 V_2. \tag{5.38}$$

Using the notation in (5.37), let us define

$$\begin{bmatrix} P_1 \\ P_2 \end{bmatrix} := \begin{bmatrix} V_1 B \\ V_2 B \end{bmatrix} = X^{-1} B, \qquad \begin{bmatrix} Q_1 & Q_2 \end{bmatrix} := \begin{bmatrix} C U_1 & C U_2 \end{bmatrix} = C X.$$

Then

$$M = \begin{bmatrix} A & B \\ C & D \end{bmatrix} = \begin{bmatrix} U_1 & U_2 & 0 \\ 0 & 0 & I \end{bmatrix} \begin{bmatrix} \Sigma & 0 & P_1 \\ 0 & N & P_2 \\ Q_1 & Q_2 & D \end{bmatrix} \begin{bmatrix} V_1 & 0 \\ V_2 & 0 \\ 0 & I \end{bmatrix} \tag{5.39}$$

and

$$\begin{bmatrix} I & 0 & 0 \\ 0 & 0 & I \\ 0 & I & 0 \end{bmatrix} \begin{bmatrix} \Sigma & 0 & P_1 \\ 0 & N & P_2 \\ Q_1 & Q_2 & D \end{bmatrix} \begin{bmatrix} I & 0 & 0 \\ 0 & 0 & I \\ 0 & I & 0 \end{bmatrix} = \begin{bmatrix} \Sigma & P_1 & 0 \\ Q_1 & D & Q_2 \\ 0 & P_2 & N \end{bmatrix}. \tag{5.40}$$

Next we define

$$M_1 := \begin{bmatrix} \Sigma & P_1 & 0 \\ Q_1 & D & Q_2 \\ 0 & P_2 & N \end{bmatrix}.$$

From (5.39) and (5.40), we have that

$$M = \begin{bmatrix} U_1 & 0 & U_2 \\ 0 & I & 0 \end{bmatrix} M_1 \begin{bmatrix} V_1 & 0 \\ 0 & I \\ V_2 & 0 \end{bmatrix}.$$

Notice that both M and M_1 are square matrices of the same order and that the product $\begin{bmatrix} V_1 & 0 \\ 0 & I \\ V_2 & 0 \end{bmatrix} \times \begin{bmatrix} U_1 & 0 & U_2 \\ 0 & I & 0 \end{bmatrix}$ is the identity matrix. Consequently, it follows from Lemma 5.2 that

$$\text{Ind}(M) = \text{Ind}(M_1), \qquad M^D = \begin{bmatrix} U_1 & 0 & U_2 \\ 0 & I & 0 \end{bmatrix} M_1^D \begin{bmatrix} V_1 & 0 \\ 0 & I \\ V_2 & 0 \end{bmatrix}. \tag{5.41}$$

In order to get an explicit formula for M_1^D by an application of Lemma 5.2, we repartition M_1 as a 2 × 2 block matrix, i.e.,

$$M_1 = \begin{bmatrix} \Sigma & P_1 & 0 \\ Q_1 & D & Q_2 \\ 0 & P_2 & N \end{bmatrix} = \begin{bmatrix} F & \tilde{Q}_2 \\ \tilde{P}_2 & N \end{bmatrix}, \tag{5.42}$$

where

$$F = \begin{bmatrix} \Sigma & P_1 \\ Q_1 & D \end{bmatrix}, \quad \tilde{P}_2 = [0 \;\; P_2], \quad \tilde{Q}_2 = \begin{bmatrix} 0 \\ Q_2 \end{bmatrix}. \tag{5.43}$$

Now, we can prove one of the main results of this section.

Theorem 5.8 *Let M be given by (5.30). If*

$$C(I - AA^D)B = 0, \qquad A(I - AA^D)B = 0, \tag{5.44}$$

and $Z = D - CA^DB$ is invertible, then $\text{Ind}(M) \leq \text{Ind}(A) + 1$ and

$$M^D = \left(I + \begin{bmatrix} 0 & (I - AA^D)B \\ 0 & 0 \end{bmatrix} R \right) R \left(I + \sum_{i=0}^{k-1} R^{i+1} \begin{bmatrix} 0 & 0 \\ C(I - AA^D)A^i & 0 \end{bmatrix} \right), \tag{5.45}$$

where $k = \text{Ind}(A)$ and

$$R = \begin{bmatrix} A^D + A^D BZ^{-1}CA^D & -A^D BZ^{-1} \\ -Z^{-1}CA^D & Z^{-1} \end{bmatrix}. \tag{5.46}$$

Proof We first convert (5.44) into an equivalent form in terms of matrices $\tilde{P}_2$, $\tilde{Q}_2$ and N. Using the identities in (5.38), we obtain

$$C(I - AA^D)B = CU_2V_2B = Q_2P_2,$$
$$A(I - AA^D)B = (U_1\Sigma V_1 + U_2NV_2)U_2V_2B = U_2NP_2.$$

Then (5.44) is equivalent to

$$\tilde{Q}_2 \tilde{P}_2 = 0, \qquad N\tilde{P}_2 = 0.$$

Recall that N is nilpotent. Applying Corollary 5.2 to M_1 in (5.42) yields

$$\text{Ind}(M_1) \le \text{Ind}(F) + \text{Ind}(N) + 1$$

and

$$M_1^{\mathrm{D}} = \begin{bmatrix} I \\ S \end{bmatrix} F^{\mathrm{D}} \left[I \sum_{i=0}^{k-1} (F^{\mathrm{D}})^i T N^i \right], \tag{5.47}$$

where $k = \text{Ind}(A)$, F is given by (5.43) and

$$S = \tilde{P}_2 F^{\mathrm{D}} = [0 \; P_2] F^{\mathrm{D}}, \qquad T = F^{\mathrm{D}} \tilde{Q}_2 = F^{\mathrm{D}} \begin{bmatrix} 0 \\ Q_2 \end{bmatrix}.$$

Substituting (5.47) into (5.41), we have

$$M^{\mathrm{D}} = \left(\begin{bmatrix} U_1 & 0 \\ 0 & I \end{bmatrix} + \begin{bmatrix} U_2 \\ 0 \end{bmatrix} S \right) F^{\mathrm{D}} \left(\begin{bmatrix} V_1 & 0 \\ 0 & I \end{bmatrix} + \sum_{i=0}^{k-1} (F^{\mathrm{D}})^i T N^i [V_2 \; 0] \right)$$

$$= \left(I + \begin{bmatrix} U_2 \\ 0 \end{bmatrix} S \begin{bmatrix} V_1 & 0 \\ 0 & I \end{bmatrix} \right) R \left(I + \begin{bmatrix} U_1 & 0 \\ 0 & I \end{bmatrix} \sum_{i=0}^{k-1} (F^{\mathrm{D}})^i T N^i [V_2 \; 0] \right) \tag{5.48}$$

where

$$R = \begin{bmatrix} U_1 & 0 \\ 0 & I \end{bmatrix} F^{\mathrm{D}} \begin{bmatrix} V_1 & 0 \\ 0 & I \end{bmatrix}. \tag{5.49}$$

It is easy to verify that

$$\begin{bmatrix} U_2 \\ 0 \end{bmatrix} S \begin{bmatrix} V_1 & 0 \\ 0 & I \end{bmatrix} = \begin{bmatrix} 0 & (I - AA^{\mathrm{D}})B \\ 0 & 0 \end{bmatrix} R,$$

$$\begin{bmatrix} U_1 & 0 \\ 0 & I \end{bmatrix} T [V_2 \; 0] = R \begin{bmatrix} 0 & 0 \\ C(I - AA^{\mathrm{D}}) & 0 \end{bmatrix},$$

$$\begin{bmatrix} U_2 \\ 0 \end{bmatrix} N^i [V_2 \; 0] = \begin{bmatrix} (I - AA^{\mathrm{D}})A^i & 0 \\ 0 & 0 \end{bmatrix}. \tag{5.50}$$

Equation (5.45) immediately follows from (5.48) and (5.50), where R is given by (5.49).

It suffices to show that $\text{Ind}(F) = 0$ and that (5.46) follows from (5.49). As a matter of fact, the Schur complement of F in (5.43), $Z = D - Q_1 \Sigma^{-1} P_1 = D - CA^{\mathrm{D}}B$ is invertible by hypothesis. Then F is invertible and its index is zero. Applying Theorem 5.7 to F, we have

$$F^{-1} = \begin{bmatrix} \Sigma^{-1} + \Sigma^{-1} P_1 Z^{-1} Q_1 \Sigma^{-1} & -\Sigma^{-1} P_1 Z^{-1} \\ -Z^{-1} Q_1 \Sigma^{-1} & Z^{-1} \end{bmatrix}. \tag{5.51}$$

Substituting (5.51) into (5.49), we obtain

$$R = \begin{bmatrix} U_1(\Sigma^{-1} + \Sigma^{-1} P_1 Z^{-1} Q_1 \Sigma^{-1}) V_1 & -U_1 \Sigma^{-1} P_1 Z^{-1} \\ -Z^{-1} Q_1 \Sigma^{-1} V_1 & Z^{-1} \end{bmatrix}.$$

Now (5.46) follows from (5.38). □

Remark 5.2.1 We remark that (5.44) is weaker than (5.33). To see this in the case when $k = \text{Ind}(A) \geq 1$, one could consider the following geometric interpretations of these two conditions. Equation (5.33) in Theorem 5.7 means that $\mathscr{R}(B) \subset \mathscr{R}(A^k)$ and $\mathscr{N}(A^k) \subset \mathscr{N}(C)$ while (5.44) in Theorem 5.8 means that $\mathscr{R}(B) \subset \mathscr{R}(A^{k-1})$ and that the projection of $\mathscr{R}(B)$ on $\mathscr{N}(A^k)$ along $\mathscr{R}(A^k)$ is contained in $\mathscr{N}(C)$.

Two corollaries are easily derived from Theorem 5.8 by taking adjoints. If $A(I - AA^D)B = 0$ in (5.44) is replaced by $C(I - AA^D)A = 0$ (which means that $\mathscr{N}(A^{k-1}) \subset \mathscr{N}(C)$) we have the following:

Corollary 5.3 *Let M be given by (5.30). If*

$$C(I - AA^D)B = 0, \qquad C(I - AA^D)A = 0,$$

and $Z = D - CA^D B$ is invertible, then $\text{Ind}(M) \leq \text{Ind}(A) + 1$ and

$$M^D = \left(I + \sum_{i=0}^{k-1} \begin{bmatrix} 0 & A^i(I - AA^D)B \\ 0 & 0 \end{bmatrix} R^{i+1} \right) R \left(I + R \begin{bmatrix} 0 & 0 \\ C(I - AA^D) & 0 \end{bmatrix} \right),$$

where $k = \text{Ind}(A)$ and R is given by (5.46).

If the extra condition $C(I - AA^D)A = 0$ is imposed in Theorem 5.8, then we have a simpler formula for M^D.

Corollary 5.4 *Let M be given by (5.30). If*

$$C(I - AA^D)B = 0, \quad A(I - AA^D)B = 0, \quad C(I - AA^D)A = 0, \tag{5.52}$$

and $Z = D - CA^D B$ is invertible, then $\text{Ind}(M) \leq \text{Ind}(A) + 1$ and

$$M^D = \begin{bmatrix} I - (I - AA^D)BZ^{-1}CA^D & (I - AA^D)BZ^{-1} \\ 0 & I \end{bmatrix}$$
$$\times \begin{bmatrix} A^D + A^D BZ^{-1}CA^D & -A^D BZ^{-1} \\ -Z^{-1}CA^D & Z^{-1} \end{bmatrix} \begin{bmatrix} I - A^D BZ^{-1}C(I - AA^D) & 0 \\ Z^{-1}C(I - AA^D) & I \end{bmatrix}. \tag{5.53}$$

It is easy to see that Theorem 5.8 and Corollaries 5.3 and 5.4 are extensions of Theorem 5.7. As consequences they entail many results related to the Drazin inverse of a partitioned matrix in [22].

Below we will extend Theorem 5.7 and establish a new representation for M^D.

Theorem 5.9 *Let M be given by (5.30). Suppose $P = 0$, $Q = 0$, $CZ^DZ = C$, and $Z^DZB = B$. Then*

$$M^D = \begin{bmatrix} A^D + KZ^D H & -KZ^D \\ -Z^D H & Z^D \end{bmatrix}. \tag{5.54}$$

Proof Denote the right-hand side of (5.54) by X. Since

$$MX = \begin{bmatrix} AA^D + AA^DCZ^DBA^D - CZ^DBA^D & -AA^DCZ^D + CZ^D \\ BA^D + (D-Z)Z^DBA^D - DZ^DBA^D & -(D-Z)Z^D + DZ^D \end{bmatrix} = \begin{bmatrix} AA^D & 0 \\ 0 & ZZ^D \end{bmatrix},$$

and

$$XM = \begin{bmatrix} A^DA + A^DCZ^DBA^DA - A^DCZ^DB & A^DC + A^DCZ^D(D-Z) - A^DCZ^DD \\ -Z^DBA^DA + Z^DB & -Z^D(D-Z) + Z^DD \end{bmatrix}$$

$$= \begin{bmatrix} A^DA & 0 \\ 0 & Z^DZ \end{bmatrix},$$

we thus obtain

$$XMX = \begin{bmatrix} A^DA(A^D + A^DCZ^DBA^D) & -A^DA A^DCZ^D \\ -Z^DZZ^DBA^D & -Z^DZZ^D \end{bmatrix} = X.$$

It is not difficult to verify that for every $m \geq \text{Ind}(A)$,

$$M^{m+1}X = M^m.$$

□

If $Z = D - BA^DC$ is invertible, then we have the following corollary:

Corollary 5.5 *Let M be given by (5.30). Suppose $P = 0$, $Q = 0$, and Z is invertible. Then*

$$M^D = \begin{bmatrix} A^D + KZ^{-1}H & -KZ^{-1} \\ -Z^{-1}H & Z^{-1} \end{bmatrix} = \begin{bmatrix} A^D & 0 \\ 0 & 0 \end{bmatrix} + SZ^{-1}T, \tag{5.55}$$

where $S = \begin{bmatrix} -K \\ I \end{bmatrix}$ and $T = \begin{bmatrix} -H & I \end{bmatrix}$.

We now turn our attention to the problem of finding representations for M^D in the case when the generalized Schur complement of M is zero, to which the remainder of this section is devoted.

Theorem 5.10 *Let M be given by (5.30). Suppose $P = 0$, $Q = 0$ and $Z = 0$, and $C(I + \Gamma)^D(I + \Gamma) = C$, $(I + \Gamma)(I + \Gamma)^D B = B$. Then*

$$
M^D = \begin{bmatrix} (I - K(I + \Gamma)^D H)A^D(I - K(I + \Gamma)^D H) & (I - K(I + \Gamma)^D H)A^D K(I + \Gamma)^D \\ (I + \Gamma)^D H A^D(I - K(I + \Gamma)^D H) & (I + \Gamma)^D H A^D K(I + \Gamma)^D \end{bmatrix}
$$

$$
= (I - S(I + \Gamma)^D T)\begin{bmatrix} A^D & 0 \\ 0 & 0 \end{bmatrix}(I - S(I + \Gamma)^D T). \tag{5.56}
$$

Proof Denote the right-hand side of (5.56) by X. By direct manipulations, we have

$$
MX = \begin{bmatrix} AA^D(I - K(I + \Gamma)^D H) & K(I + \Gamma)^D \\ (I - \Gamma(I + \Gamma)^D)H & \Gamma(I + \Gamma)^D \end{bmatrix},
$$

and

$$
XM = \begin{bmatrix} (I - K(I + \Gamma)^D H)A^D A & K(I - (I + \Gamma)^D)\Gamma) \\ (I + \Gamma)^D H & (I + \Gamma)^D \Gamma \end{bmatrix}.
$$

Since

$$
K(I - (I + \Gamma)^D \Gamma) = K(I - (I + \Gamma)^D(I + \Gamma)) + K(I + \Gamma)^D = K(I + \Gamma)^D,
$$

a similar argument leads to

$$
(I - \Gamma(I + \Gamma)^D)H = (I + \Gamma)^D H.
$$

The relation $\Gamma(I + \Gamma)^D = (I + \Gamma)^D \Gamma$ follows from the fact that $(I + \Gamma)^D$ is a polynomial of Γ [2, Theorem 7.5.1]. Thus $MX = XM$. Furthermore, we have

$$
\begin{aligned}
(XMX)_{11} &= ([I - K(I + \Gamma)^D H][I - K(I + \Gamma)^D H] \\
&\quad + K(I + \Gamma)^D(I + \Gamma)^D H])A^D[I - K(I + \Gamma)^D H] \\
&= [I - 2K(I + \Gamma)^D H + K(I + \Gamma)^D(I + \Gamma - I)(I + \Gamma)^D H \\
&\quad + K(I + \Gamma)^D(I + \Gamma)^D H]A^D[I - K(I + \Gamma)^D H] \\
&= [I - K(I + \Gamma)^D H]A^D[I - K(I + \Gamma)^D H] \\
&= (X)_{11},
\end{aligned}
$$

and similarly $(XMX)_{12} = (X)_{12}$. Notice that

$$
\begin{aligned}
&(I + \Gamma)^D H - (I + \Gamma)^D \Gamma(I + \Gamma)^D H + \Gamma(I + \Gamma)^D(I + \Gamma)^D H \\
&= (I + \Gamma)^D H - (I + \Gamma)^D(I + \Gamma - I)(I + \Gamma)^D H + \Gamma(I + \Gamma)^D(I + \Gamma)^D H \\
&= (I + \Gamma)^D H,
\end{aligned}
$$

Hence, we obtain

$$
(XMX)_{21} = (X)_{21}, \qquad (XMX)_{22} = (X)_{22},
$$

Thus

$$XMX = X.$$

Finally from $P = 0$, $Q = 0$ and $Z = 0$, we get $M = \begin{bmatrix} I \\ H \end{bmatrix} A \begin{bmatrix} I & K \end{bmatrix}$, and

$$M^m = \begin{bmatrix} I \\ H \end{bmatrix} (AE)^{m-1} A \begin{bmatrix} I & K \end{bmatrix}, \quad m = 2, 3, \dots$$

where $E = I + KH$. Now we have

$$
\begin{aligned}
M^{k+2}X &= \begin{bmatrix} I \\ H \end{bmatrix} (AE)^{k+1} A \begin{bmatrix} I & K \end{bmatrix} X \\
&= \left[\begin{bmatrix} I \\ H \end{bmatrix} (AE)^{k+1} AA^D[I - K(I + \Gamma)^D H] \quad \begin{bmatrix} I \\ H \end{bmatrix} (AE)^{k+1} AA^D K(I + \Gamma)^D \right] \\
&= \left[\begin{bmatrix} I \\ H \end{bmatrix} (AE)^k AA^D A \quad \begin{bmatrix} I \\ H \end{bmatrix} (AE)^k AK(I + \Gamma)(I + \Gamma)^D \right] \\
&= \left[\begin{bmatrix} I \\ H \end{bmatrix} (AE)^k A \quad \begin{bmatrix} I \\ H \end{bmatrix} (AE)^k AK \right] \\
&= M^{k+1}.
\end{aligned}
$$

□

Combining Corollary 5.5 and Theorem 5.10, we get the following:

Theorem 5.11 *Let M be given by (5.30). Suppose $P = 0$, $Q = 0$, $\mathrm{Ind}(Z) = 1$, $C(I + \Gamma)^D(I + \Gamma) = C$, $(I + \Gamma)(I + \Gamma)^D B = B$, $D(I + \Gamma)^D(I + \Gamma) = D$ and $\Gamma Z = Z\Gamma$. Then*

$$M^D = (I - S(I - Z^\# Z)(I + \Gamma)^D T) \begin{bmatrix} A^D & 0 \\ 0 & 0 \end{bmatrix} (I - S(I + \Gamma)^D(I - ZZ^\#)T)$$
$$+ SZ^\# T. \tag{5.57}$$

Proof Denote the right-hand side of (5.57) by X. We can verify that

$$
MX = \begin{bmatrix} AA^D[I - K(I + \Gamma)^D(I - ZZ^\#)H] & K(I + \Gamma)^D(I - ZZ^\#) \\ (I - (I + \Gamma)(I + \Gamma)^D)(I - ZZ^\#)H & \Gamma(I + \Gamma)^D(I - ZZ^\#) \end{bmatrix}
$$
$$
+ \begin{bmatrix} 0 & 0 \\ (I + \Gamma)^D(I - ZZ^\#)H & ZZ^\# \end{bmatrix}
$$

and

$$
XM = \begin{bmatrix} [I - K(I - ZZ^\#)(I + \Gamma)^D H A^D A & K(I - ZZ^\#)[I - (I + \Gamma)^D(I + \Gamma)] \\ (I - ZZ^\#)(I + \Gamma)^D)H & (I - ZZ^\#)(I + \Gamma)^D \Gamma \end{bmatrix}
$$
$$
+ \begin{bmatrix} 0 & K(I - ZZ^\#)(I + \Gamma)^D \\ 0 & Z^\# Z \end{bmatrix}.
$$

Since $(I + \Gamma)^D$ is a polynomial of Γ, $I - ZZ^\#$ is a polynomial of Z, and from $Z\Gamma = \Gamma Z$, we get

$$(I + \Gamma)^D(I - ZZ^\#) = (I - Z^\#Z)(I + \Gamma)^D$$

and

$$[I - (I + \Gamma)(I + \Gamma)^D](I - ZZ^\#) = (I - Z^\#Z)[I - (I + \Gamma)^D(I + \Gamma)],$$
$$\Gamma(I + \Gamma)^D(I - ZZ^\#) = (I - Z^\#Z)(I + \Gamma)^D\Gamma.$$

Thus

$$MX = XM,$$

The proofs of $XMX = X$ and $M^{k+1}X = M^k$ are analogous to that of $MX = XM$. □

Remark 5.2.2 Using Theorem 1 of [12], we can replace the condition $P = 0$ and $Q = 0$ with $P = 0$ or $Q = 0$ in Theorems 5.9–5.11.

In Theorem 5.7, if the invertibility of the generalized Schur complement is replaced by the requirement that it is equal to zero, it is again possible to give a representation of the Drazin inverse of M as follows.

Theorem 5.12 ([23]) *Let M be given by (5.30). If*

$$C(I - AA^D) = 0, \qquad (I - AA^D)B = 0,$$

and the generalized Schur complement $Z = D - CA^DB$ is equal to 0, then

$$M^D = \begin{bmatrix} I \\ CA^D \end{bmatrix}[(AW)^D]^2A\begin{bmatrix} I & A^DB \end{bmatrix}, \qquad (5.58)$$

where $W = AA^D + A^DBCA^D$.

In order to mention one more expression for M^D, we will first introduce the notion of the weighted Drazin inverse defined by Cline and Greville in [25].

Definition 5.2 For $G \in \mathbb{C}^{m \times n}$ and $W \in \mathbb{C}^{n \times m}$, the weighted Drazin inverse of G with weight W, denoted by $G_{d,W}$, is defined as

$$G_{d,W} = [(GW)^D]^2G. \qquad (5.59)$$

It is well-known that $G_{d,W}$ is the unique matrix satisfying the following three equations:

$$(GW)^k = (GW)^{k+1}G_{d,W}W, \quad G_{d,W} = G_{d,W}WGWG_{d,W} \quad GWG_{d,W} = G_{d,W}WG,$$

where $k = \text{Ind}(GW)$.

If $G \in \mathbb{C}^{n \times n}$ and $W = I_n$ then the *weighted* Drazin inverse of G with weight I is the Drazin inverse of G, i.e.,

$$G_{d,I} = G^D.$$

Thus, the weighted Drazin inverse of G with weight W is a generalization of the notion of the Drazin inverse of a square matrix.

Theorem 5.13 *Let M be given by (5.30) and let*

$$W = AA^D + A^D BCA^D.$$

If the generalized Schur complement $Z = D - CA^D B$ is equal to 0, and

$$C(I - AA^D)B = 0, \qquad A(I - AA^D)B = 0, \tag{5.60}$$

then

$$\mathrm{Ind}(M) \le \mathrm{Ind}(AW) + \mathrm{Ind}(A) + 2, \tag{5.61}$$

and

$$M^D = \left(I + \begin{bmatrix} 0 & (I - AA^D)B \\ 0 & 0 \end{bmatrix} R_1 \right) R_1 \left(I + \sum_{i=0}^{k-1} R_1^{i+1} \begin{bmatrix} 0 & 0 \\ C(I - AA^D)A^i & 0 \end{bmatrix} \right), \tag{5.62}$$

where $k = \mathrm{Ind}(A)$,

$$R_1 = \begin{bmatrix} I \\ CA^D \end{bmatrix} A_{d,W} \begin{bmatrix} I & A^D B \end{bmatrix}, \tag{5.63}$$

and $A_{d,w} = [(AW)^D]^2 A$ is the weighted Drazin inverse of A with weight W.

Proof Let

$$M_1 = \begin{bmatrix} F & \tilde{Q}_2 \\ \tilde{P}_2 & N \end{bmatrix},$$

where

$$F = \begin{bmatrix} \Sigma & P_1 \\ Q_1 & D \end{bmatrix}, \quad \tilde{P}_2 = \begin{bmatrix} 0 & P_2 \end{bmatrix}, \quad \tilde{Q}_2 = \begin{bmatrix} 0 \\ Q_2 \end{bmatrix}$$

are as in (5.43). Using the same arguments as in the proof of Theorem 5.8 following (5.50), we know that (5.60) implies

$$\mathrm{Ind}(M) \le \mathrm{Ind}(F) + \mathrm{Ind}(A) + 1, \tag{5.64}$$

and that (5.62) is valid for R_1 given by

$$R_1 = \begin{bmatrix} U_1 & 0 \\ 0 & I \end{bmatrix} F^D \begin{bmatrix} V_1 & 0 \\ 0 & I \end{bmatrix}. \tag{5.65}$$

It suffices to derive (5.61) and (5.63) under the assumption $Z = D - CA^D B = 0$.

We first show that (5.63) follows from (5.65). Note that Σ is invertible and rank$(F) =$ rank(Σ) since $D - Q_1\Sigma^{-1}P_1 = D - .CA^D B = 0$ by our hypothesis. It follows from ([2, Theorems 7.7.5 and 7.7.6]) that

$$\text{Ind}(F) = \text{Ind}\left[\Sigma(I + \Sigma^{-1}P_1 Q_1 \Sigma^{-1})\right] + 1, \tag{5.66}$$

and

$$F^D = \begin{bmatrix} I \\ Q_1\Sigma^{-1} \end{bmatrix}\left[(\Sigma H)^2\right]^D \Sigma\left[I \ \ \Sigma^{-1}P_1\right], \tag{5.67}$$

where $H = I + \Sigma^{-1}P_1 Q_1 \Sigma^{-1}$.

Notice that

$$\begin{bmatrix} U_1 & 0 \\ 0 & I \end{bmatrix}\begin{bmatrix} I \\ Q_1\Sigma^{-1} \end{bmatrix} = \begin{bmatrix} I \\ CA^D \end{bmatrix}U_1,$$

$$\begin{bmatrix} I & \Sigma^{-1}P_1 \end{bmatrix}\begin{bmatrix} V_1 & 0 \\ 0 & I \end{bmatrix} = V_1\begin{bmatrix} I & A^D B \end{bmatrix}.$$

Applying (5.38) yields

$$\begin{aligned} U_1\Sigma H V_1 &= U_1\Sigma V_1 + U_1 V_1 BCU_1\Sigma^{-1}V_1 \\ &= (U_1\Sigma V_1 + U_2 N V_2)U_1 V_1 + U_1 V_1 BCU_1 V_1 A^D U_1 V_1 \\ &= A(AA^D + A^D BCA^D) = AW. \end{aligned}$$

By an application of Lemma 5.2, we obtain

$$U_1\left[(\Sigma H)^2\right]^D V_1 = \left[U_1(\Sigma H)^D V_1\right]^2 = \left[(U_1\Sigma H V_1)^D\right]^2 = \left[(AW)^D\right]^2. \tag{5.68}$$

Then it follows from (5.65) and (5.67) that

$$R_1 = \begin{bmatrix} I \\ CA^D \end{bmatrix}\left[(AW)^D\right]^2 A^2 A^D\left[I \ \ A^D B\right].$$

From (5.68) and Lemma 5.2 we have that

$$\left[(AW)^D\right]^2 AA^D = U_1\left[(\Sigma H)^2\right]^D V_1 U_1 V_1 = U_1\left[(\Sigma H)^2\right]^D V_1 = \left[(AW)^D\right]^2.$$

Thus (5.63) follows immediately.

Finally, we show (5.61). By (5.64) and (5.66), we have

$$\text{Ind}(M) \leq \text{Ind}\left[\Sigma(I + \Sigma^{-1}P_1 Q_1 \Sigma^{-1})\right] + \text{Ind}(A) + 2.$$

By an application of Lemma 5.2, we obtain

$$\text{Ind}[\Sigma(I + \Sigma^{-1}P_1 Q_1 \Sigma^{-1})] \leq \text{Ind}\left[U_1 \Sigma V_1 U_1 (I + \Sigma^{-1}P_1 Q_1 \Sigma^{-1})V_1\right]$$
$$= \text{Ind}\left[A(AA^D + A^D BCA^D)\right] = \text{Ind}(AW),$$

which immediately leads to (5.61).   □

Remark 5.2.3 The constraint on $Z = D - CA^D B$ in Theorem 5.8 or 5.13 (i.e., Z is either invertible or zero) is essential in this section. It is an open problem to find an explicit formula for M^D in terms of A, B, C, D, A^{-1} and D^D in the case where A is invertible and $Z = D - CA^{-1}B$ is a non-zero singular matrix.

If we replace $A(I - AA^D)B = 0$ in (5.60) in Theorem 5.13 by $C(I - AA^D)A = 0$, we obtain the following corollary by taking adjoints.

Corollary 5.6 *Let M be given by (5.30) and let $W = AA^D + A^D BCA^D$. If for the generalized Schur complement we have $Z = D - CA^D B = 0$, and*

$$C(I - AA^D)B = 0, \qquad C(I - AA^D)A = 0,$$

then $\text{Ind}(M) \leq \text{Ind}(AW) + \text{Ind}(A) + 2$ *and*

$$M^D = \left(I + \sum_{i=0}^{k-1}\begin{bmatrix} 0 & A^i(I - AA^D)B \\ 0 & 0 \end{bmatrix} R_1^{i+1}\right) R_1 \left(I + R_1 \begin{bmatrix} 0 & 0 \\ C(I - AA^D) & 0 \end{bmatrix}\right),$$

where $k = \text{Ind}(A)$ and R_1 is given by (5.63) in Theorem 5.13.

If the condition $C(I - AA^D)A = 0$ is added in Theorem 5.13, then we have a simpler formula for M^D.

Corollary 5.7 *Let M be given by (5.30) and let $W = AA^D + A^D BCA^D$. If $Z = D - CA^D B = 0$ and*

$$C(I - AA^D)B = 0, \quad A(I - AA^D)B = 0, \quad C(I - AA^D)A = 0,$$

then $\text{Ind}(M) \leq \text{Ind}(AW) + \text{Ind}(A) + 2$ *and*

$$M^D = \begin{bmatrix} I + (I - AA^D)BCA^D A_{d,w} & (I - AA^D)BCA^D A_{d,w} A^D B \\ 0 & I \end{bmatrix}$$
$$\times R_1 \begin{bmatrix} I + A_{d,w} A^D BC(I - AA^D) & 0 \\ CA^D A_{d,w} A^D BC(I - AA^D) & I \end{bmatrix},$$

where R_1 is given by (5.63) in Theorem 5.13.

It is evident that Theorem 5.13 and Corollaries 5.4 and 5.7 are extensions of Theorem 5.12.

Example 5.2 [12] Consider a 2 × 2 block matrix $M = \begin{bmatrix} A & B \\ C & D \end{bmatrix}$, where

$$A = \begin{bmatrix} 0 & 1 & -1 & 1 \\ 0 & 1 & -1 & 1 \\ 0 & 1 & -1 & 1 \\ 0 & 0 & 0 & 1 \end{bmatrix}, \quad B = \begin{bmatrix} 4 & 5 \\ 3 & 4 \\ 3 & 4 \\ 2 & 3 \end{bmatrix}, \quad C = \begin{bmatrix} -1 & 4 & -2 & 0 \\ -2 & 3 & 1 & 0 \end{bmatrix}, \quad D = \begin{bmatrix} 3 & 3 \\ 5 & 7 \end{bmatrix}.$$

It is calculated that

$$A^D = \begin{bmatrix} 0 & 0 & 0 & 1 \\ 0 & 0 & 0 & 1 \\ 0 & 0 & 0 & 1 \\ 0 & 0 & 0 & 1 \end{bmatrix}, \quad I - AA^D = \begin{bmatrix} 1 & 0 & 0 & -1 \\ 0 & 1 & 0 & -1 \\ 0 & 0 & 1 & -1 \\ 0 & 0 & 0 & 0 \end{bmatrix}.$$

Then $D - CA^D B = \begin{bmatrix} 1 & 0 \\ 1 & 1 \end{bmatrix}$ is nonsingular and (5.44) is satisfied. It is verified that $\text{Ind}(A) = 2$ and $\text{Ind}(M) = 3$ satisfying Theorem 5.8. Equation (5.45) gives the exact value of M^D,

$$M^D = \begin{bmatrix} -106 & 734 & -522 & -126 & -5 & 11 \\ -37 & 255 & -181 & -44 & -2 & 4 \\ -37 & 255 & -181 & -44 & -2 & 4 \\ 32 & -224 & 160 & 38 & 1 & -3 \\ -6 & 47 & -35 & -7 & 1 & 0 \\ -5 & 31 & -21 & -6 & -1 & 1 \end{bmatrix}.$$

5.3 Generalized Drazin Inverse of Operator Matrices on Banach Spaces

In this section we will consider the generalized Drazin inverse of a 2 × 2 operator matrix

$$M = \begin{pmatrix} A & B \\ C & D \end{pmatrix}, \tag{5.69}$$

where $A \in \mathcal{B}(\mathcal{X})$ and $D \in \mathcal{B}(\mathcal{Y})$ are generalized Drazin invertible and $\mathcal{X}, \mathcal{Y}$ are complex Banach spaces.

Generalized Drazin inverses of operator matrices have various applications in singular differential and difference equations, Markov chains, and iterative methods (see [1, 4, 8, 13, 17, 26–38]). Various formulae for M^d appear frequently in connection with many problems arising in diverse areas of research and have thus long been studied [3, 12, 22, 39], but it is still an open problem to find an explicit formula for M^d in the general case. Here, we list some explicit generalized Drazin inverse formulae for a 2 × 2 operator matrix M under a number of different conditions.

Throughout the section if $T \in \mathscr{B}(\mathscr{X})$ is generalized Drazin invertible, then the spectral idempotent T^{π} of T corresponding to $\{0\}$ will be denoted by $T^{\pi} = I - TT^{d}$. The operator matrix form of T with respect to the space decomposition $X = \mathscr{N}(T^{\pi}) \oplus \mathscr{R}(T^{\pi})$ is given by $T = T_1 \oplus T_2$, where T_1 is invertible and T_2 is quasi-nilpotent.

First we present an additive result for the generalized Drazin inverse of $P + Q$, which is closely connected with the generalized Drazin inverse of a 2×2 operator matrix.

Lemma 5.3 *Let P and $Q \in \mathscr{B}(\mathscr{X})$ be generalized Drazin invertible.*

(1) ([40]) If $PQ = QP$, then $P + Q$ is generalized Drazin invertible if and only if $I + P^{d}Q$ is generalized Drazin invertible. In this case we have

$$(P + Q)^{d} = P^{d}(I + P^{d}Q)^{d}QQ^{d} + (I - QQ^{d})\left[\sum_{n=0}^{\infty}(-Q)^{n}(P^{d})^{n}\right]P^{d}$$

$$+ Q^{d}\left[\sum_{n=0}^{\infty}(Q^{d})^{n}(-P)^{n}\right](I - PP^{d}).$$

(2) ([41]) If $PQ = 0$, then $P + Q$ is generalized Drazin invertible and

$$(P + Q)^{d} = (I - QQ^{d})\left[\sum_{n=0}^{\infty}Q^{n}(P^{d})^{n}\right]P^{d} + Q^{d}\left[\sum_{n=0}^{\infty}(Q^{d})^{n}P^{n}\right](I - PP^{d}).$$

Throughout this section, we will use the following notation

$$S_0 = \sum_{n=0}^{\infty}(A^{d})^{n+2}BD^{n}(I - DD^{d}), \quad T_0 = (I - AA^{d})\sum_{n=0}^{\infty}A^{n}B(D^{d})^{n+2}, \quad (5.70)$$

$$S = (I - DD^{d})\sum_{n=0}^{\infty}D^{n}C(A^{d})^{n+2}, \quad T = \sum_{n=0}^{\infty}(D^{d})^{n+2}CA^{n}(I - AA^{d}). \quad (5.71)$$

Using different splittings of the operator matrix M as $M = P + Q$, i.e., carefully choosing the operators P and Q (the operator Q being fully determined by the choice of the operator Q of course), and then imposing the one-sided condition $PQ = 0$ or the two-sided condition $PQ = QP$, we will then apply Lemma 5.3 to obtain various explicit expressions for $M^{d} = (P + Q)^{d}$. This will be carried out in Theorems 5.14–5.18.

Theorem 5.14 *Let $A \in \mathscr{B}(\mathscr{X})$ and $D \in \mathscr{B}(\mathscr{Y})$ be generalized Drazin invertible and let M be given by (5.69).*

(1) *If $BC = 0$ and $BD = 0$, then M is generalized Drazin invertible, and*

$$M^d = \begin{pmatrix} A^d & (A^d)^2 B \\ S + T - D^d C A^d & D^d + (D^d T + S A^d) B - D^d (D^d C + C A^d) A^d B \end{pmatrix}.$$

(2) *If $AB = 0$ and $CB = 0$, then M is generalized Drazin invertible and*

$$M^d = \begin{pmatrix} A^d + B(D^d T + S A^d) - B D^d (D^d C + C A^d) A^d & B(D^d)^2 \\ S + T - D^d C A^d & D^d \end{pmatrix}.$$

Proof (1) Let

$$P = \begin{pmatrix} A & B \\ 0 & 0 \end{pmatrix}, \qquad Q = \begin{pmatrix} 0 & 0 \\ C & D \end{pmatrix}.$$

For $n \geq 1$, then

$$P^n = \begin{pmatrix} A^n & A^{n-1} B \\ 0 & 0 \end{pmatrix}, \qquad Q^n = \begin{pmatrix} 0 & 0 \\ D^{n-1} C & D^n \end{pmatrix}.$$

Since A and D are generalized Drazin invertible, using the assumptions from (1) and Corollary 7.7.2 from [2], we have

$$P^d = \begin{pmatrix} A^d & (A^d)^2 B \\ 0 & 0 \end{pmatrix}, \qquad Q^d = \begin{pmatrix} 0 & 0 \\ (D^d)^2 C & D^d \end{pmatrix}.$$

Now, for $n \geq 1$,

$$(P^d)^n = \begin{pmatrix} (A^d)^n & (A^d)^{n+1} B \\ 0 & 0 \end{pmatrix}, \qquad (Q^d)^n = \begin{pmatrix} 0 & 0 \\ (D^d)^{n+1} C & (D^d)^n \end{pmatrix},$$

so

$$(I - Q Q^d) \left[\sum_{n=0}^{\infty} Q^n (P^d)^n \right] P^d = \begin{pmatrix} A^d & (A^d)^2 B \\ -D^d C A^d + S & -D^d C (A^d)^2 B + S A^d B \end{pmatrix},$$

and

$$Q^d \left[\sum_{n=0}^{\infty} (Q^d)^n P^n \right] (I - P P^d) = \begin{pmatrix} 0 & 0 \\ T & D^d - (D^d)^2 C A^d B + D^d T B \end{pmatrix}.$$

Since $PQ = 0$, by (2) from Lemma 5.3, we obtain

$$M^d = (I - Q Q^d) \left[\sum_{n=0}^{\infty} Q^n (P^d)^n \right] P^d + Q^d \left[\sum_{n=0}^{\infty} (Q^d)^n P^n \right] (I - P P^d)$$
$$= \begin{pmatrix} A^d & (A^d)^2 B \\ S + T - D^d C A^d & D^d + (D^d T + S A^d) B - D^d (D^d C + C A^d) A^d B \end{pmatrix}.$$

(2) Let $P = \begin{pmatrix} A & 0 \\ C & 0 \end{pmatrix}$ and $Q = \begin{pmatrix} 0 & B \\ 0 & D \end{pmatrix}$. Since $AB = 0$ and $CB = 0$ we have $PQ = 0$. The proof then follows a similar argument as previously and is therefore omitted. $\square$

It is obvious that although the two previous formulae for M^D in general apply under different conditions, they coincide when $B = 0$. Next, we will present some formulae valid if the blocks satisfy certain new requirements.

Theorem 5.15 *Let $A \in \mathscr{B}(\mathscr{X})$ and $D \in \mathscr{B}(\mathscr{Y})$ be generalized Drazin invertible and let M be given by (5.69).*

(1) *If $CA = 0$ and $CB = 0$, then M is generalized Drazin invertible, and*

$$M^d = \begin{pmatrix} A^d + (T_0 D^d + A^d S_0)C - A^d(BD^d + A^d B)D^d C & S_0 + T_0 - A^d B D^d \\ (D^d)^2 C & D^d \end{pmatrix}.$$

(2) *If $BC = 0$ and $DC = 0$, then M is generalized Drazin invertible, and*

$$M^d = \begin{pmatrix} A^d & S_0 + T_0 - A^d B D^d \\ C(A^d)^2 & D^d + C(T_0 D^d + A^d S_0) - CA^d(BD^d + A^d B)D^d \end{pmatrix}.$$

Proof (1) Observe that $CA = 0$ and $CB = 0$ implies $A^*C^* = 0$ and $B^*C^* = 0$. Hence, M^* is generalized Drazin invertible and by (2) in Theorem 5.14,

$$(M^*)^d = \begin{pmatrix} [A^d + (T_0 D^d + A^d S_0)C - A^d(BD^d + A^d B)D^d C]^* & [(D^d)^2 C]^* \\ (S_0 + T_0 - A^d B D^d)^* & (D^d)^* \end{pmatrix}.$$

Consequently, we deduce that M is generalized Drazin invertible and obtain the corresponding expression for M^d.
Item (2) can be proved in a similar manner. $\square$

From Theorems 5.14 and 5.15, a number of results obtained in literature follow as can be seen from the corollary below.

Corollary 5.8 *Let $A \in \mathscr{B}(\mathscr{X})$ and $D \in \mathscr{B}(\mathscr{Y})$ be generalized Drazin invertible and let M be given by (5.69).*

(1) ([3]) *If $BC = 0$, $DC = 0$ and $BD = 0$, then*

$$M^d = \begin{pmatrix} A^d & (A^d)^2 B \\ C(A^d)^2 & D^d + C(A^d)^3 B \end{pmatrix}.$$

(2) ([12, Lemma 2.2]) *If $BC = 0$, $DC = 0$ and D is quasi-nilpotent, then*

$$M^d = \begin{pmatrix} A^d & \sum_{n=0}^{\infty}(A^d)^{n+2} B D^n \\ C(A^d)^2 & CA^d \sum_{n=0}^{\infty}(A^d)^{n+2} B D^n \end{pmatrix}.$$

(3) ([3, 4, 42]) If $C = 0$, then

$$M^d = \begin{pmatrix} A^d & S_0 + T_0 - A^d B D^d \\ 0 & D^d \end{pmatrix}$$

(4) ([3, 4, 42]) If $B = 0$, then

$$M^d = \begin{pmatrix} A^d & 0 \\ T + S - D^d C A^d & D^d \end{pmatrix}.$$

(5) ([30]) If $BC = 0$ and $D = 0$, then

$$M^d = \begin{pmatrix} A^d & (A^d)^2 B \\ C(A^d)^2 & C(A^d)^3 B \end{pmatrix}.$$

□

In the next three theorems to derive explicit expressions for M^d, the two-sided condition is used.

Theorem 5.16 *Let $A \in \mathscr{B}(\mathscr{X})$ and $D \in \mathscr{B}(\mathscr{Y})$ be generalized Drazin invertible and M be given by (5.69).*
(1) If $BC = 0$, $CB = 0$ and $DC = CA$, then M is generalized Drazin invertible, and

$$M^d = \begin{pmatrix} A^d & S_0 + T_0 - A^d B D^d \\ -C(A^d)^2 & D^d + C A^d (A^d B + B D^d) D^d - C (A^d S_0 + T_0 D^d) \end{pmatrix}.$$

(2) If $BC = 0$, $CB = 0$ and $AB = BD$, then M is generalized Drazin invertible, and

$$M^d = \begin{pmatrix} A^d + B D^d (C A^d + D^d C) A^d - B(S A^d + D^d T) & -B(D^d)^2 \\ S + T - D^d C A^d & D^d \end{pmatrix}.$$

Proof (1) Let

$$P = \begin{pmatrix} A & B \\ 0 & D \end{pmatrix}, \quad Q = \begin{pmatrix} 0 & 0 \\ C & 0 \end{pmatrix}.$$

By (3) of Corollary 5.8, we have

$$P^d = \begin{pmatrix} A^d & S_0 + T_0 - A^d B D^d \\ 0 & D^d \end{pmatrix}$$

and $Q^d = Q^2 = 0$. From $BC = 0$, $CB = 0$ and $DC = CA$, we know that $PQ = QP$. By Lemma 5.3, we obtain

$$(P + Q)^d = P^d - Q(P^d)^2.$$

Note that $A^d T_0 = 0$ and $S_0 D^d = 0$. Direct computations now leads to

$$
(P + Q)^d = P^d - Q(P^d)^2
$$
$$
= \begin{pmatrix} A^d & S_0 + T_0 - A^d B D^d \\ -C(A^d)^2 & D^d + C A^d (A^d B + B D^d) D^d - C(A^d S_0 + T_0 D^d) \end{pmatrix}.
$$

(2) This is done similarly as in the proof of (1). The details are omitted.   □

Theorem 5.17 Let $A \in \mathcal{B}(\mathcal{X})$ and $D \in \mathcal{B}(\mathcal{Y})$ be generalized Drazin invertible and M be given by (5.69).

(1) If $BC = 0$, $CB = 0$, $CA(I - A^\pi) = (I - D^\pi)DC$ and $AA^\pi B = BDD^\pi$, then M is generalized Drazin invertible, and

$$
M^d = \sum_{n=0}^{\infty} \begin{pmatrix} -AA^\pi & 0 \\ -C & -DD^\pi \end{pmatrix}^n \begin{pmatrix} A^d & (A^d)^2 B D^\pi + A^\pi B(D^d)^2 - A^d B D^d \\ 0 & D^d \end{pmatrix}^{n+1}.
$$

(2) If $CB = 0$, $CA(I - A^\pi) = 0$, and $AA^\pi B = 0$, then M is generalized Drazin invertible, and

$$
M^d = \sum_{n=0}^{\infty} \begin{pmatrix} A^d & (A^d)^2 B D^\pi + A^\pi B(D^d)^2 - A^d B D^d \\ 0 & D^d \end{pmatrix}^{n+1} \begin{pmatrix} AA^\pi & 0 \\ C & DD^\pi \end{pmatrix}^n.
$$

(3) If $BC = 0$, $(I - D^\pi)DC = 0$, and $BDD^\pi = 0$, then M is generalized Drazin invertible, and

$$
M^d = \sum_{n=0}^{\infty} \begin{pmatrix} AA^\pi & 0 \\ C & DD^\pi \end{pmatrix}^n \begin{pmatrix} A^d & (A^d)^2 B D^\pi + A^\pi B(D^d)^2 - A^d B D^d \\ 0 & D^d \end{pmatrix}^{n+1}.
$$

Proof (1) Let

$$
P = \begin{pmatrix} A^2 A^d & B \\ 0 & D^2 D^d \end{pmatrix}, \qquad Q = \begin{pmatrix} AA^\pi & 0 \\ C & DD^\pi \end{pmatrix}.
$$

Then $M = P + Q$. We can show that $Q^d = 0$ by proving that Q is quasi-nilpotent or using (4) in Corollary 5.8. By [2, Theorem 7.7.3], we have

$$
P^d = \begin{pmatrix} A^d & (A^d)^2 B D^\pi + A^\pi B(D^d)^2 - A^d B D^d \\ 0 & D^d \end{pmatrix}.
$$

Since $PQ = QP$, by Lemma (2) in 5.3 it follows that

$$(P + Q)^d = \sum_{n=0}^{\infty} (-Q)^n (P^d)^{n+1}$$

$$= \sum_{n=0}^{\infty} \begin{pmatrix} -AA^\pi & 0 \\ -C & -DD^\pi \end{pmatrix}^n \begin{pmatrix} A^d & (A^d)^2 BD^\pi + A^\pi B(D^d)^2 - A^d BD^d \\ 0 & D^d \end{pmatrix}^{n+1}.$$

(2) Choosing the same P and Q as in (1) we have that $QP = 0$. Hence,

$$(P + Q)^d = \sum_{n=0}^{\infty} (P^d)^{n+1} Q^n$$

$$= \sum_{n=0}^{\infty} \begin{pmatrix} A^d & (A^d)^2 BD^\pi + A^\pi B(D^d)^2 - A^d BD^d \\ 0 & D^d \end{pmatrix}^{n+1} \begin{pmatrix} AA^\pi & 0 \\ C & DD^\pi \end{pmatrix}^n.$$

(3) Choosing the same P and Q as in (1) we have that $PQ = 0$. Hence,

$$(P + Q)^d = \sum_{n=0}^{\infty} Q^n (P^d)^{n+1}$$

$$= \sum_{n=0}^{\infty} \begin{pmatrix} AA^\pi & 0 \\ C & DD^\pi \end{pmatrix}^n \begin{pmatrix} A^d & (A^d)^2 BD^\pi + A^\pi B(D^d)^2 - A^d BD^d \\ 0 & D^d \end{pmatrix}^{n+1}.$$

$\square$

Theorem 5.18 Let $A \in \mathcal{B}(\mathcal{X})$ and $D \in \mathcal{B}(\mathcal{Y})$ be generalized Drazin invertible and M be given by (5.69).
(1) If $BC = 0$, $CB = 0$, $CA(I - A^\pi) = D^\pi DC$, and $A^\pi AB = BD(I - D^\pi)$, then M is generalized Drazin invertible and

$$M^d = \begin{pmatrix} A^d & S_0 \\ T - D^\pi C(A^d)^2 & D^d - D^\pi CA^d S_0 + \sum_{n=0}^{\infty} (-D^d)^{n+1} TB(DD^\pi)^n \end{pmatrix}.$$

(2) If $BC = 0$, $BD(I - D^\pi) = 0$, and $D^\pi DC = 0$, then M is generalized Drazin invertible and

$$M^d = \begin{pmatrix} A^d & S_0 \\ T + D^\pi C(A^d)^2 & D^d + D^\pi CA^d S_0 + \sum_{n=0}^{\infty} (D^d)^{n+1} TB(DD^\pi)^n \end{pmatrix}.$$

Proof Let

$$P = \begin{pmatrix} A(I - A^\pi) & B \\ 0 & DD^\pi \end{pmatrix}, \qquad Q = \begin{pmatrix} AA^\pi & 0 \\ C & D(I - D^\pi) \end{pmatrix}.$$

Evidently, $M = P + Q$. By (3) in Corollary 5.8,

$$P^d = \begin{pmatrix} A^d & \sum_{n=0}^{\infty} (A^d)^{n+2} B D^n D^\pi \\ 0 & 0 \end{pmatrix} = \begin{pmatrix} A^d & S_0 \\ 0 & 0 \end{pmatrix}, \quad I - PP^d = \begin{pmatrix} A^\pi & -AS_0 \\ 0 & I \end{pmatrix}$$

and by (4) in Corollary 5.8

$$Q^d = \begin{pmatrix} 0 & 0 \\ \sum_{n=0}^{\infty} (D^d)^{n+2} CA^n A^\pi & D^d \end{pmatrix} = \begin{pmatrix} 0 & 0 \\ T & D^d \end{pmatrix}, \quad I - QQ^d = \begin{pmatrix} I & 0 \\ -DT & D^\pi \end{pmatrix}.$$

(1) It is easy to verify that $PQ = QP$. Also $S_0 DT = 0, TAS_0 = 0, S_0 D^d = 0$ and $TA^d = 0$, so

$$M^d = P^d(I + P^d Q)^d QQ^d + (I - QQ^d)\left[\sum_{n=0}^{\infty}(-Q)^n (P^d)^n\right] P^d$$

$$+ Q^d \left[\sum_{n=0}^{\infty}(Q^d)^n(-P)^n\right](I - PP^d)$$

$$= \begin{pmatrix} A^d & S_0 \\ 0 & 0 \end{pmatrix}\begin{pmatrix} (I + S_0 C)^d & 0 \\ 0 & I \end{pmatrix}\begin{pmatrix} 0 & 0 \\ DT & D^d D \end{pmatrix}$$

$$+ \begin{pmatrix} I & 0 \\ -DT & D^\pi \end{pmatrix}\sum_{n=0}^{\infty}\begin{pmatrix} -AA^\pi & 0 \\ -C & -D^2 D^d \end{pmatrix}^n\begin{pmatrix} (A^d)^{n+1} & (A^d)^n S_0 \\ 0 & 0 \end{pmatrix}$$

$$+ \sum_{n=0}^{\infty}\begin{pmatrix} 0 & 0 \\ (D^d)^n T & (D^d)^{n+1} \end{pmatrix}\begin{pmatrix} -A^2 A^d & -B \\ 0 & -DD^\pi \end{pmatrix}^n\begin{pmatrix} A^\pi & -AS_0 \\ 0 & I \end{pmatrix}$$

$$= 0 + \begin{pmatrix} A^d & S_0 \\ -D^\pi C(A^d)^2 & -D^\pi CA^d S_0 \end{pmatrix} + \begin{pmatrix} 0 & 0 \\ T D^d + \sum_{n=0}^{\infty}(-D^d)^{n+1} TB(DD^\pi)^n \end{pmatrix}$$

$$= \begin{pmatrix} A^d & S_0 \\ T - D^\pi C(A^d)^2 & D^d - D^\pi CA^d S_0 + \sum_{n=0}^{\infty}(-D^d)^{n+1} TB(DD^\pi)^n \end{pmatrix}.$$

(2) Since $PQ = 0$, we have

$$
\begin{aligned}
M^{\mathsf{d}} &= (I - QQ^{\mathsf{d}})\left[\sum_{n=0}^{\infty} Q^{n}(P^{\mathsf{d}})^{n}\right] P^{\mathsf{d}} + Q^{\mathsf{d}}\left[\sum_{n=0}^{\infty}(Q^{\mathsf{d}})^{n} P^{n}\right](I - PP^{\mathsf{d}}) \\
&= \begin{pmatrix} I & 0 \\ -DT & D^{\pi} \end{pmatrix} \sum_{n=0}^{\infty} \begin{pmatrix} AA^{\pi} & 0 \\ C & D^{2}D^{\mathsf{d}} \end{pmatrix}^{n} \begin{pmatrix} (A^{D})^{n+1} & (A^{D})^{n} S_{0} \\ 0 & 0 \end{pmatrix} \\
&\quad + \sum_{n=0}^{\infty} \begin{pmatrix} 0 & 0 \\ (D^{\mathsf{d}})^{n}T & (D^{\mathsf{d}})^{n+1} \end{pmatrix} \begin{pmatrix} A^{2}A^{\mathsf{d}} & B \\ 0 & DD^{\pi} \end{pmatrix}^{n} \begin{pmatrix} A^{\pi} & -AS_{0} \\ 0 & I \end{pmatrix} \\
&= \begin{pmatrix} A^{\mathsf{d}} & S_{0} \\ D^{\pi}C(A^{\mathsf{d}})^{2} & D^{\pi}CA^{\mathsf{d}}S_{0} \end{pmatrix} + \begin{pmatrix} 0 & 0 \\ TD^{\mathsf{d}} + \sum_{n=0}^{\infty}(D^{\mathsf{d}})^{n+1}TB(DD^{\pi})^{n} \end{pmatrix} \\
&= \begin{pmatrix} A^{\mathsf{d}} & S_{0} \\ T + D^{\pi}C(A^{\mathsf{d}})^{2} & D^{\mathsf{d}} + D^{\pi}CA^{\mathsf{d}}S_{0} + \sum_{n=0}^{\infty}(D^{\mathsf{d}})^{n+1}TB(DD^{\pi})^{n} \end{pmatrix}.
\end{aligned}
$$

$\square$

Assuming that $A \in \mathcal{B}(\mathcal{X})$ is generalized Drazin invertible and that the generalized Schur complement $S = D - CA^{\mathsf{d}}B$ is invertible, the next theorem gives a representation for M^{d} under some additional conditions. Denote $X_1 = \mathcal{N}(A^{\pi})$ and $X_2 = \mathcal{R}(A^{\pi})$. The operator

$$
I_0 = I \oplus \begin{pmatrix} 0 & I \\ I & 0 \end{pmatrix}
$$

from $X_1 \oplus X_2 \oplus Y$ onto $X_1 \oplus Y \oplus X_2$ is invertible. Then M as an operator on $X_1 \oplus X_2 \oplus Y$ has the following operator matrix form

$$
M = \begin{pmatrix} A_1 & 0 & B_1 \\ 0 & A_2 & B_2 \\ C_1 & C_2 & D \end{pmatrix} = I_0^{-1} \begin{pmatrix} A_1 & B_1 & 0 \\ C_1 & D & C_2 \\ 0 & B_2 & A_2 \end{pmatrix} I_0 = I_0^{-1} \begin{pmatrix} A_0 & B_0 \\ C_0 & D_0 \end{pmatrix} I_0, \qquad (5.72)
$$

where A_1 is invertible and $D_0 = A_2$ is quasi-nilpotent.

Let $A = \begin{pmatrix} A_1 & 0 \\ 0 & A_2 \end{pmatrix}$, $B = \begin{pmatrix} B_1 \\ B_2 \end{pmatrix}$, $C = \begin{pmatrix} C_1 & C_2 \end{pmatrix}$, $A_0 = \begin{pmatrix} A_1 & B_1 \\ C_1 & D_1 \end{pmatrix}$, $B_0 = \begin{pmatrix} 0 \\ C_2 \end{pmatrix}$ and $C_0 = \begin{pmatrix} 0 & B_2 \end{pmatrix}$. With these notations the following can be proved.

Theorem 5.19 Let $A \in \mathcal{B}(\mathcal{X})$ be generalized Drazin invertible, $D \in \mathcal{B}(\mathcal{Y})$ and M be given by (5.69). If

$$
A^{\pi}BC = 0, \quad CA^{\pi}B = 0, \quad A^{\pi}AB = A^{\pi}BD
$$

and $S = D - CA^dB$ is invertible, then M is generalized Drazin invertible,

$$M^d = \left[R - \begin{pmatrix} 0 & (I - AA^d)B \\ 0 & 0 \end{pmatrix} R^2 \right] \left[I + \sum_{n=0}^{\infty} R^{n+1} \begin{pmatrix} 0 & 0 \\ C(I - AA^d)A^n & 0 \end{pmatrix} \right],$$

where

$$R = \begin{pmatrix} A^d + A^dBS^{-1}CA^d & -A^dBS^{-1} \\ -S^{-1}CA^d & S^{-1} \end{pmatrix}. \tag{5.73}$$

Proof Note that the Schur complement S can be expressed as

$$S = D - CA^dB = D - \begin{pmatrix} C_1 & C_2 \end{pmatrix} \begin{pmatrix} A_1^{-1} & 0 \\ 0 & 0 \end{pmatrix} \begin{pmatrix} B_1 \\ B_2 \end{pmatrix} = D - C_1 A_1^{-1} B_1.$$

Since S and A_1 are invertible, we conclude that A_0 is invertible and that

$$\begin{aligned}
A_0^{-1} &= \begin{pmatrix} A_1^{-1} + A_1^{-1}B_1S^{-1}C_1A_1^{-1} & -A_1^{-1}B_1S^{-1} \\ -S^{-1}C_1A_1^{-1} & S^{-1} \end{pmatrix} \\
&= \begin{pmatrix} I & 0 & 0 \\ 0 & 0 & I \end{pmatrix} \begin{pmatrix} A_1^{-1} + A_1^{-1}B_1S^{-1}C_1A_1^{-1} & 0 & -A_1^{-1}B_1S^{-1} \\ 0 & 0 & 0 \\ -S^{-1}C_1A_1^{-1} & 0 & S^{-1} \end{pmatrix} \begin{pmatrix} I & 0 \\ 0 & 0 \\ 0 & I \end{pmatrix} \\
&= \begin{pmatrix} I & 0 & 0 \\ 0 & 0 & I \end{pmatrix} R \begin{pmatrix} I & 0 \\ 0 & 0 \\ 0 & I \end{pmatrix}.
\end{aligned}$$

Also,

$$A_0^{-(n+2)} = \begin{pmatrix} I & 0 & 0 \\ 0 & 0 & I \end{pmatrix} R^{n+2} \begin{pmatrix} I & 0 \\ 0 & 0 \\ 0 & I \end{pmatrix}.$$

From $A^{\pi}BC = 0$, we have

$$\begin{pmatrix} 0 & 0 \\ 0 & I \end{pmatrix} \begin{pmatrix} B_1 \\ B_2 \end{pmatrix} \begin{pmatrix} C_1 & C_2 \end{pmatrix} = \begin{pmatrix} 0 & 0 \\ B_2C_1 & B_2C_2 \end{pmatrix} = 0.$$

Similarly, $CA^{\pi}B = 0$ implies $C_2B_2 = 0$ and $A^{\pi}AB = A^{\pi}BD$ implies $A_2B_2 = B_2D$. Hence

$$B_0C_0 = 0, \quad C_0B_0 = 0, \quad D_0C_0 = C_0A_0.$$

By Theorem 5.18 (1), we get

$$M^d = I_0^{-1} \begin{pmatrix} A_0 & B_0 \\ C_0 & D_0 \end{pmatrix}^d I_0$$

$$= I_0^{-1} \begin{pmatrix} A_0^{-1} & \sum_{n=0}^{\infty} A_0^{-(n+2)} B_0 D_0^n \\ -C_0 A_0^{-2} & -C_0 A_0^{-1} \sum_{n=0}^{\infty} A_0^{-(n+2)} B_0 D_0^n \end{pmatrix} I_0.$$

Now we express B_0, C_0 and D_0 in terms of A, B, C and A^d. We have

$$B_0 = \begin{pmatrix} 0 \\ I \end{pmatrix} C(I - AA^d) \begin{pmatrix} 0 \\ I \end{pmatrix}, \quad D_0^n = \begin{pmatrix} 0 & I \end{pmatrix} A^n \begin{pmatrix} 0 \\ I \end{pmatrix}$$

and

$$C_0 = \begin{pmatrix} 0 & I \end{pmatrix} (I - AA^d) B \begin{pmatrix} 0 & I \end{pmatrix}.$$

Direct calculation shows that

$$M^d = \begin{pmatrix} I & 0 & 0 \\ 0 & 0 & I \\ 0 & I & 0 \end{pmatrix} \begin{pmatrix} A_0^{-1} & \sum_{n=0}^{\infty} A_0^{-(n+2)} B_0 D_0^n \\ -C_0 A_0^{-2} & -C_0 A_0^{-1} \sum_{n=0}^{\infty} A_0^{-(n+2)} B_0 D_0^n \end{pmatrix} \begin{pmatrix} I & 0 & 0 \\ 0 & 0 & I \\ 0 & I & 0 \end{pmatrix}$$

$$= \begin{pmatrix} I & 0 & 0 \\ 0 & 0 & 0 \\ 0 & 0 & I \end{pmatrix} R \begin{pmatrix} I & 0 & 0 \\ 0 & 0 & 0 \\ 0 & 0 & I \end{pmatrix} - \begin{pmatrix} 0 & 0 \\ 0 & I \\ 0 & 0 \end{pmatrix} (I - AA^d) B \begin{pmatrix} 0 & 0 & I \end{pmatrix} R^2 \begin{pmatrix} I & 0 & 0 \\ 0 & 0 & 0 \\ 0 & 0 & I \end{pmatrix}$$

$$+ \left[I - \begin{pmatrix} 0 & 0 \\ 0 & I \\ 0 & 0 \end{pmatrix} (I - AA^d) B \begin{pmatrix} 0 & 0 & I \end{pmatrix} R \right]$$

$$\times \sum_{n=0}^{\infty} \begin{pmatrix} I & 0 & 0 \\ 0 & 0 & 0 \\ 0 & 0 & I \end{pmatrix} R^{n+2} \begin{pmatrix} 0 \\ 0 \\ I \end{pmatrix} C(I - AA^d) A^n \begin{pmatrix} 0 & 0 & 0 \\ 0 & I & 0 \end{pmatrix}$$

$$= R - \begin{pmatrix} 0 & (I - AA^d) B \\ 0 & 0 \end{pmatrix} R^2 + \left[I - \begin{pmatrix} 0 & (I - AA^d) B \\ 0 & 0 \end{pmatrix} R \right]$$

$$\times \sum_{n=0}^{\infty} R^{n+2} \begin{pmatrix} 0 & 0 \\ C(I - AA^d) A^n & 0 \end{pmatrix}$$

$$= \left[R - \begin{pmatrix} 0 & (I - AA^d) B \\ 0 & 0 \end{pmatrix} R^2 \right] \left[I + \sum_{n=0}^{\infty} R^{n+1} \begin{pmatrix} 0 & 0 \\ C(I - AA^d) A^n & 0 \end{pmatrix} \right].$$

□

Using (2) from Theorem 5.18 and the notations from above, we can derive the following result.

Theorem 5.20 *Let $A \in \mathcal{B}(\mathcal{X})$ be generalized Drazin invertible, $D \in \mathcal{B}(\mathcal{Y})$ and M be given by (5.69). Let R be defined by (5.73). If*

$$BCA^{\pi} = 0, \quad CA^{\pi}B = 0, \quad CAA^{\pi} = DCA^{\pi},$$

and $S = D - CA^{d}B$ invertible, then M is generalized Drazin invertible, and

$$M^{d} = R + \sum_{n=0}^{\infty} \begin{pmatrix} 0 & A^{n}(I - AA^{d})B \\ 0 & 0 \end{pmatrix} R^{n+2} - \sum_{n=0}^{\infty} \begin{pmatrix} 0 & 0 \\ 0 & CA^{n}(I - AA^{d})B \end{pmatrix} R^{n+3}.$$

Proof Using the notations from the proof of Theorem 5.19, we can show that

$$B_0 C_0 = 0, \quad C_0 B_0 = 0, \quad A_0 B_0 = B_0 D_0.$$

By (2) from Theorem 5.18, we get

$$M^{d} = \begin{pmatrix} I & 0 & 0 \\ 0 & 0 & I \\ 0 & I & 0 \end{pmatrix} \begin{pmatrix} A_0^{-1} - B_0 \sum_{n=0}^{\infty} D_0^n C_0 A_0^{-(n+3)} & 0 \\ \sum_{n=0}^{\infty} D_0^n C_0 A_0^{-(n+2)} & 0 \end{pmatrix} \begin{pmatrix} I & 0 & 0 \\ 0 & 0 & I \\ 0 & I & 0 \end{pmatrix}$$

$$= \begin{pmatrix} I & 0 & 0 \\ 0 & 0 & 0 \\ 0 & 0 & I \end{pmatrix} R \begin{pmatrix} I & 0 & 0 \\ 0 & 0 & 0 \\ 0 & 0 & I \end{pmatrix} - \begin{pmatrix} 0 \\ 0 \\ I \end{pmatrix} C \sum_{n=0}^{\infty} A^{n}(I - AA^{d})B \begin{pmatrix} 0 & 0 & I \end{pmatrix} R^{n+3} \begin{pmatrix} I & 0 & 0 \\ 0 & 0 & 0 \\ 0 & 0 & I \end{pmatrix}$$

$$+ \sum_{n=0}^{\infty} \begin{pmatrix} 0 & 0 \\ 0 & I \\ 0 & 0 \end{pmatrix} A^{n}(I - AA^{d})B \begin{pmatrix} 0 & 0 & I \end{pmatrix} R^{n+2} \begin{pmatrix} I & 0 & 0 \\ 0 & 0 & 0 \\ 0 & 0 & I \end{pmatrix}$$

$$= R + \sum_{n=0}^{\infty} \begin{pmatrix} 0 & A^{n}(I - AA^{d})B \\ 0 & 0 \end{pmatrix} R^{n+2} - \sum_{n=0}^{\infty} \begin{pmatrix} 0 & 0 \\ 0 & CA^{n}(I - AA^{d})B \end{pmatrix} R^{n+3}.$$

□

Now, we consider the case when the generalized Schur complement $S = D - CA^{d}B$ is equal to zero. We recall one result from [12].

Lemma 5.4 ([12, Theorem 1.2]) *Let $A \in \mathcal{B}(\mathcal{X})$ be generalized Drazin invertible, $D \in \mathcal{B}(\mathcal{Y})$ and let M be given by (5.69). Let $W = AA^{d} + A^{d}BCA^{d}$. If AW is generalized Drazin invertible,*

$$C(I - AA^{d}) = 0, \quad (I - AA^{d})B = 0$$

and if for the generalized Schur complement we have $S = D - CA^{d}B = 0$, then M is generalized Drazin invertible, and

$$M^{d} = \begin{pmatrix} I \\ CA^{d} \end{pmatrix} [(AW)^{d}]^{2} A \begin{pmatrix} I & A^{d}B \end{pmatrix}.$$

In the following result, we present an expression for M^{d} using the weighted Drazin inverse $A_{d,W}$ of A with weight $W = AA^{d} + A^{d}BCA^{d}$.

Theorem 5.21 Let $A \in \mathcal{B}(\mathcal{X})$ be generalized Drazin invertible, $D \in \mathcal{B}(\mathcal{Y})$ and M be given by (5.69). Let $W = AA^d + A^d BCA^d$. If AW is generalized Drazin invertible,

$$A^\pi BC = 0, \quad CA^\pi B = 0, \quad A^\pi AB = A^\pi BD$$

and the generalized Schur complement $S = D - CA^d B = 0$, then M is generalized Drazin invertible, and

$$M^d = \left[R_0 - \begin{pmatrix} 0 & (I - AA^d)B \\ 0 & 0 \end{pmatrix} R_0^2 \right] \left[I + \sum_{n=0}^{\infty} R_0^{n+1} \begin{pmatrix} 0 & 0 \\ C(I - AA^d)A^n & 0 \end{pmatrix} \right],$$

where

$$R_0 = \begin{pmatrix} I \\ CA^d \end{pmatrix} A_{d,w} \begin{pmatrix} I & A^d B \end{pmatrix}, \quad W = AA^d + A^d BCA^d. \tag{5.74}$$

Proof Following the proof of Theorem 5.18, notice that for the Schur complement we have

$$S = D - CA^d B = D - C_1 A_1^{-1} B_1 = 0.$$

Let $W_1 = I + A_1^{-1} B_1 C_1 A_1^{-1}$. Since AW is generalized Drazin invertible, so is $A_1 W_1$. Let $(A_1)_{d,W_1} = \left[(A_1 W_1)^d \right]^2 A_1$. By Lemma 5.4, we obtain

$$A^d_0 = \begin{pmatrix} A_1 & B_1 \\ C_1 & D_1 \end{pmatrix}^d = \begin{pmatrix} I \\ C_1 A_1^d \end{pmatrix} A_1^{d,W_1} \begin{pmatrix} I & A_1^d B_1 \end{pmatrix}$$

$$= \begin{pmatrix} I & 0 & 0 \\ 0 & 0 & I \end{pmatrix} R_0 \begin{pmatrix} I & 0 \\ 0 & 0 \\ 0 & I \end{pmatrix}$$

and

$$(A_0^d)^{n+2} = \begin{pmatrix} I & 0 & 0 \\ 0 & 0 & I \end{pmatrix} R_0^{n+2} \begin{pmatrix} I & 0 \\ 0 & 0 \\ 0 & I \end{pmatrix}.$$

From $A^\pi BC = 0, CA^\pi B = 0$, and $A^\pi AB = A^\pi BD$, we have $B_0 C_0 = 0, C_0 B_0 = 0$ and $D_0 C_0 = C_0 A_0$. By (1) from Theorem 5.16, we get

$$M^d = I_0^{-1} \begin{pmatrix} A_0 & B_0 \\ C_0 & D_0 \end{pmatrix}^d I_0 = I_0^{-1} \begin{pmatrix} A_0^d & \sum_{n=0}^{\infty} (A_0^d)^{n+2} B_0 D_0^n \\ -C_0(A_0^d)^2 & -C_0 A_0^d \sum_{n=0}^{\infty} (A_0^d)^{n+2} B_0 D_0^n \end{pmatrix} I_0.$$

$$= \begin{pmatrix} I & 0 & 0 \\ 0 & 0 & I \\ 0 & I & 0 \end{pmatrix} \begin{pmatrix} A_0^d & \sum_{n=0}^{\infty} (A_0^d)^{n+2} B_0 D_0^n \\ -C_0(A_0^d)^2 & -C_0 A_0^d \sum_{n=0}^{\infty} (A_0^d)^{n+2} B_0 D_0^n \end{pmatrix} \begin{pmatrix} I & 0 & 0 \\ 0 & 0 & I \\ 0 & I & 0 \end{pmatrix}$$

$$= \left[R_0 - \begin{pmatrix} 0 & (I - AA^d)B \\ 0 & 0 \end{pmatrix} R_0^2 \right] \left[I + \sum_{n=0}^{\infty} R_0^{n+1} \begin{pmatrix} 0 & 0 \\ C(I - AA^d)A^n & 0 \end{pmatrix} \right].$$

□

Using Lemma 5.21 and (2) in Theorem 5.8, we can prove the following theorem.

Theorem 5.22 *Let $A \in \mathscr{B}(\mathscr{X})$ be generalized Drazin invertible, $D \in \mathscr{B}(\mathscr{Y})$ and M be given by (5.69). Let $W = AA^d + A^d BCA^d$. If AW is generalized Drazin invertible,*

$$BCA^\pi = 0, \quad CA^\pi B = 0, \quad CAA^\pi = DCA^\pi$$

and for the generalized Schur complement we have $S = D - CA^d B = 0$, then M is generalized Drazin invertible, and

$$M^d = R_0 + \sum_{n=0}^{\infty} \begin{pmatrix} 0 & A^n(I - AA^d)B \\ 0 & 0 \end{pmatrix} R_0^{n+2} - \sum_{n=0}^{\infty} \begin{pmatrix} 0 & 0 \\ 0 & CA^n(I - AA^d)B \end{pmatrix} R_0^{n+3},$$

where R_0 is defined by (5.74).

Proof Similarly as in the proof of Theorem 5.20, we have

$$B_0 C_0 = 0, \quad C_0 B_0 = 0, \quad A_0 B_0 = B_0 D_0.$$

By (2) of Theorem 5.8 and Lemma 5.21, we get

$$M^d = \begin{pmatrix} I & 0 & 0 \\ 0 & 0 & I \\ 0 & I & 0 \end{pmatrix} \begin{pmatrix} A_0^d - B_0 \sum_{n=0}^{\infty} D_0^n C_0 (A_0^d)^{n+3} & 0 \\ \sum_{n=0}^{\infty} D_0^n C_0 A_0^{-(n+2)} & 0 \end{pmatrix} \begin{pmatrix} I & 0 & 0 \\ 0 & 0 & I \\ 0 & I & 0 \end{pmatrix}$$

$$= R_0 + \sum_{n=0}^{\infty} \begin{pmatrix} 0 & A^n(I - AA^d)B \\ 0 & 0 \end{pmatrix} R_0^{n+2} - \sum_{n=0}^{\infty} \begin{pmatrix} 0 & 0 \\ 0 & CA^n(I - AA^d)B \end{pmatrix} R_0^{n+3}.$$

□

5.4   Representations for the Drazin Inverse of the Sum $P + Q + R + S$

A representation for the Drazin inverse of a complex upper triangular block-matrix

$$\begin{bmatrix} A & B \\ 0 & C \end{bmatrix},$$

which is known as the Hartwig-Shoaf-Meyer-Rose formula was given in [2, Theorem 7.7.1] and [11, 12] independantly.

In the literature, we can find numerous applications of the Hartwig-Shoaf-Meyer-Rose formula, since it is an effective and basic tool for finding various explicit representations for the Drazin inverse of block matrices and modified matrices in particular. In this section we will extend the Hartwig-Shoaf-Meyer-Rose formula in several ways.

For a 2×2 block matrix $M = \begin{bmatrix} A & B \\ D & C \end{bmatrix}$, where A and C are square matrices of different sizes in general, we consider the decomposition of $M = P + Q + R + S$, where

$$P = \begin{bmatrix} A & 0 \\ 0 & 0 \end{bmatrix}, \quad Q = \begin{bmatrix} 0 & 0 \\ 0 & C \end{bmatrix}, \quad R = \begin{bmatrix} 0 & B \\ 0 & 0 \end{bmatrix}, \quad S = \begin{bmatrix} 0 & 0 \\ D & 0 \end{bmatrix}.$$

We can check that P, Q, R and S satisfy the following relations:

$$PQ = QP = 0, \quad PS = SQ = QR = RP = 0, \quad R^D = S^D = 0. \qquad (5.75)$$

Motivated by this we introduce the following definition:

Definition 5.3 Let $M \in \mathbb{C}^{n \times n}$. If there exist $P, Q, R, S \in \mathbb{C}^{n \times n}$ satisfying (5.75) such that $M = P + Q + R + S$, then the quadruple (P, Q, R, S) is called a pseudo-block decomposition of M. Moreover, M is called the pseudo-block matrix corresponding to (P, Q, R, S).

It is obvious that a 2×2 block-matrix $\begin{bmatrix} A & B \\ 0 & C \end{bmatrix}$ has the pseudo-block decomposition $(P_0, Q_0, R_0, 0)$, where $P_0 = \begin{bmatrix} A & 0 \\ 0 & 0 \end{bmatrix}$, $Q_0 = \begin{bmatrix} 0 & 0 \\ 0 & C \end{bmatrix}$, $R_0 = \begin{bmatrix} 0 & B \\ 0 & 0 \end{bmatrix}$ and that then

$$R_0^2 = 0, \quad \text{rank}(P_0 + Q_0) = \text{rank}(P_0) + \text{rank}(Q_0).$$

This is a very special kind of a pseudo-block decomposition of M.

One of the main goals of this section will be to find explicit formulae for $(P + Q + R)^D$, where $(P, Q, R, 0)$ is a pseudo-block decomposition of the matrix $P + Q + R$ and to extend the Hartwig-Shoaf-Meyer-Rose formula in three aspects. Firstly, the condition

$$\text{rank}(P + Q) = \text{rank}(P) + \text{rank}(Q)$$

will be removed. Secondly, the condition $\text{Ind}(P + Q + R) \geq \max\{\text{Ind}(P), \text{Ind}(Q)\}$ will also be relaxed. Thirdly, the condition $R^2 = 0$ will be replaced by $R^D = 0$, i.e., that R is nilpotent. Also, we will find a formula for $(P + Q + R + S)^D$ under some restrictions, where (P, Q, R, S) is a pseudo-block decomposition of the matrix $P + Q + R + S$ and generalize the Hartwig-Shoaf-Meyer-Rose formula to the case $S \neq 0$.

We begin by recalling an additive result for the sum of two matrices which will be used later.

Lemma 5.5 ([12, Lemma 4], [4, Corollary 2.1]) *Let* $P, Q, R \in \mathbb{C}^{n \times n}$ *and* $R^k = 0$.

(1) *If* $RP = 0$, *then* $(R + P)^D = P^D + \displaystyle\sum_{i=1}^{k-1} (P^D)^{i+1} R^i$,

(2) *If* $QR = 0$, *then* $(Q + R)^D = Q^D + \displaystyle\sum_{i=1}^{k-1} R^i (Q^D)^{i+1}$.

Now, we present a representation for the Drazin inverse of the pseudo-block matrix $P + Q + R$ corresponding to $(P, Q, R, 0)$ under some conditions.

Theorem 5.23 *Let* $P, Q, R \in \mathbb{C}^{m \times m}$ *and let* $(P, Q, R, 0)$ *be a pseudo-block decomposition of* M. *Suppose that* $\mathrm{Ind}(P) \leq 1$, $\mathrm{Ind}(Q) \leq 1$ *and* $k = \mathrm{Ind}(R) \geq 2$. *Then the following hold:*
(1) $\mathrm{Ind}(M) \leq \mathrm{Ind}(R)$,
(2) $\mathrm{Ind}(M) \leq 1$ *if and only if*

$$\sum_{i=1}^{k-2} (P^{\#})^i R^{i+1} (I - QQ^{\#}) + \sum_{j=1}^{k-2} (I - PP^{\#}) R^{j+1} (Q^{\#})^j$$

$$= (I - PP^{\#}) R (I - QQ^{\#}) + \sum_{i=1}^{k-3} \sum_{j=1}^{k-i-2} (P^{\#})^i R^{i+j+1} (Q^{\#})^j. \qquad (5.76)$$

Furthermore, M^D *is given by*

$$M^D = \sum_{i=1}^{k} (P^{\#})^i R^{i-1} (I - QQ^{\#}) + \sum_{j=1}^{k} (I - PP^{\#}) R^{j-1} (Q^{\#})^j$$

$$- \sum_{i=1}^{k-1} \sum_{j=1}^{k-i} (P^{\#})^i R^{i+j-1} (Q^{\#})^j. \qquad (5.77)$$

Proof Denote the right-hand side of (5.77) by X. First, we will show that $X = M^D$. Since

$$
MX = \left[(PP^\# + \sum_{i=1}^{k-1}(P^\#)^i R^i(I - QQ^\#)) - (\sum_{j=1}^{k-1} PP^\# R^j(Q^\#)^j \right.
$$

$$
\left. + \sum_{i=1}^{k-2}\sum_{j=1}^{k-i-1} (P^\#)^i R^{i+j}(Q^\#)^j) \right] + QQ^\# + \sum_{j=1}^{k-1} R^j(Q^\#)^j \tag{5.78}
$$

$$
= PP^\# + QQ^\# + \sum_{i=1}^{k-1}(P^\#)^i R^i(I - QQ^\#) + \sum_{j=1}^{k-1}(I - PP^\#)R^j(Q^\#)^j
$$

$$
- \sum_{i=1}^{k-2}\sum_{j=1}^{k-i-1} (P^\#)^i R^{i+j}(Q^\#)^j.
$$

and

$$
XM = PP^\# + QQ^\# + \sum_{i=1}^{k-1}(P^\#)^i R^i(I - QQ^\#) + \sum_{j=1}^{k-1}(I - PP^\#)R^j(Q^\#)^j
$$

$$
- \sum_{i=1}^{k-2}\sum_{j=1}^{k-i-1} (P^\#)^i R^{i+j}(Q^\#)^j, \tag{5.79}
$$

we have that

$$
MX = XM. \tag{5.80}
$$

We now prove that $XMX = X$. Let us denote the first, second and third term on the right-hand side of (5.77) by X_1, X_2 and X_3, respectively. Then $X = X_1 + X_2 - X_3$. Expanding XM as (5.79), we obtain that

$$
XMX_2 = \left(Q^\# + \sum_{j=1}^{k-1}(I - PP^\#)R^j(Q^\#)^{j+1} - \sum_{i=1}^{k-2}\sum_{j=1}^{k-i-1}(P^\#)^i R^{i+j}(Q^\#)^{j+1} \right)
$$

$$
+ \sum_{i=1}^{k-2}\sum_{j=1}^{k-i-1} (P^\#)^i R^{i+j}(Q^\#)^{j+1}
$$

$$
= \sum_{j=1}^{k}(I - PP^\#)R^{j-1}(Q^\#)^j.
$$

Hence

$$XMX = \sum_{i=1}^{k}(P^{\#})^i R^{i-1}(I - QQ^{\#}) + \sum_{j=1}^{k}(I - PP^{\#})R^{j-1}(Q^{\#})^j$$

$$- \sum_{i=1}^{k-1}\sum_{j=1}^{k-i}(P^{\#})^i R^{i+j-1}(Q^{\#})^j.$$

Thus, $XMX = X$. Now we claim that

$$M^{k+1}X = M^k. \tag{5.81}$$

In fact, by induction on $l > 1$, one can see that

$$M^l = P^l + Q^l + R^l + \sum_{i=1}^{l-1}\sum_{j=0}^{l-i}P^j R^i Q^{l-i-j}. \tag{5.82}$$

Notice that $P(I - PP^{\#}) = 0$, $Q(I - QQ^{\#}) = 0$, and $R^k = R^{k+1}R^D = 0$. Combining (5.82) and (5.78), we obtain

$$M^{k+1}X = \left(P^k + \sum_{i=1}^{k-1} P^{k-i} R^i(I - QQ^{\#}) - \sum_{i=1}^{k-2}\sum_{j=1}^{k-i-1} P^{k-i} R^{i+j}(Q^{\#})^j\right)$$

$$+Q^k + \left(\sum_{i=1}^{k-1}\sum_{j=0}^{k-i} P^j R^i Q^{k-i-j} QQ^{\#} + \sum_{i=1}^{k-1}\sum_{j=1}^{k-1} P^{k-i} R^{i+j}(Q^{\#})^j\right) \tag{5.83}$$

$$=P^k + Q^k + \sum_{i=1}^{k-1}\sum_{j=0}^{k-i} P^j R^i Q^{k-i-j}$$

$$=M^k.$$

Hence, $\mathrm{Ind}(M) \le \mathrm{Ind}(R)$ and (5.77) is satisfied. Since

$$M^2 X = \left[P + \left(PP^{\#}R(I - QQ^{\#}) + \sum_{i=1}^{k-2}(P^{\#})^i R^{i+1}(I - QQ^{\#})\right)\right.$$

$$- \left(\sum_{j=1}^{k-2} PP^{\#}R^{j+1}(Q^{\#})^j + \sum_{i=1}^{k-3}\sum_{j=1}^{k-i-2}(P^{\#})^i R^{i+j+1}(Q^{\#})^j\right)\right] + Q \tag{5.84}$$

$$+ \left[(R - R(I - QQ^{\#})) + \sum_{j=1}^{k-1} R^{j+1}(Q^{\#})^j\right]$$

$$= (P + Q + R) + \left(\sum_{i=1}^{k-2} (P^{\#})^i R^{i+1}(I - QQ^{\#}) + \sum_{j=1}^{k-2} (I - PP^{\#}) R^{j+1}(Q^{\#})^j \right)$$

$$- \left((I - PP^{\#}) R(I - QQ^{\#}) + \sum_{i=1}^{k-3} \sum_{j=1}^{k-i-2} (P^{\#})^i R^{i+j+1}(Q^{\#})^j \right),$$

it is evident that $\text{Ind}(M) \le 1$ if and only if (5.76) holds.   □

Remark 5.4.1 The importance of Theorem 5.23 lies not only in representing the Drazin inverse, but can also be very useful for the existence of the group inverse. If we assume that $\text{Ind}(R) < 2$ (equivalently that $R = 0$, if $(P, Q, R, 0)$ is a pseudo-block decomposition of M), and keep the remaining assumptions of Theorem 5.23, then (5.77) still holds and gives an expression for the group inverse of M.

Finally, we present an explicit representation for the Drazin inverse of the pseudo-block matrix $P + Q + R$ corresponding to $(P, Q, R, 0)$.

Theorem 5.24 *Let* $P, Q, R \in \mathbb{C}^{m \times m}$ *and* $(P, Q, R, 0)$ *be a pseudo-block decomposition of* M. *Let* $k = \text{Ind}(R)$, $l_P = \text{Ind}(P)$, *and* $l_Q = \text{Ind}(Q)$. *Then*

$$M^D = P^D + Q^D - \sum_{i=1}^{k-1} \sum_{j=1}^{k-i} (P^D)^i R^{i+j-1}(Q^D)^j \tag{5.85}$$

$$+ \left(\sum_{i=1}^{k-1} \sum_{j=0}^{i_Q} (P^D)^{i+j+1} R^i Q^j + \sum_{i=1}^{k-2} \sum_{j=1}^{k-i-1} (P^D)^{m+i+1} R^{i+j} Q^{m-j} \right)(I - QQ^D)$$

$$+ (I - PP^D) \left(\sum_{i=1}^{k-1} \sum_{j=0}^{i_P} P^j R^i (Q^D)^{i+j+1} + \sum_{i=1}^{k-2} \sum_{j=1}^{k-i-1} P^{m-i} R^{i+j} (Q^D)^{m+j+1} \right),$$

where $i_P = \min\{l_P - 1, m - i\}$ *and* $i_Q = \min\{l_Q - 1, m - i\}$, *and also*

$$M^D = P^D + Q^D - \sum_{i=1}^{m-1} \sum_{j=1}^{m-i} (P^D)^i R^{i+j-1}(Q^D)^j \tag{5.86}$$

$$+ \left(\sum_{i=1}^{m-1} \sum_{j=0}^{m-i} (P^D)^{i+j+1} R^i Q^j + \sum_{i=1}^{m-2} \sum_{j=1}^{m-i-1} (P^D)^{m+i+1} R^{i+j} Q^{m-j} \right)(I - QQ^D)$$

$$+ (I - PP^D) \left(\sum_{i=1}^{m-1} \sum_{j=0}^{m-i} P^j R^i (Q^D)^{i+j+1} + \sum_{i=1}^{m-2} \sum_{j=1}^{m-i-1} P^{m-i} R^{i+j} (Q^D)^{m+j+1} \right).$$

Proof Using Lemma 5.5, we have

$$M^D = [P + (R + Q)]^D$$

$$= \sum_{t=0}^{l-1} (P^D)^{t+1}(R + Q)^t \Pi + \sum_{t=0}^{l_P-1}(I - PP^D)P^t[(R + Q)^D]^{t+1}, \quad (5.87)$$

where $\Pi = I - (R + Q)(R + Q)^D$ and $l = \text{Ind}(R + Q)$.

Similarly, we have

$$(R + Q)^D = \sum_{j=0}^{k-1} R^j (Q^D)^{j+1}, \qquad \Pi = I - QQ^D + \sum_{j=0}^{k-1} R^{j+1}(Q^D)^{j+1}.$$

Since

$$(R + Q)^t = \sum_{i=0}^{t} R^{t-i} Q^i, \qquad [(R + Q)^D]^{t+1} = \sum_{j=0}^{k-1} R^i (Q^D)^{t+j+1},$$

substituting the above expressions in (5.87), by computing we obtain the expressions from the theorem.   □

The following is an obvious corollary of Theorem 5.24 and contains as a special case the Hartwig-Shoaf-Meyer-Rose formula.

Corollary 5.9 *Let* $P, Q, R \in \mathbb{C}^{m \times m}$ *and* $(P, Q, R, 0)$ *be a pseudo-block decomposition of* M. *Let* $l_P = \text{Ind}(P)$ *and* $l_Q = \text{Ind}(Q)$. *If* $\text{Ind}(R) \leq 2$ *(i.e.,* $R^2 = 0$*), then*

$$(P + Q + R)^D = P^D + Q^D - P^D R Q^D + \left(\sum_{j=0}^{m-1}(P^D)^{j+2} R Q^j\right)(I - QQ^D)$$

$$+ (I - PP^D)\left(\sum_{j=0}^{m-1} P^j R(Q^D)^{j+2}\right) \qquad (5.88)$$

$$= P^D + Q^D - P^D R Q^D + \left(\sum_{j=0}^{l_Q-1}(P^D)^{j+2} R Q^j\right)(I - QQ^D)$$

$$+ (I - PP^D)\left(\sum_{j=0}^{l_P-1} P^j R(Q^D)^{j+2}\right).$$

Remark 5.4.2 For $P = \left[\begin{smallmatrix} A & 0 \\ 0 & 0 \end{smallmatrix}\right]$, $Q = \left[\begin{smallmatrix} 0 & 0 \\ 0 & C \end{smallmatrix}\right]$ and $R = \left[\begin{smallmatrix} 0 & B \\ 0 & 0 \end{smallmatrix}\right]$, we have that $(P, Q, R, 0)$ is a pseudo-block decomposition of $M = \left[\begin{smallmatrix} A & B \\ 0 & C \end{smallmatrix}\right]$. So the Hartwig-Shoaf-Meyer-Rose formula follows immediately from Corollary 5.9 taking into account that $P^D = \left[\begin{smallmatrix} A^D & 0 \\ 0 & 0 \end{smallmatrix}\right]$ and $Q^D = \left[\begin{smallmatrix} 0 & 0 \\ 0 & C^D \end{smallmatrix}\right]$.

Corollary 5.10 (Hartwig-Shoaf-Meyer-Rose formula [2, Theorem 7.7.1], [11, 12])
Let $M = \begin{bmatrix} A & B \\ 0 & C \end{bmatrix} \in \mathbb{C}^{(m+n)\times(m+n)}$, $A \in \mathbb{C}^{n\times n}$, $C \in \mathbb{C}^{m\times m}$, $l_A=\text{Ind}(A)$ and $l_C=\text{Ind}(C)$.
Then
$$M^D = \begin{bmatrix} A^D & X \\ 0 & C^D \end{bmatrix},$$

where

$$X = \left(\sum_{j=0}^{l_C-1} (A^D)^{j+2} BC^j \right)(I - CC^D) + (I - AA^D)\left(\sum_{j=0}^{l_A-1} A^j B(C^D)^{j+2} \right) - A^D BC^D.$$

Our next goal is to find some explicit representations for the Drazin inverse of a pseudo-block matrix $P + Q + R + S$ corresponding to (P, Q, R, S) under certain assumptions. First, we need to list some auxiliary results for the case $S = 0$.

Lemma 5.6 *Let $P, Q, R \in \mathbb{C}^{m\times m}$ and $(P, Q, R, 0)$ be a pseudo-block decomposition of M. Then*

$$
\begin{aligned}
\left(M^D\right)^l =& \left((P + Q + R)^D\right)^l \\
=& (P^D)^l + (Q^D)^l + \sum_{k=1}^{l-1}\sum_{i=1}^{m-2}\sum_{j=1}^{m-i-1} (P^D)^{k+i} R^{i+j} (Q^D)^{l-k+j} \\
& - \sum_{k=1}^{l}\sum_{i=1}^{m-1}\sum_{j=1}^{m-i} (P^D)^{k+i-1} R^{i+j-1} (Q^D)^{l-k+j} \\
& + \left(\sum_{i=1}^{m-1}\sum_{j=0}^{m-i} (P^D)^{l+i+j} R^i Q^j + \sum_{i=1}^{m-2}\sum_{j=1}^{m-i-1} (P^D)^{l+m+i} R^{i+j} Q^{m-j} \right)(I - QQ^D) \\
& + (I - PP^D)\left(\sum_{i=1}^{m-1}\sum_{j=0}^{m-i} P^j R^i (Q^D)^{l+i+j} + \sum_{i=1}^{m-2}\sum_{j=1}^{m-i-1} P^{m-i} R^{i+j} (Q^D)^{l+m+j} \right),
\end{aligned}
$$

for any $l \geq 2$.

Proof By (5.86) in Theorem 5.24 and induction on l, the desired result follows after careful verification. □

Theorem 5.25 *Let $P, Q, R \in \mathbb{C}^{m\times m}$ and (P, Q, R, S) be a pseudo-block decomposition of M such that $SP = SR = 0$. Then*

$$M^D = P^D + Q^D + \sum_{l=2}^{l_S}(Q^D)^l S^{l-1} + \sum_{l=2}^{l_S}\sum_{k=1}^{l-1}\sum_{i=1}^{m-1}\sum_{j=1}^{m-i}(P^D)^{k+i}R^{i+j}(Q^D)^{l-k+j}S^{l-1}$$

$$+ \sum_{l=1}^{l_S}\sum_{i=1}^{m-1}\sum_{j=0}^{m-i}(P^D)^{l+i+j}R^i Q^j(I - QQ^D)S^{l-1}$$

$$+ \sum_{l=1}^{l_S}\sum_{i=1}^{m-1}\sum_{j=0}^{m-i}(I - PP^D)P^j R^i(Q^D)^{l+i+j}S^{l-1}$$

$$+ \sum_{l=1}^{l_S}\sum_{i=1}^{m-2}\sum_{j=1}^{m-i-1}(P^D)^{l+m+i}R^{i+j}Q^{m-j}(I - QQ^D)S^{l-1} \qquad (5.89)$$

$$+ \sum_{l=1}^{l_S}\sum_{i=1}^{m-2}\sum_{j=1}^{m-i-1}(I - PP^D)P^{m-i}R^{i+j}(Q^D)^{l+m+j}S^{l-1}$$

$$- \sum_{l=1}^{l_S}\sum_{k=1}^{l}\sum_{i=1}^{m-1}\sum_{j=1}^{m-i}(P^D)^{k+i-1}R^{i+j-1}(Q^D)^{l-k+j}S^{l-1},$$

where $l_S = \text{Ind}(S)$. *Replacing l_S by an integer n ($m \geq n \geq l_S$) in (5.89), the above explicit representation still holds.*

Proof From Lemma 5.5 and $SP = SQ = SR = 0$, it follows that

$$M^D = (P + Q + R + S)^D = \sum_{l=1}^{l_S}\left((P + Q + R)^D\right)^l S^{l-1}. \qquad (5.90)$$

Combining (3.8) with Lemma 5.6, we get (5.89).   □

Replacing the condition $SP = SR = 0$ in the above theorem by $RS = QS = 0$, we can obtain a similar result.

Theorem 5.26 *Let $P, Q, R \in \mathbb{C}^{m \times m}$ and (P, Q, R, S) be a pseudo-block decomposition of M such that $RS = QS = 0$. Then*

$$M^D = P^D + Q^D + \sum_{l=2}^{l_S}S^{l-1}(P^D)^l + \sum_{l=1}^{l_S}\sum_{i=1}^{m-1}\sum_{j=0}^{m-i}S^{l-1}(P^D)^{l+i+j}R^i Q^j(I - QQ^D)$$

$$+ \sum_{l=2}^{l_S}\sum_{k=1}^{l-1}\sum_{i=1}^{m-2}\sum_{j=1}^{m-i-1}S^{l-1}(P^D)^{k+i}R^{i+j}(Q^D)^{l-k+j}$$

$$+ \sum_{l=1}^{l_S}\sum_{i=1}^{m-1}\sum_{j=0}^{m-i}S^{l-1}(I - PP^D)P^j R^i(Q^D)^{l+i+j}$$

$$+\sum_{l=1}^{l_S}\sum_{i=1}^{m-2}\sum_{j=1}^{m-i-1} S^{l-1}(P^D)^{l+m+i}R^{i+j}Q^{m-j}(I-QQ^D) \tag{5.91}$$

$$+\sum_{l=1}^{l_S}\sum_{i=1}^{m-2}\sum_{j=1}^{m-i-1} S^{l-1}(I-PP^D)P^{m-i}R^{i+j}(Q^D)^{l+m+j}$$

$$-\sum_{l=1}^{l_S}\sum_{k=1}^{l}\sum_{i=1}^{m-1}\sum_{j=1}^{m-i} S^{l-1}(P^D)^{k+i-1}R^{i+j-1}(Q^D)^{l-k+j},$$

where $l_S = \mathrm{Ind}(S)$. *Replacing l_S by any integer n ($m \ge n \ge l_S$) in (5.91), the above explicit representation still holds.*

Proof The proof is similar to that of Theorem 5.25. $\square$

Specializing Theorem 5.25 to the case $\mathrm{Ind}(S) \le 2$ and $\mathrm{Ind}(R) \le 2$ (i.e., $R^2 = S^2 = 0$), we have the following corollary.

Corollary 5.11 *Let $P, Q, R \in \mathbb{C}^{m \times m}$ and (P, Q, R, S) be a pseudo-block decomposition of M such that $SP = SR = 0$ and $S^2 = R^2 = 0$. Then*

$$M^D = P^D + Q^D - P^D R Q^D + ((Q^D)^2 - P^D R(Q^D)^2 - (P^D)^2 R Q^D)S$$

$$+\sum_{j=0}^{m-1}(P^D)^{j+2}R Q^j(I-QQ^D) + \sum_{j=0}^{m-1}(I-PP^D)P^j R(Q^D)^{j+2} \tag{5.92}$$

$$+\sum_{j=0}^{m-1}\left((P^D)^{j+3}R Q^j(I-QQ^D) + (I-PP^D)P^j R(Q^D)^{j+3}\right)S,$$

or alternatively,

$$M^D = P^D + Q^D - P^D R Q^D + ((Q^D)^2 - P^D R(Q^D)^2 - (P^D)^2 R Q^D)S$$

$$+\sum_{j=0}^{l_Q-1}(P^D)^{j+2}R Q^j(I-QQ^D) + \sum_{j=0}^{l_P-1}(I-PP^D)P^j R(Q^D)^{j+2} \tag{5.93}$$

$$+\sum_{j=0}^{l_Q-1}(P^D)^{j+3}R Q^j(I-QQ^D)S + \sum_{j=0}^{l_P-1}(I-PP^D)P^j R(Q^D)^{j+3}S,$$

where $l_P = \mathrm{Ind}(P)$ and $l_Q = \mathrm{Ind}(Q)$. Furthermore,

$$\mathrm{Ind}(M) \le \mathrm{Ind}(P) + \mathrm{Ind}(Q) + 2. \tag{5.94}$$

Proof The formulae (5.92) and (5.93) are immediate from Theorem 5.25. All we need to do now is to prove (5.94).

By induction on l, we have

$$M^{l+1} = P^{l+1} + Q^{l+1} + Q^l S + \sum_{i=0}^{l} P^{l-i} R Q^i + \sum_{i=0}^{l-1} P^{l-i-1} R Q^i S,$$

for $l \geq 1$. Denote the five terms on the right-hand side of the above equality by Y_i $(i = 1, 2, \ldots, 5)$, respectively, and put $X = M^D$. Combining the above equality with (5.93), for $l \geq l_P + l_Q + 2$, we obtain

$$M^{l+1} M^D = Y_1 X + Y_2 X + Y_3 X + Y_4 X + Y_5 X = Y_1 X + Y_2 X + Y_4 X \qquad (5.95)$$

$$= \Big[P^l - P^l R Q^D - P^l R Q^D - (P^l R (Q^D)^2 + P^{l-1} R Q^D) S$$

$$+ \sum_{j=0}^{l-1} P^{l-j-1} R Q^j (I - Q Q^D) + \sum_{j=0}^{l-2} P^{l-j-2} R Q^j (I - Q Q^D) S \Big]$$

$$+ (Q^l + Q^{l-1} S) + \Big[\Big(P^l R Q^D + \sum_{i=0}^{l-1} P^{l-i+1} R Q^i \cdot Q Q^D \Big)$$

$$+ \Big(P^l R (Q^D)^2 S + P^{l-1} R Q^D S + \sum_{i=0}^{l-2} P^{l-i-2} R Q^i \cdot Q Q^D S \Big) \Big]$$

$$= M^l,$$

since $SX_3 = 0$, $SX_5 = 0$, $\sum_{j=l_Q}^{l-1} P^{l-j-1} R Q^j (I - Q Q^D) = 0$, $P^l (P^D)^j = P^{l-j}$ for $l - \max\{\text{Ind}(P), 1\} \geq j > 0$, and $\sum_{j=l_Q}^{l-2} P^{l-j-2} R Q^j (I - Q Q^D) S = 0$. Clearly, (5.94) follows from (5.95).   □

We now pay our attention to finding expressions for the Drazin inverse of a 2 × 2 block matrix, $M = \begin{bmatrix} A & B \\ 0 & C \end{bmatrix} \in \mathbb{C}^{m \times m}$, where A and C are square matrices.

Let $P = \begin{bmatrix} A & 0 \\ 0 & 0 \end{bmatrix}$, $Q = \begin{bmatrix} 0 & 0 \\ 0 & C \end{bmatrix}$, $R = \begin{bmatrix} 0 & B \\ 0 & 0 \end{bmatrix}$ and $S = \begin{bmatrix} 0 & 0 \\ D & 0 \end{bmatrix}$. Considering the block decomposition (P, Q, R, S) of $M = P + Q + R + S = \begin{bmatrix} A & B \\ D & C \end{bmatrix}$, the desired results are derived from Corollary 5.11 and Theorem 5.26, respectively.

Corollary 5.12 *Let* $M = \begin{bmatrix} A & B \\ D & C \end{bmatrix} \in \mathbb{C}^{m \times m}$, $l_A = \text{Ind}(A)$ *and* $l_C = \text{Ind}(C)$, *where* A *and* C *are square matrices. If* $DA = 0$ *and* $DB = 0$, *then*

$$M^D = \begin{bmatrix} A^D + X_2 D & X_1 \\ (C^D)^2 D & C^D \end{bmatrix},$$

where

$$X_i = \sum_{j=0}^{m-1} \left((A^D)^{i+j+1} BC^j (I - CC^D) + (I - AA^D) A^j B(C^D)^{i+j+1} \right)$$

$$- \sum_{j=0}^{i-1} (A^D)^{j+1} B(C^D)^{i-j} \tag{5.96}$$

$$= \left(\sum_{j=0}^{l_C-1} (A^D)^{i+j+1} BC^j \right)(I - CC^D) + (I - AA^D)\left(\sum_{j=0}^{l_A-1} A^j B(C^D)^{i+j+1} \right)$$

$$- \sum_{j=0}^{i-1} (A^D)^{j+1} B(C^D)^{i-j}, \quad (i = 1, 2)$$

Furthermore,

$$\mathrm{Ind}(M) \le \mathrm{Ind}(A) + \mathrm{Ind}(C) + 2.$$

Corollary 5.13 ([13, Theorem 2.1]) *Let* $M = \begin{bmatrix} A & B \\ D & C \end{bmatrix} \in \mathbb{C}^{m \times m}$, $l_A = \mathrm{Ind}(A)$ *and* $l_C = \mathrm{Ind}(C)$, *where A and C are square matrices. If* $BD = 0$ *and* $CD = 0$, *then*

$$M^D = \begin{bmatrix} A^D & X_1 \\ D(A^D)^2 & C^D + DX_2 \end{bmatrix},$$

where X_1 and X_2 are defined in (5.96). Furthermore,

$$\mathrm{Ind}(M) \le \mathrm{Ind}(A) + \mathrm{Ind}(C) + 2.$$

We conclude this section with the following example in which none of the results on representations of M^D using block decompositions of M can be applied to compute M^D but choosing a special pseudo-block decomposition of M does the job.

Example 5.3 Let

$$M = \begin{bmatrix} 8 & 0 & 8 & 8 \\ 0 & 0 & -6 & 6 \\ 4 & -4 & -7 & 15 \\ 4 & -4 & -7 & 15 \end{bmatrix} = \begin{bmatrix} A & B \\ D & C \end{bmatrix}.$$

There are three block forms for M given as follows:

Case (1): $A = [\, 8\,], B = [\, 0,\ 8,\ 8], C = \begin{bmatrix} 0 & -6 & 6 \\ -4 & -7 & 15 \\ -4 & -7 & 15 \end{bmatrix}, D = \begin{bmatrix} 0 \\ 4 \\ 4 \end{bmatrix}.$

Case (2): $A = \begin{bmatrix} 8 & 0 \\ 0 & 0 \end{bmatrix}, B = \begin{bmatrix} 8 & 8 \\ -6 & 6 \end{bmatrix}, C = \begin{bmatrix} -7 & 15 \\ -7 & 15 \end{bmatrix}, D = \begin{bmatrix} 4 & -4 \\ 4 & -4 \end{bmatrix}.$

Case (3): $A = \begin{bmatrix} 8 & 0 & 8 \\ 0 & 0 & -6 \\ 4 & -4 & -7 \end{bmatrix}, B = \begin{bmatrix} 8 \\ 6 \\ 15 \end{bmatrix}, C = [15], D = [4, \; -4, \; -7].$

In each case, one can verify that $DB \neq 0$ and $BD \neq 0$. Thus the block matrix version results in Corollaries 5.12 and 5.13 cannot apply to yield M^D. Similarly, let (P_1, Q_1, R_1, S_1) be the block decompositions of M, where

$$P_1 = \begin{bmatrix} A & 0 \\ 0 & 0 \end{bmatrix}, \quad Q_1 = \begin{bmatrix} 0 & 0 \\ 0 & C \end{bmatrix}, \quad R_1 = \begin{bmatrix} 0 & B \\ 0 & 0 \end{bmatrix}, \quad S_1 = \begin{bmatrix} 0 & 0 \\ D & 0 \end{bmatrix}.$$

We see that Theorems 5.25, 5.26 and Corollary 5.11, can neither be used to produce M^D, since $S_1 R_1 \neq 0$ and $R_1 S_1 \neq 0$. If we take a special pseudo-block decomposition (P_2, Q_2, R_2, S_2) of M as follows:

$$P_2 = \begin{bmatrix} 8 & -8 & 8 & 8 \\ 0 & 0 & 0 & 0 \\ 4 & -4 & 2 & 6 \\ 4 & -4 & 2 & 6 \end{bmatrix}, Q_2 = \begin{bmatrix} 0 & 0 & 2 & -2 \\ 0 & 0 & -2 & 2 \\ 0 & 0 & -2 & 2 \\ 0 & 0 & -2 & 2 \end{bmatrix}, R_2 = \begin{bmatrix} 0 & 8 & 0 & 0 \\ 0 & 0 & -4 & 4 \\ 0 & 0 & -8 & 8 \\ 0 & 0 & -8 & 8 \end{bmatrix}, S_2 = \begin{bmatrix} 0 & 0 & -2 & 2 \\ 0 & 0 & 0 & 0 \\ 0 & 0 & 1 & -1 \\ 0 & 0 & 1 & -1 \end{bmatrix},$$

then we can compute M^D using Theorem 5.25. Indeed, one can verify that (P_2, Q_2, R_2, S_2) is a pseudo-block decomposition of M, as $S_2 P_2 = 0$ and $S_2 R_2 = 0$.

Let $\alpha_1^T = (0, 0, 1, 1)$, $\alpha_2^T = (1, 0, 0, 0)$, $\alpha_3^T = (-1, 1, 1, 1)$, $\beta_1^T = (4, -4, 2, 6)$, $\beta_2^T = (8, -8, 8, 8)$, and $\beta_3^T = (0, 0, -2, 2)$. We have

$$P_2^D = (\alpha_1 \beta_1^T + \alpha_2 \beta_2^T)^D = \left([\alpha_1, \alpha_2] \begin{bmatrix} \beta_1^T \\ \beta_2^T \end{bmatrix} \right)^D$$

$$= [\alpha_1, \alpha_2] \left[\left(\begin{bmatrix} \beta_1^T \\ \beta_2^T \end{bmatrix} [\alpha_1, \alpha_2] \right)^D \right]^2 \begin{bmatrix} \beta_1^T \\ \beta_2^T \end{bmatrix} = \frac{1}{256} \begin{bmatrix} 8 & -8 & 6 & 10 \\ 0 & 0 & 0 & 0 \\ 4 & -4 & 3 & 5 \\ 4 & -4 & 3 & 5 \end{bmatrix},$$

and $Q_2^D = (\alpha_3 \beta_3^T)^D = \alpha_3 \left[(\beta_3^T \alpha_3)^D \right]^2 \beta_3^T = 0$. Hence by Theorem 5.25, we obtain

$$M^D = P_2^D + \sum_{l=1}^{l_{S_2}} \sum_{i=1}^{m-1} \sum_{j=0}^{m-i} (P_2^D)^{l+i+j} R_2^i Q_2^j S_2^{l-1} + \sum_{l=1}^{l_{S_2}} \sum_{i=1}^{m-2} \sum_{j=1}^{m-i-1} (P_2^D)^{l+m+i} R_2^{i+j} Q_2^{m-j} S_2^{l-1}$$

$$= P_2^D + (P_2^D)^2 R_2 + (P_2^D)^3 (R_2 Q_2 + R_2^2)$$

$$= \frac{1}{256} \begin{bmatrix} 8 & -8 & 6 & 10 \\ 0 & 0 & 0 & 0 \\ 4 & -4 & 3 & 5 \\ 4 & -4 & 3 & 5 \end{bmatrix} + \frac{1}{256} \begin{bmatrix} 0 & 4 & -6 & 6 \\ 0 & 0 & 0 & 0 \\ 0 & 2 & -3 & 3 \\ 0 & 2 & -3 & 3 \end{bmatrix} + \frac{3}{1024} \begin{bmatrix} 0 & 0 & -2 & 2 \\ 0 & 0 & 0 & 0 \\ 0 & 0 & -1 & 1 \\ 0 & 0 & -1 & 1 \end{bmatrix}$$

$$= \frac{1}{1024} \begin{bmatrix} 32 & -16 & -6 & 70 \\ 0 & 0 & 0 & 0 \\ 16 & -8 & -3 & 35 \\ 16 & -8 & -3 & 35 \end{bmatrix},$$

since $Q_2^D = 0$, $l_{S_2} = \mathrm{Ind}(S_2) = 2$ (since $S_2^2 = 0$ and $S_2 \neq 0$), $m = 4$, and $R_2^3 = 0$, $Q_2 S_2 = 0$, $R_2 S_2 = 0$, $Q_2^2 = 0$, and $R_2^2 Q_2 = 0$.

References

1. Campbell, S.L.: The Drazin inverse and systems of second order linear differential equations. Linear Multilinear Algebra **14**, 195–198 (1983)
2. Campbell, S.L., Meyer, C.D.: Generalized Inverse of Linear Transformations, Pitman, London (1979). Dover, New York (1991)
3. Djordjević, D.S., Stanimirović, P.S.: On the generalized Drazin inverse and generalized resolvent. Czechoslovak Math. J. **51**(126), 617–634 (2001)
4. Hartwig, R.E., Wang, G., Wei, Y.: Some additive results on Drazin inverse. Linear Algebra Appl. **322**, 207–217 (2001)
5. Li, X., Wei, Y.: An improvement on perturbation of thr group inverse and oblique projection. Linear Algebra Appl. **338**, 53–66 (2001)
6. Wei, Y., Li, X.: An improvement on perturbation bounds for the Drazin inverse. Numer. Linear Algebra Appl. **10**, 563–575 (2003)
7. Wei, Y., Li, X., Bu, F.: A perturbation bound of the Drazin inverse of a matrix by separation of simple invariant subspaces. SIAM J. Matrix Anal. Appl. **27**, 72–81 (2005)
8. Campbell, S.L.: Singular Systems of Differential Equations. Pitman, London (1980)
9. Zhang, N., Wei, Y.: Solving EP singular linear systems. Int. J. Comput. Math. **81**, 1395–1405 (2004)
10. Hartwig, R.E., Shoaf, J.M.: Group inverse and Drazin inverse of bidiagonal and triangular Toeplitz matrices. Austr. J. Math. **24A**, 10–34 (1977)
11. Meyer, C.D., Rose, N.J.: The index and the Drazin inverse of block triangular matrices. SIAM J. Appl. Math. **33**, 1–7 (1977)
12. Hartwig, R.E., Li, X., Wei, Y.: Representations for the Drazin inverse of 2 × 2 block matrix. SIAM J. Matrix Anal. Appl. **27**, 757–771 (2006)
13. Cvetković-Ilić, D.S.: A note on the representation for the Drazin inverse of 2 × 2 block matrices. Linear Algebra Appl. **429**, 242–248 (2008)
14. Cvetković-Ilić, D.S., Chen, J., Xu, Z.: Explicit representation of the Drazin inverse of block matrix and modified matrix. Linear Multilinear Algebra **57**(4), 355–364 (2009)
15. Yang, H., Liu, X.: The Drazin inverse of the sum of two matrices and its applications. J. Comput. Appl. Math. **235**, 1412–1417 (2011)
16. Ljubisavljević, J., Cvetković-Ilić, D.S.: Additive results for the Drazin inverse of block matrices and applications. J. Comput. Appl. Math. **235**, 3683–3690 (2011)

17. Castro-González, N., Dopazo, E., Robles, J.: Formulas for the Drazin inverse of special block matrices. Appl. Math. Comput. **174**, 252–270 (2006)
18. Cvetković, A.S., Milovanović, G.V.: On Drazin inverse of operator matrices. J. Math. Anal. Appl. **375**, 331–335 (2011)
19. Cvetković-Ilić, D.S.: New additive results on Drazin inverse and its applications. Appl. Math. Comput. **218**(7), 3019–3024 (2011)
20. Bu, C., Zhang, K.: The explicit representations of the Drazin inverses of a class of block matrices. Electron. J. Linear Algebra **20**, 406–418 (2010)
21. Ljubisavljević, J., Cvetković-Ilić, D.S.: Representations for the Drazin inverse of block matrix. J. Comput. Anal. Appl. **15**(3), 481–497 (2013)
22. Wei, Y.: Expressions for the Drazin inverse of a 2 × 2 block matrix. Linear Multilinear Algebra **45**, 131–146 (1998)
23. Miao, J.: Results of the Drazin inverse of block matrices. J. Shanghai Normal Univ. **18**, 25–31 (1989). (in Chinese)
24. Ben-Isreal, A., Greville. T.N.E.: Generalized Inverse: Theory and Applications, 2nd edn. Springer, New York (2003)
25. Cline, R.E., Greville, T.N.E.: A Drazin inverse for rectangular matrices. Linear Algebra Appl. **29**, 53–62 (1980)
26. Benítez, J., Thome, N.: The generalized Schur complement in group inverses and $(k+1)$-potent matrices. Linear Multilinear Algebra **54**, 405–413 (2006)
27. Bru, R., Climent, J., Neumann, M.: On the index of block upper triangular matrices. SIAM J. Matrix Anal. Appl. **16**, 436–447 (1995)
28. Campbell, S.L., Meyer, C.D.: Continuality properties of the Drazin inverse. Linear Algebra Appl. **10**, 77–83 (1975)
29. Castro-González, N., Koliha, J., Rakočević, V.: Continuity and general perturbation of the Drazin inverse for closed linear operators. Abstr. Appl. Anal. **7**, 335–347 (2002)
30. Castro-González, N., Dopazo, E.: Representations of the Drazin inverse for a class of block matrices. Linear Algebra Appl. **400**, 253–269 (2005)
31. Chen, J., Xu, Z., Wei, Y.: Representations for the Drazin inverse of the sum $P + Q + R + S$ and its applications. Linear Algebra Appl. **430**, 438–454 (2009)
32. Chen, X., Hartwig, R.E.: The group inverse of a triangular matrix. Linear Algebra Appl. **237**(238), 97–108 (1996)
33. Deng, C., Du, H.: The reduced minimum modulus of Drazin inverses of linear operators on Hilbert spaces. Proc. Am. Math. Soc. **134**, 3309–3317 (2006)
34. Djordjević, D.S.: Further results on the reverse order law for generalized inverses. SIAM J. Matrix Anal. Appl. **29**(4), 1242–1246 (2007)
35. Du, H., Deng, C.: The representation and characterization of Drazin inverses of operators on a Hilbert space. Linear Algebra Appl. **407**, 117–124 (2005)
36. Wei, Y., Diao, H.: On group inverse of singular Toeplitz matrices. Linear Algebra Appl. **399**, 109–123 (2005)
37. Wei, Y., Diao, H., Ng, M.K.: On Drazin inverse of singular Toeplitz matrix. Appl. Math. Comput. **172**, 809–817 (2006)
38. Wei, Y., Li, X., Bu, F., Zhang, F.: Relative perturbation bounds for the eigenvalues of diagonalizable and singular matrices-application of perturbation theory for simple invariant subspaces. Linear Algebra Appl. **419**, 765–771 (2006)
39. Li, X., Wei, Y.: A note on the representations for the Drazin inverse of 2 × 2 block matrices. Linear Algebra Appl. **423**, 332–338 (2007)
40. Deng, C.Y., Wei, Y.: New additive results for the generalized Drazin inverse. J. Math. Anal. Appl. **370**, 313–321 (2009)
41. Djordjević, D.S., Wei, Y.: Additive results for the generalized Drazin inverse. J. Austr. Math. Soc. **73**, 115–125 (2002)
42. Castro-González, N., Koliha, J.J.: New additive results for the g-Drazin inverse. Proc. R. Soc. Edinburgh **134A**, 1085–1097 (2004)

Chapter 6
Additive Results for the Drazin Inverse

The Drazin inverse, introduced in [1], named after Michael P. Drazin in 1958 in the setting of an abstract ring, is a kind of generalized inverse of a matrix. Many interesting spectral properties of the Drazin inverse make it as a concept that is extremely useful in various considerations in topics such as Markov chains, multibody system dynamics, singular difference and differential equations, differential-algebraic equations and numerical analysis ([1–6]).

In this chapter we will focus our attention on the behavior of the Drazin inverse of a sum of two Drazin invertible elements in the setting of matrices as well as in Banach algebras, where we will also consider the concept of the generalized Drazin inverse. In 1958, while considering the question of Drazin invertibility of a sum of two Drazin invertible elements of a ring Drazin proved that

$$(A + E)^{\mathrm{D}} = A^{\mathrm{D}} + E^{\mathrm{D}}$$

provided that $AE = EA = 0$. After that this topic received considerable interest with many authors working on this problem [4, 7–10], which in turn lead to a number of different formulae for the Drazin inverse $(A + E)^{\mathrm{D}}$ as a function of A, E, A^{D} and E^{D}.

6.1 Additive Results for the Drazin Inverse

Although it was already even in 1958 that Drazin [1] pointed out that computing the Drazin inverse of a sum of two elements in a ring was not likely to be easy, this problem remains open to this day even for matrices. It is precisely this problem when considered in rings of matrices that will be the subject of our interest in this section, i.e., under various conditions we will compute $(P + Q)^{\mathrm{D}}$ as a function of P, Q, P^{D} and Q^{D}. We will extend Drazin's result in the sense that only one of the conditions

© Springer Nature Singapore Pte Ltd. 2017

D. Cvetković Ilić and Y. Wei, *Algebraic Properties of Generalized Inverses*,
Developments in Mathematics 52, DOI 10.1007/978-981-10-6349-7_6

$PQ = 0$ or $PQ = QP$ is assumed. The results obtained will be then used to analyze a special class of perturbations of the type $A - X$.

Throughout the section, we shall assume familiarity with the theory of Drazin inverses (see [11]). Also, for $A \in \mathbb{C}^{n \times n}$, we denote $Z_A = I - AA^D$.

First, we will give a representation of $(P + Q)^D$ under the condition $PQ = 0$ which was considered in [10, Theorem 2.1]:

Theorem 6.1 *Let* $P, Q \in \mathbb{C}^{n \times n}$. *If* $PQ = 0$, *then*

$$(P + Q)^D = (I - QQ^D)[I + QP^D + \cdots + Q^{k-1}(P^D)^{k-1}]P^D$$
$$+ Q^D[I + Q^D P + \cdots + (Q^D)^{k-1} P^{k-1}](I - PP^D), \qquad (6.1)$$

and

$$(P + Q)(P + Q)^D = (I - QQ^D)[I + QP^D + \cdots + Q^{k-1}(P^D)^{k-1}]PP^D$$
$$+ QQ^D[I + Q^D P + \cdots + (Q^D)^{k-1} P^{k-1}](I - PP^D) + QQ^D PP^D, \qquad (6.2)$$

where $\max\{\mathrm{Ind}(P), \mathrm{Ind}(Q)\} \le k \le \mathrm{Ind}(P) + \mathrm{Ind}(Q)$.

Proof Under the assumption $PQ = 0$, we have

$$P^D Q = PQ^D = 0, \quad Z_P Q = Q \quad \text{and} \quad PZ_Q = P. \qquad (6.3)$$

Using Cline's Formula [12], $(AB)^D = A[(BA)^D]^2 B$, we have

$$(P + Q)^D = \left([I, Q] \begin{bmatrix} P \\ I \end{bmatrix}\right)^D = [I, Q] \left(\begin{bmatrix} P & PQ \\ I & Q \end{bmatrix}^D\right)^2 \begin{bmatrix} P \\ I \end{bmatrix}.$$

Now, by Theorem 1 of [4], we have that

$$\begin{bmatrix} P & 0 \\ I & Q \end{bmatrix}^D = \begin{bmatrix} P^D & 0 \\ R & Q^D \end{bmatrix},$$

for

$$R = -Q^D P^D + Z_Q Y_k (P^D)^{k+1} + (Q^D)^{k+1} Y_k Z_P$$

and

$$Y_k = Q^{k-1} + Q^{k-2} P + \cdots + QP^{k-2} + P^{k-1},$$

where $\max\{\mathrm{Ind}(P), \mathrm{Ind}(Q)\} \le k \le \mathrm{Ind}(P) + \mathrm{Ind}(Q)$.

Hence

$$(P + Q)^D = [I, Q] \left(\begin{bmatrix} P^D & 0 \\ R & Q^D \end{bmatrix}\right)^2 \begin{bmatrix} P \\ I \end{bmatrix} = P^D + QRPP^D + QQ^D RP + Q^D.$$

Substituting R in the above equality, we get (6.1). It is straightforward to prove (6.2) from (6.1) and (6.3).   □

Now we list some special cases of the previous result:

Corollary 6.1 *Let* $P, Q \in \mathbb{C}^{n \times n}$ *be such that* $PQ = 0$ *and let* k *be such that* $\max\{\text{Ind}(P), \text{Ind}(Q)\} \leq k \leq \text{Ind}(P) + \text{Ind}(Q)$.

(i) *If* Q *is nilpotent, then* $(P + Q)^D = P^D + Q(P^D)^2 + \cdots + Q^{k-1}(P^D)^k$.

(ii) *If* $Q^2 = 0$, *then* $(P + Q)^D = P^D + Q(P^D)^2$.

(iii) *If* P *is nilpotent, then* $(P + Q)^D = Q^D + (Q^D)^2 P + \cdots + (Q^D)^k P^{k-1}$.

(iv) *If* $P^2 = 0$, *then* $(P + Q)^D = Q^D + (Q^D)^2 P$.

(v) *If* $P^2 = P$, *then* $(P + Q)^D = (I - QQ^D)(I + Q + \cdots + Q^{k-1})P + Q^D(I - P)$, *and*
$(P + Q)^D(I - P) = Q^D(I - P)$.

(vi) *If* $Q^2 = Q$, *then* $(P + Q)^D = (I - Q)P^D + Q(I + P + \cdots + P^{k-1})(I - PP^D)$, *and*
$(I - Q)(P + Q)^D = (I - Q)P^D$.

(vii) *If* $PR = 0$, *then* $(P + Q)^D R = (I - QQ^D)P^D R + Q^D R = Q^D R$.

Theorem 6.1 may be used to obtain several additional perturbation results concerning the matrix $\Gamma = A - X$. Needless to say these are rather special, since addition and inversion rarely mix. First a useful result.

Lemma 6.1 *Let* $A, F, X \in \mathbb{C}^{n \times n}$. *If* $AF = FA$ *and* $FX = X$, *then*

$$(AF - X)^k X = (A - X)^k X, \quad \text{for all} \quad k \in \mathbb{N}. \tag{6.4}$$

Proof Since $AF = FA$ and $(I - F)X = 0$, we have that

$$(I - F)(A - X)^k X = 0. \tag{6.5}$$

Now the assertion is proved by induction. The case $k = 1$ is trivial. Suppose $(AF - X)^k X = (A - X)^k X$. Then by (6.5),

$$(AF - X)^{k+1} X = (AF - X)(A - X)^k X = AF(A - X)^k X - X(A - X)^k X$$
$$= A(A - X)^k X - X(A - X)^k X = (A - X)^{k+1} X. \square$$

Now we present a perturbation result.

Corollary 6.2 *Let* $A, F, X \in \mathbb{C}^{n \times n}$ *and let* F *be an idempotent matrix which commutes with* A. *Let* $\Gamma = A - X$ *and let* $\max\{\text{Ind}(A), \text{Ind}(X)\} \leq k \leq \text{Ind}(A) + \text{Ind}(X)$. *If* $FX = X$ *and* $R = \Gamma F = AF - XF$, *then*

$$(A - X)^D = R^D - \sum_{i=0}^{k-1} (R^D)^{i+2} X (I - F) A^i (I - AA^D)$$

$$+ (I + R^D X)(I - F) A^D - (I - RR^D) \sum_{i=0}^{k-2} (A - X)^i X (I - F)(A^D)^{i+2}. \quad (6.6)$$

Proof Let $\Gamma = A - X = P + Q$, where $P = A(I - F)$ and $Q = AF - FX$. Since $F^2 = F$ we have that $(I - F)^2 = I - F$ and $(I - F)^D = (I - F)$. Since

$$PQ = A(I - F)(AF - FX) = A(I - F)AF = A^2[(I - F)F] = 0,$$

after applying Theorem 6.1 we get

$$(P + Q)^D = (I - QQ^D)V + W(I - PP^D) = T_1 + T_2,$$

where $V = [P^D + Q(P^D)^2 + \cdots + Q^{k-1}(P^D)^k]$ and $W = [Q^D + (Q^D)^2 P + \cdots + (Q^D)^k P^{k-1}]$. Put $T_1 = (I - QQ^D)V$ and $T_2 = W(I - PP^D)$. So we see that we need to compute Q^D and P^D. The latter is easily found because A and F commute:

$$P^D = [A(I - F)]^D = (I - F)A^D, \quad PP^D = (I - F)AA^D.$$

On the other hand, in order to compute Q^D, we split Q further as

$$Q = R - S,$$

where $R = (A - X)F = AF - FXF$ and $S = FX(I - F)$. Since

$$SR = FX(I - F)(FA - FXF) = FX[(I - F)F](A - XF) = 0,$$

and $S^2 = FX[(I - F)F]X(I - F) = 0$, by (iv) of Corollary 6.1 we get

$$Q^D = (-S + R)^D = R^D - (R^D)^2 S, \quad QQ^D = (R - S)[R^D - (R^D)^2 S].$$

Since $SR^D = SR = 0$, it follows that $QQ^D = RR^D - R^D S$. Also, $R^D P = 0$, because

$$RP = (AF - FXF)A(I - F) = (A - FX)[F(I - F)]A = 0.$$

So $Q^D P = -(R^D)^2 SP = -(R^D)^2 XP$. Similarly, since $SR^D = 0$, we get

$$(Q^D)^2 P = [R^D - (R^D)^2 S][-(R^D)^2 XP] = -(R^D)^3 XP.$$

Repeating the process, we obtain

$$(Q^D)^{t+1} P^t = -(R^D)^{t+2} X P^t, \quad t = 1, 2, \ldots$$

which when substituted yields the second term:

$$
\begin{aligned}
T_2 &= W(I - PP^D) = [R^D - (R^D)^2 S - (R^D)^3 XP - \cdots - (R^D)^{k+1} X P^{k-1}](I - PP^D) \\
&= [R^D - (R^D)^2 X(I - F) - (R^D)^3 XA(I - F) \cdots - (R^D)^{k+1} X A^{k-1}(I - F)] \\
&\quad - [R^D - (R^D)^2 X(I - F) - (R^D)^3 XA(I - F) \cdots - (R^D)^{k+1} X A^{k-1}(I - F)](I - F)AA^D \\
&= R^D - \sum_{i=0}^{k-1} (R^D)^{i+2} X(I - F) A^i (I - AA^D).
\end{aligned}
$$

Let us next examine the first term

$$T_1 = (I - QQ^D)V = [I - (RR^D - R^D S)][P^D + Q(P^D)^2 + \cdots + Q^{k-1}(P^D)^k].$$

We compute the powers $Q^i (P^D)^{i+1} = (AF - X)^i (I - F)(A^D)^{i+1}$. For $i = 1$, this becomes $(AF - X)(I - F)(A^D)^2 = -X(I - F)(A^D)^2$, while for higher powers of i we may use Lemma 6.1 to obtain

$$
\begin{aligned}
Q^i (P^D)^{i+1} &= (AF - X)^{i-1}(AF - X)(I - F)(A^D)^{i+1} \\
&= -(AF - X)^{i-1} X(I - F)(A^D)^{i+1} = -(A - X)^{i-1} X(I - F)(A^D)^{i+1}.
\end{aligned}
$$

Now

$$
\begin{aligned}
S(A - X)^{i-1} X &= X(I - F)(A - X)(A - X)^{i-2} X \\
&= XA(I - F)(A - X)^{i-2} X = \cdots = XA^{i-1}(I - F)X = 0
\end{aligned}
$$

for all i, and $R^D(I - F) = (R^D)^2 R(I - F) = (R^D)^2 (A - X)[F(I - F)] = 0$, so

$$
\begin{aligned}
T_1 &= (I - RR^D + R^D S)(I - F)A^D + (I - RR^D + R^D S)[Q(P^D)^2 + \cdots + Q^{k-1}(P^D)^k] \\
&= [I + R^D X(I - F)](I - F)A^D - (I - RR^D) \sum_{i=1}^{k-1} (A - X)^{i-1} X(I - F)(A^D)^{i+1} \\
&= (I + R^D X)(I - F)A^D - (I - RR^D) \sum_{i=0}^{k-2} (A - X)^i X(I - F)(A^D)^{i+2},
\end{aligned}
$$

completing the proof.   □

Using the previous result we will analyze some special types of perturbations of the matrix $A - X$. We shall thereby extend earlier work by several authors [13–16] and partially solve a problem posed in 1975 by Campbell and Meyer [17], who considered it difficult to establish norm estimates for the perturbation of the Drazin inverse.

In the following five special cases, we assume $FX = X$ and $R = AF - XF$.
Case (1) $XF = 0$.

Clearly $(R^D)^i = (A^D)^i F$ and $S = X$. Moreover $(A - X)^i FX = A^i X$ for $i \geq 0$.
Thus (6.6) reduces to

$$(A - X)^D = A^D F - \sum_{i=0}^{k-1} (A^D)^{i+2} X A^i (I - AA^D)$$

$$+ (I - F + A^D X)A^D - \sum_{i=0}^{k-2} A^i (I - AA^D) X (A^D)^{i+2}. \quad (6.7)$$

Case (1a) $XF = 0$ and $F = AA^D$.

If we in addition assume that $F = AA^D$, then $XA^D = 0$ and (6.7) is reduced to

$$(A - X)^D = A^D - \sum_{i=0}^{k-1} (A^D)^{i+2} X A^i. \quad (6.8)$$

Case (1b) $XF = 0$ and $F = I - AA^D$.
In this case, $A^D X = 0$ and (6.7) becomes

$$(A - X)^D = A^D - \sum_{i=0}^{k-2} A^i X (A^D)^{i+2}. \quad (6.9)$$

Case (2) $F = AA^D$.

Now $AA^D X = X$, $R = A^2 A^D (I - A^D X AA^D)$ and (6.6) simplifies to

$$(A - X)^D = R^D - \sum_{i=0}^{k-1} (R^D)^{i+2} X A^i (I - AA^D). \quad (6.10)$$

If we set $U = I - A^D X AA^D$ and $V = I - AA^D X A^D$, then $UA^D = A^D V$ and $R = A^2 A^D U = V A^2 A^D$. Now if we assume that U is invertible, then so will be V and $U^{-1} A^D = A^D V^{-1}$. It is now easily verified that $R^\#$ exists and

$$R^\# = U^{-1} A^D = A^D V^{-1}.$$

In fact $RR^\# = A^2 A^D U U^{-1} A^D = AA^D = A^D V^{-1} V A^2 A^D = R^\# R$ and $R^2 R^\# = RAA^D = R$ and $R^\# RR^\# = U^{-1} A^D AA^D = U^{-1} A^D = R^\#$. We then have two subcases.

Case (2a) $F = AA^D$, and $U = I - A^D X AA^D$ is invertible.

In this case, (6.10) is just

$$(A - X)^D = R^\# - \sum_{i=0}^{k-1}(R^\#)^{i+2}XA^i(I - AA^D), \tag{6.11}$$

where $R = A^2A^DU$, $R^\# = U^{-1}A^D$. In general, $(R^\#)^i \neq U^{-i}(A^D)^i$.

Remark 6.1.1 The matrix $U = I - A^DXAA^D$ is invertible if and only if $I - A^DX$ is invertible. This result generalizes the main results from [13–16].

Case (2b) $F = I - AA^D$.
We have $A^DX = A^DF = 0$, and (6.6) becomes

$$(A - X)^D = R^D + (I + R^DX)A^D - (I - RR^D)\sum_{i=0}^{k-2}(A - X)^iX(A^D)^{i+2}, \tag{6.12}$$

where $R = A(I - AA^D) - (I - AA^D)X(I - AA^D)$.
  Case (3) $AA^DXF = XFAA^D = XF$, $U = I - A^DXF$ is invertible and $(AF)^\#$ exists.
  Now $R = AF - XF = AF - AA^DFXF = AF(I - A^DXF) = AFU = VFA$, where $V = I - XFA^D$. Furthermore $A^DFV = UA^DF$. We may now conclude that U is invertible exactly when V is, in which case $Y = U^{-1}A^DF = A^DFV^{-1}$.
  We then have $RY = AFU(U^{-1}A^DF) = AA^DF = A^DFV^{-1}(VFA) = YR$.
Lastly,
$$Y^2R = U^{-1}A^DF(AA^DF) = U^{-1}A^DF = Y$$

and $R^2Y = RAA^DF = A^2A^DF - AA^DFXFAA^D = A^2A^DF - XF$.
  If $(AF)^\#$ exists then $AF = AF(AF)^\#AF = AFF^\#A^DAF = AFF^\#FAA^D = A^2A^DF$, so $R^2Y = AF - XF = R$, i.e., $Y = R^\#$ and (6.6) becomes

$$(A - X)^D = R^\# - \sum_{i=0}^{k-1}(R^\#)^{i+2}X(I - AA^D)A^i. \tag{6.13}$$

Case (4) $FX = XF = X$.
In this case, (6.6) reduces to

$$(A - X)^D = R^D + (I - F)A^D. \tag{6.14}$$

If in addition to $F = AA^D$, the matrix $U = I - A^DX$ is invertible, this reduces further to [15]
$$(A - X)^D = R^D = U^{-1}A^D. \tag{6.15}$$

Case (5) If $X = A^2A^D$ then Γ is nilpotent and $\Gamma^D = 0$.

Although Theorem 6.1 solves our problem under the assumption that $PQ = 0$, the condition can be relaxed and the result therefore generalized as follows: Since

$$\left(\begin{bmatrix} P & PQ \\ I & Q \end{bmatrix}^{D} \right)^{k} = \left(\begin{bmatrix} P & PQ \\ I & Q \end{bmatrix}^{k} \right)^{D} = \begin{bmatrix} P(P+Q)^{k-1} & P(P+Q)^{k-1}Q \\ (P+Q)^{k-1} & (P+Q)^{k-1}Q \end{bmatrix}^{D}, \text{ for all } k \in \mathbb{N},$$

we may extend the considerations above to the case when $P(P+Q)^{k-1}Q = 0$.
    In fact

$$(P+Q)^{D} = [I, Q] \left(\begin{bmatrix} P & PQ \\ I & Q \end{bmatrix}^{k} \right)^{D} \begin{bmatrix} P & PQ \\ I & Q \end{bmatrix}^{k-2} \begin{bmatrix} P \\ I \end{bmatrix} =$$

$$[I, Q] \begin{bmatrix} P(P+Q)^{k-1} & 0 \\ (P+Q)^{k-1} & (P+Q)^{k-1}Q \end{bmatrix}^{D} \begin{bmatrix} P(P+Q)^{k-3} & P(P+Q)^{k-3}Q \\ (P+Q)^{k-3} & (P+Q)^{k-3}Q \end{bmatrix} \begin{bmatrix} P \\ I \end{bmatrix}.$$

This requires computation of $[P(P+Q)^{k-1}]^{D}$ and $[(P+Q)^{k-1}Q]^{D}$, which may actually be easier than that of $(P+Q)^{D}$.
    A second attempt to generalize Theorem 6.1 would be to assume only that $P^2Q = 0$. Needless to say, this is the best attempted via the block form, which in turn should give a suitable formula.
    Now, we will investigate explicit representations for the Drazin inverse $(A + E)^{D}$ in the case when $AE = EA$, which was considered in [18, Theorem 2]. For $A \in \mathbb{C}^{n \times n}$ with $\text{Ind}(A) = k$ and $\text{rank}(A^k) = r$, there exists an nonsingular matrix $P \in \mathbb{C}^{n \times n}$ such that

$$A = P \begin{bmatrix} C & 0 \\ 0 & N \end{bmatrix} P^{-1}, \tag{6.16}$$

where $C \in \mathbb{C}^{r \times r}$ is a nonsingular matrix, N is nilpotent of index k and $\text{Ind}(N) = \text{Ind}(A) = k$. In that case

$$A^{D} = P \begin{bmatrix} C^{-1} & 0 \\ 0 & 0 \end{bmatrix} P^{-1}, \tag{6.17}$$

If $P = I$, then the block-diagonal matrices A and A^{D} are written as $A = C \oplus N$ and $A^{D} = C^{-1} \oplus 0$.
    Now we state the following result which was obtained by Hartwig and Shoaf [19] and Meyer and Rose [20], since it will be used in the theorem to follow.

Theorem 6.2 *If* $M = \begin{bmatrix} A & C \\ 0 & B \end{bmatrix}$, *where* $A \in \mathbb{C}^{n \times n}$ *and* $B \in \mathbb{C}^{m \times m}$ *with* $\text{Ind}(A) = k$ *and* $\text{Ind}(B) = l$, *then* $M^{D} = \begin{bmatrix} A^{D} & X \\ 0 & B^{D} \end{bmatrix}$,

where

$$X = \left[\sum_{n=0}^{l-1}(A^{\mathrm{D}})^{n+2}CB^n\right](I - BB^{\mathrm{D}}) + (I - AA^{\mathrm{D}})\left[\sum_{n=0}^{k-1}A^nC(B^{\mathrm{D}})^{n+2}\right] - A^{\mathrm{D}}CB^{\mathrm{D}}.$$

Theorem 6.3 *If $A, E \in \mathbb{C}^{n\times n}$, $AE = EA$ and $\mathrm{Ind}(A) = k$, then*

$$(A + E)^{\mathrm{D}} = (I + A^{\mathrm{D}}E)^{\mathrm{D}}A^{\mathrm{D}} + (I - AA^{\mathrm{D}})\sum_{i=0}^{k-1}(E^{\mathrm{D}})^{i+1}(-A)^i$$

$$= (I + A^{\mathrm{D}}E)^{\mathrm{D}}A^{\mathrm{D}} + (I - AA^{\mathrm{D}})E^{\mathrm{D}}[I + A(I - AA^{\mathrm{D}})E^{\mathrm{D}}]^{-1},$$

and

$$(A + E)^{\mathrm{D}}(A + E) = (I + A^{\mathrm{D}}E)^{\mathrm{D}}A^{\mathrm{D}}(A + E) + (I - AA^{\mathrm{D}})EE^{\mathrm{D}}.$$

Proof Let $A \in \mathbb{C}^{n\times n}$ be given by (6.16). Without loss of generality, we assume that $P = I$ and $A = C \oplus N$, where C is invertible and N is nilpotent with $N^k = 0$. From $AE = EA$, we have $A^kE = EA^k$. Now $E = E_1 \oplus E_2$, $CE_1 = E_1C$ and $NE_2 = E_2N$. Hence

$$(A + E)^{\mathrm{D}} = (C + E_1)^{\mathrm{D}} \oplus (N + E_2)^{\mathrm{D}}.$$

Since C and $I + C^{-1}E_1$ commute, we get

$$(C + E_1)^{\mathrm{D}} \oplus 0 = (I + C^{-1}E_1)^{\mathrm{D}}C^{-1} \oplus 0 = (I + A^{\mathrm{D}}E)^{\mathrm{D}}A^{\mathrm{D}}.$$

Notice that $(I + T)^{-1} = \sum_{i=0}^{k-1}(-T)^i$ if $T^k = 0$. Applying Lemma 4 [19] we get $(N + E_2)^{\mathrm{D}} = E_2^{\mathrm{D}}(I + E_2^{\mathrm{D}}N)^{-1}$ and

$$0 \oplus E_2^{\mathrm{D}}(I + E_2^{\mathrm{D}}N)^{-1} = 0 \oplus \sum_{i=0}^{k-1}(E_2^{\mathrm{D}})^{i+1}(-N)^i = (I - AA^{\mathrm{D}})\sum_{i=0}^{k-1}(E^{\mathrm{D}})^{i+1}(-A)^i$$

$$= (I + A^{\mathrm{D}}E)^{\mathrm{D}}A^{\mathrm{D}} + (I - AA^{\mathrm{D}})E^{\mathrm{D}}[I + A(I - AA^{\mathrm{D}})E^{\mathrm{D}}]^{-1}.$$

Hence

$$(A + E)^{\mathrm{D}} = (I + A^{\mathrm{D}}E)^{\mathrm{D}}A^{\mathrm{D}} + (I - AA^{\mathrm{D}})\sum_{i=0}^{k-1}(E^{\mathrm{D}})^{i+1}(-A)^i$$

$$= (I + A^{\mathrm{D}}E)^{\mathrm{D}}A^{\mathrm{D}} + (I - AA^{\mathrm{D}})E^{\mathrm{D}}[I + A(I - AA^{\mathrm{D}})E^{\mathrm{D}}]^{-1},$$

and

$$(A + E)^D(A + E)$$

$$= \left\{ (I + A^D E)^D A^D + (I - A A^D) \sum_{i=0}^{k-1} (E^D)^{i+1}(-A)^i \right\} (A + E)$$

$$= (I + A^D E)^D (A + E) A^D + (I - A A^D) E^D A \sum_{i=0}^{k-1} (E^D)^i (-A)^i$$

$$+ (I - A A^D) E^D E \sum_{i=0}^{k-1} (E^D)^i (-A)^i$$

$$= (I + A^D E)^D (A + E) A^D + (I - A A^D) \left(-\sum_{i=1}^{k} (E^D)^i (-A)^i \right)$$

$$+ (I - A A^D) \left(E^D E + \sum_{i=1}^{k-1} (E^D)^i (-A)^i \right)$$

$$= (I + A^D E)^D (A + E) A^D + (I - A A^D) E E^D. \qquad \square$$

From Theorem 6.3, we can see that the generalized Schur complement $I + A^D E$ [21] plays an important role in the representation of the Drazin inverse $(A + E)^D$. In some special cases, it is possible to give an expression for $(I + A^D E)^D$.

Theorem 6.4 *Let* $A, E \in \mathbb{C}^{n \times n}$ *be such that* $AE = EA$ *and let* $\mathrm{Ind}(A) = k$ *and* $\mathrm{Ind}(E) = l$.
(1) *If* $A^D E^D = 0$, *then*

$$(A + E)^D = (I - A A^D) \sum_{i=0}^{k-1} (E^D)^{i+1}(-A)^i + \sum_{i=0}^{l-1} (-E)^i (A^D)^{i+1}(I - E E^D).$$

(2) *If* $A^D E = 0$, *then* $(A + E)^D = A^D + (I - A A^D) \sum_{i=0}^{k-1} (E^D)^{i+1}(-A)^i$.
(3) *If* $\mathrm{Ind}(A) = 1$, *then* $(A + E)^D = (I + A^\# E)^D A^\# + (I - A A^\#) E^D$.

Proof We use the notations from the proof of Theorem 6.3.
(1) If $A^D E^D = 0$, then E_1 is nilpotent with $E_1^l = 0$. So we have

$$(I + A^D E)^D A^D = (I + C^{-1} E_1)^{-1} C^{-1} \oplus 0 = \sum_{i=0}^{l-1} (-E_1)^i (C)^{-(i+1)} \oplus 0$$

$$= \sum_{i=0}^{l-1} (-E)^i (A^D)^{i+1}(I - E E^D).$$

The result now follows from Theorem 6.3.

(2)–(3) Note that if $A^D E = 0$, then $E_1 = 0$; if $\text{Ind}(A) = 1$, then $N = 0$. The results follow directly from the proof of Theorem 6.3. $\qquad\qquad\square$

Let $A, E \in \mathbb{C}^{n \times n}$. If there exists a nonzero idempotent matrix $P = P^2$ such that $AEP = EAP$ (or $PAE = PEA$), then A and E are partially commutative. For $A, E \in \mathbb{C}^{n \times n}$, let $A^\pi = I - AA^D$ and $\text{Ind}(A) = k$ and suppose $E^2 = 0$. In [22], Castro-González proved that if $A^\pi E = E$ and $AEA^\pi = 0$, then

$$(A + E)^D = A^D + \sum_{i=0}^{k} A^i E (A^D)^{i+2} + \sum_{i=0}^{k-1} E A^i E (A^D)^{i+3}.$$

But no representations of $(A + E)^D$ assuming only partial commutativity are known. Under the conditions $A^\pi E = E$ and $AEA^\pi = EAA^\pi$, we are able to give an expression for $(A + E)^D$.

Theorem 6.5 *Let $A \in \mathbb{C}^{n \times n}$ with $\text{Ind}(A) = k$ and $E \in \mathbb{C}^{n \times n}$ be nilpotent of index l. If $E A^D = 0$ and $A^\pi A E = A^\pi E A$, then*

$$(A + E)^D = A^D + \sum_{i=0}^{k+l-2} (A^D)^{i+2} E T(i),$$

where $T(i) = (I - AA^D) \sum_{j=0}^{i} \binom{j}{i} A^j E^{i-j}$.

Proof Similarly as in the proof of Theorem 6.3, let $A = C \oplus N$, where C is invertible and N is nilpotent with $N^k = 0$. It follows from $E A^D = 0$ that E can be written as $E = \begin{bmatrix} 0 & E_1 \\ 0 & E_2 \end{bmatrix}$ with $E_2^l = 0$. Also by $A^\pi A E = A^\pi E A$, we get $E_2 N = N E_2$. Thus

$$(N + E_2)^i \oplus 0 = \sum_{j=0}^{i} \binom{j}{i} N^j E_2^{i-j} \oplus 0 = (I - AA^D) \left(\sum_{j=0}^{i} \binom{j}{i} A^j E^{i-j} \right) = T(i).$$

We observe that $N + E_2$ is nilpotent of index $k + l - 1$. From Theorem 6.2, we further obtain

$$(A + E)^D = \begin{bmatrix} C & E_1 \\ 0 & N + E_2 \end{bmatrix}^D = \begin{bmatrix} C^{-1} & X \\ 0 & 0 \end{bmatrix},$$

where

$$X = \sum_{i=0}^{k+l-2} C^{-(i+2)} E_1 (N + E_2)^i = \sum_{i=0}^{k+l-2} C^{-(i+2)} E_1 \left(\sum_{j=0}^{i} \binom{j}{i} N^j E_2^{i-j} \right). \qquad (6.18)$$

Hence

$$A^D + \sum_{i=0}^{k+l-2} (A^D)^{i+2} ET(i) = \begin{bmatrix} C^{-1} \sum_{i=0}^{k+l-2} C^{-(i+2)} E_1 \left(\sum_{j=0}^{i} \binom{j}{i} N^j E_2^{i-j} \right) \\ 0 \end{bmatrix} = (A + E)^D.$$

□

The following result generalizes Theorems 6.3 and 6.5 to the case of partial commutativity.

Theorem 6.6 *Let* $A, E \in \mathbb{C}^{n \times n}$ *and* $\text{Ind}(A) = k$. *Also let* $Q \in \mathbb{C}^{n \times n}$ *be an idempotent matrix such that* $QA = AQ$ *and* $EQ = 0$. *If* $(I - Q)AE = (I - Q)EA$, *then*

$$(A + E)^D = QA^D + (I - Q)\Psi - QA^DE\Psi + Q(I - AA^D) \left[\sum_{i=0}^{k-1} A^i E\Psi^{i+2} \right]$$

$$+ Q \left[\sum_{i=0}^{h-1} (A^D)^{i+2} E(A + E)^i \right] (I - Q)[I - (A + E)\Psi],$$

(6.19)

where $\Psi = (I + A^DE)^D A^D + (I - AA^D) \sum_{i=0}^{k-1} (E^D)^{i+1} (-A)^i$ *and* $h = \text{Ind}[(I - Q)(A + E)]$.

Proof Suppose that $Q = I_{r \times r} \oplus 0_{(n-r) \times (n-r)}$, where $r \le n$. If $QA = AQ$, $EQ = 0$ and $(I - Q)AE = (I - Q)EA$, then $A = A_1 \oplus A_2$ and $E = \begin{bmatrix} 0 & E_1 \\ 0 & E_2 \end{bmatrix}$ with $A_2 E_2 = E_2 A_2$. Using Theorems 6.2 and 6.3, we have

$$(A + E)^D = \begin{bmatrix} A_1^D & X \\ 0 & (I + A_2^D E_2)^D A_2^D + (I - A_2 A_2^D) \sum_{i=0}^{k-1} (E_2^D)^{i+1} (-A_2)^i \end{bmatrix},$$

where

$$X = \left[\sum_{i=0}^{h-1} (A_1^D)^{i+2} E_1 (A_2 + E_2)^i \right] [I - (A_2 + E_2)(A_2 + E_2)^D]$$

$$+ (I - A_1 A_1^D) \left[\sum_{i=0}^{k-1} A_1^i E_1 ((A_2 + E_2)^D)^{i+2} \right] - A_1^D E_1 (A_2 + E_2)^D,$$

and $\text{Ind}(A_2 + E_2) = h$.

If we write $\Psi = (I + A^DE)^D A^D + (I - AA^D) \sum_{i=0}^{k-1} (E^D)^{i+1} (-A)^i$, then $(I - Q)\Psi = 0 \oplus (A_2 + E_2)^D$. We can simplify the expression for $(A + E)^D$ using the block decomposition above. We deduce

$$\Sigma_1 = Q\left[\sum_{i=0}^{h-1}(A^D)^{i+2}E(A+E)^i\right](I-Q)[I-(A+E)\Psi]$$

$$= \begin{bmatrix} 0 & \left[\sum_{i=0}^{h-1}(A_1^D)^{i+2}E_1(A_2+E_2)^i\right][I-(A_2+E_2)(A_2+E_2)^D] \\ 0 & 0 \end{bmatrix},$$

$$\Sigma_2 = QA^\pi\left[\sum_{i=0}^{k-1}A^iE\Psi^{i+2}\right] = \begin{bmatrix} 0 & A_1^\pi\left[\sum_{i=0}^{k-1}A_1^iE_1((A_2+E_2)^D)^{i+2}\right] \\ 0 & 0 \end{bmatrix}$$

and $\Sigma_3 = QA^DE\Psi = \begin{bmatrix} 0 & A_1^DE_1(A_2+E_2)^D \\ 0 & 0 \end{bmatrix}$.

Thus

$$(A+E)^D = QA^D + (I-Q)\Psi + \Sigma_1 + \Sigma_2 - \Sigma_3.$$

□

Now a few special cases follow immediately.

Corollary 6.3 *Let* $A, E \in \mathbb{C}^{n\times n}$ *with* $\text{Ind}(A) = k$ *and* $\text{Ind}(E) = l$.
(1) *If* $EA^\pi = 0$ *and* $(I-A^\pi)AE = (I-A^\pi)EA$, *then*

$$(A+E)^D = AA^D\Psi + (I-AA^D)\left[\sum_{i=0}^{k-1}A^iE\Psi^{i+2}\right],$$

where $\Psi = (I+A^DE)^DA^D + (I-AA^D)\sum_{i=0}^{k-1}(E^D)^{i+1}(-A)^i$.
(2) *If* E *is nilpotent,* $EA^\pi = E$ *and* $A^\pi AE = A^\pi EA$, *then*

$$(A+E)^D = A^D + \sum_{i=0}^{k+l-2}(A^D)^{i+2}E(A+E)^i.$$

Proof We adopt the notations from Theorem 6.6.
(1) Let $Q = I - AA^D$ in Theorem 6.6 and apply $QA^D = 0$ to (6.19).
(2) Let $Q = AA^D$ in Theorem 6.6. Since $EA^\pi = E$, we obtain $EA^D = EA^\pi$ $A^D = 0$. Thus $(A^DE)^2 = A^DEA^DE = 0$ and $(I+A^DE)^DA^D = (I+A^DE)^{-1}A^D = A^D$. Note that E is nilpotent so that $\Psi = A^D$. Hence

$$E(A+E)^i(I-Q)[I-(A+E)\Psi] = E(A+E)^iA^\pi = E(A+E)^i, \quad for \ i \geq 0.$$

The result follows directly from (6.19).   □

Let A be an $n \times n$ complex matrix and $B = A + E$ be a perturbation of A. The classical Bauer-Fike theorem on eigenvalue perturbation gives a bound on the distance between an eigenvalue μ of B and the closest eigenvalue λ of A, which is required to be diagonalizable.

Let $A = X \Sigma X^{-1}$ be an eigendecomposition, where Σ is a diagonal matrix, and X is an eigenvector matrix. The Bauer-Fike theorem [23, Theorem IIIa] states that for any eigenvalue μ of B, there exists an eigenvalue λ of A such that $|\mu - \lambda| \le \kappa(X)\|E\|$, where $\kappa(X) = \|X\|\,\|X^{-1}\|$ is the condition number of X.

The relative perturbation version of the Bauer-Fike theorem [24, Corollary 2.2] below requires, in addition, that A be invertible. That is, if A is diagonalizable and invertible, then for any eigenvalue μ of B, there exists an eigenvalue λ of A such that

$$\frac{|\mu - \lambda|}{|\lambda|} \le \kappa(X)\|A^{-1}E\|. \tag{6.20}$$

Without the assumption of diagonalizability and invertibility of A, we refine the bound (6.20) under the condition that $AE = EA$.

Theorem 6.7 *Let $B = A + E \in \mathbb{C}^{n \times n}$ be such that A is not nilpotent and $AE = EA$. For any eigenvalue μ of B, there exists a nonzero eigenvalue λ of A such that*

$$\frac{|\mu - \lambda|}{|\lambda|} \le \rho(A^D E), \tag{6.21}$$

where $\rho(A^D E)$ is the spectral radius of $A^D E$.

Proof Assume that $AE = EA$ and that A is not nilpotent. Then for any nonzero eigenvalue λ of A, there exits a common eigenvector x [25, p.250] such that

$$Ax = \lambda x, \qquad (A + E)x = \mu x.$$

Therefore

$$A^D x = \frac{1}{\lambda} x, \qquad A^D E x = A^D(\mu x - Ax) = (\mu - \lambda)A^D x = \frac{\mu - \lambda}{\lambda} x,$$

whence

$$\frac{|\mu - \lambda|}{|\lambda|} \le \rho(A^D E).$$

$\square$

Recently, the perturbation of the Drazin inverse has been studied by several authors ([6, 9, 22, 26–33]). As one application of our results in Theorem 6.3, we can establish upper bounds for the relative error $\|B^D\|$ and $\|B^D - A^D\|/\|A^D\|$ under the assumption that $AE = EA$.

Theorem 6.8 *If $B = A + E \in \mathbb{C}^{n \times n}$, $AE = EA$ and $\max\{\|A^D E\|, \|A^\pi A E^D\|\} < 1$, then*

$$\|B^D\| \le \frac{\|A^D\|}{1 - \|A^D E\|} + \frac{\|A^\pi E^D\|}{1 - \|A^\pi A E^D\|}$$

and

$$\frac{\| B^D - A^D \|}{\| A^D \|} \leq \frac{\| A^D E \|}{1 - \| A^D E \|} + \frac{\| A \| \| E^D \|}{1 - \| A^\pi A E^D \|}.$$

Proof Note that the assumption·max$\{\| A^D E \|, \| A^\pi A E^D \|\} < 1$ implies invertibility of $I + A^D E$ and $I + A^\pi A E^D$. It follows directly from Theorem 6.3 that

$$\| B^D \| \leq \| (I + A^D E)^{-1} A^D \| + \| A^\pi E^D [I + A^\pi A E^D]^{-1} \|$$
$$\leq \frac{\| A^D \|}{1 - \| A^D E \|} + \frac{\| A^\pi E^D \|}{1 - \| A^\pi A E^D \|},$$

and

$$\| B^D - A^D \| \leq \| (I + A^D E)^{-1} A^D - A^D \| + \| A^\pi E^D [I + A^\pi A E^D]^{-1} \|$$
$$\leq \| (I + A^D E)^{-1} A^D E A^D \| + \frac{\| A^\pi \| \| E^D \|}{1 - \| A^\pi A E^D \|}$$
$$\leq \frac{\| A^D E \| \| A^D \|}{1 - \| A^D E \|} + \frac{\| A A^D \| \| E^D \|}{1 - \| A^\pi A E^D \|}$$
$$\leq \left(\frac{\| A^D E \|}{1 - \| A^D E \|} + \frac{\| A \| \| E^D \|}{1 - \| A^\pi A E^D \|} \right) \| A^D \|.$$

$\square$

Remark 6.1.2 For any non-zero eigenvalue μ of the spectral set $\sigma(A + E)$, we can estimate its lower bound: let $\mu \in \sigma(A + E)$. We have $1/\mu \in \sigma[(A + E)^D]$ and $|1/\mu| \leq \rho[(A + E)^D] \leq \| (A + E)^D \|$, i.e.,

$$|\mu| \geq 1/\| (A + E)^D \| \geq 1/\left[\frac{\| A^D \|}{1 - \| A^D E \|} + \frac{\| A^\pi E^D \|}{1 - \| A^\pi A E^D \|} \right].$$

Next we will apply Theorem 6.5 to obtain a perturbation bound in terms of A^D and $\mathscr{E}_l = B^l - A^l$ for some positive integer l.

Theorem 6.9 Let $B = A + E \in \mathbb{C}^{n \times n}$ with $\mathrm{Ind}(A) = k$ and $\mathrm{Ind}(B) = s$. Denote $\mathscr{E}_l = B^l - A^l$, where $l = \max\{k, s\}$. Assume that the conditions in Theorem 6.5 hold. Then

$$\frac{\| B^D - A^D \|}{\| A^D \|} \leq \| B^\pi - A^\pi \| = \| (A^D)^l \mathscr{E}_l \|. \tag{6.22}$$

Proof Since $l = \max\{k, s\}$, using the notations in the proof of Theorem 6.5, we have

$$\mathscr{E}_l = B^l - A^l = \begin{bmatrix} 0 & \sum_{i=0}^{l-1} C^{l-1-i} E_1 \left((N + E_2)^i \right) \\ 0 & 0 \end{bmatrix} = \begin{bmatrix} 0 & \sum_{i=0}^{l-1} C^{l-1-i} E_1 \left(\sum_{j=0}^{i} \binom{j}{i} N^j E_2^{i-j} \right) \\ 0 & 0 \end{bmatrix}.$$

Then

$$
A^D + (A^D)^{l+1}\mathcal{E}_l = \begin{bmatrix} C^{-1} & 0 \\ 0 & 0 \end{bmatrix} + \begin{bmatrix} (C^{-1})^{l+1} & 0 \\ 0 & 0 \end{bmatrix} \begin{bmatrix} 0 & \sum_{i=0}^{l-1} C^{l-1-i} E_1 \left(\sum_{j=0}^{i} \binom{j}{i} N^j E_2^{i-j} \right) \\ 0 & 0 \end{bmatrix}
$$

$$
= \begin{bmatrix} C^{-1} & \sum_{i=0}^{l-1} (C^{-1})^{i+2} E_1 \left(\sum_{j=0}^{i} \binom{j}{i} N^j E_2^{i-j} \right) \\ 0 & 0 \end{bmatrix} = B^D,
$$

and

$$
A A^D + (A^D)^l \mathcal{E}_l = \begin{bmatrix} I & 0 \\ 0 & 0 \end{bmatrix} + \begin{bmatrix} (C^{-1})^l & 0 \\ 0 & 0 \end{bmatrix} \begin{bmatrix} 0 & \sum_{i=0}^{l-1} C^{l-1-i} E_1 \left(\sum_{j=0}^{i} \binom{j}{i} N^j E_2^{i-j} \right) \\ 0 & 0 \end{bmatrix}
$$

$$
= \begin{bmatrix} I & \sum_{i=0}^{l-1} (C^{-1})^{i+1} E_1 \left(\sum_{j=0}^{i} \binom{j}{i} N^j E_2^{i-j} \right) \\ 0 & 0 \end{bmatrix} = B B^D.
$$

We then have

$$
\| B^D - A^D \| = \| (A^D)^{l+1} \mathcal{E}_l \| \le \| A^D \| \| (A^D)^l \mathcal{E}_l \|,
$$

and

$$
\| B^\pi - A^\pi \| = \| B B^D - A A^D \| = \| (A^D)^l \mathcal{E}_l \|.
$$

The proof is complete.   □

   Generalizations of the results of this section to linear operators on Banach spaces can be found in [9, 34–36] while their generalizations to Banach algebra elements can be found in [37] and some will also be given in the next section where the generalized Drazin inverse will be considered.

6.2   Additive Results for the Generalized Drazin Inverse in Banach Algebra

Let $\mathscr{A}$ be a complex Banach algebra with the unit 1. By $\mathscr{A}^{-1}$, $\mathscr{A}^{nil}$, $\mathscr{A}^{qnil}$ we denote the sets of all invertible, nilpotent and quasi-nilpotent elements in $\mathscr{A}$, respectively. Let us recall that the Drazin inverse of $a \in \mathscr{A}$ [1] is the (unique) element $x \in \mathscr{A}$ (denoted by a^D) which satisfies

$$
xax = x, \quad ax = xa, \quad a^{k+1}x = a^k, \tag{6.23}
$$

for some nonnegative integer k. The least such k is the index of a, denoted by $\mathrm{ind}(a)$. When $\mathrm{ind}(a) = 1$ then the Drazin inverse a^D is called the group inverse and it is denoted by $a^\#$. The conditions (6.23) are equivalent to

$$xax = x, \quad ax = xa, \quad a - a^2x \in \mathscr{A}^{\mathrm{nil}}. \tag{6.24}$$

The concept of the generalized Drazin inverse in a Banach algebra was introduced by Koliha [38]. The condition $a - a^2x \in \mathscr{A}^{\mathrm{nil}}$ was replaced by $a - a^2x \in \mathscr{A}^{\mathrm{qnil}}$. Hence, the generalized Drazin inverse of a is the (unique) element $x \in \mathscr{A}$ (written a^d) which satisfies

$$xax = x, \quad ax = xa, \quad a - a^2x \in \mathscr{A}^{\mathrm{qnil}}. \tag{6.25}$$

We mention that an alternative definition of the generalized Drazin inverse in a ring is also given in [39–41]. These two concepts of the generalized Drazin inverse are equivalent in the case when the ring is actually a complex Banach algebra with a unit. It is well known that a^d is unique whenever it exists [38]. The set $\mathscr{A}^d$ consists of all $a \in \mathscr{A}$ such that a^d exists. For many interesting properties of the Drazin inverse see [1, 38, 42].

This section is a continuation of the previous one with the difference that here we investigate additive properties of the generalized Drazin inverse in a Banach algebra and find explicit expressions for the generalized Drazin inverse of the sum $a + b$ under various conditions.

Hartwig et al. [10] for matrices and Djordjević and Wei [9] for operators used the condition $AB = 0$ to derive a formula for $(A + B)^d$. After that Castro and Koliha [43] relaxed this hypothesis by assuming the following complimentary condition symmetric in $a, b \in \mathscr{A}^d$,

$$a^\pi b = b, \quad ab^\pi = a, \quad b^\pi aba^\pi = 0 \tag{6.26}$$

thus generalizing the results from [9]. It is easy to see that $ab = 0$ implies (6.26), but the converse is not true (see [43, Example 3.1]).

In the first part of the section we will find some new conditions, which are not equivalent with the conditions from [43], allowing for the generalized Drazin inverse of $a + b$ to be expressed in terms of a, a^d, b, b^d. It is interesting to note that in some cases the same expression for $(a + b)^d$ are obtained as in [43]. In the rest of the section we will generalize some recent results from [43].

Let $a \in \mathscr{A}$ and let $p \in \mathscr{A}$ be an idempotent ($p = p^2$). Then we can write

$$a = pap + pa(1 - p) + (1 - p)ap + (1 - p)a(1 - p)$$

and use the notations

$$a_{11} = pap, \quad a_{12} = pa(1 - p), \quad a_{21} = (1 - p)ap, \quad a_{22} = (1 - p)a(1 - p).$$

Every idempotent $p \in \mathscr{A}$ induces a representation of an arbitrary element $a \in \mathscr{A}$ given by the following matrix

$$a = \begin{bmatrix} pap & pa(1-p) \\ (1-p)ap & (1-p)a(1-p) \end{bmatrix}_p = \begin{bmatrix} a_{11} & a_{12} \\ a_{21} & a_{22} \end{bmatrix}_p. \qquad (6.27)$$

Let a^{π} be the spectral idempotent of a corresponding to $\{0\}$. It is well known that $a \in \mathscr{A}^d$ can be represented in the matrix form:

$$a = \begin{bmatrix} a_{11} & 0 \\ 0 & a_{22} \end{bmatrix}_p,$$

relative to $p = aa^d = 1 - a^{\pi}$, where a_{11} is invertible in the algebra $p\mathscr{A}p$ and a_{22} is quasi-nilpotent in the algebra $(1-p)\mathscr{A}(1-p)$. Then the generalized Drazin inverse is given by

$$a^d = \begin{bmatrix} a_{11}^{-1} & 0 \\ 0 & 0 \end{bmatrix}_p.$$

The following result is proved in [4, 20] for matrices, extended in [44] for a bounded linear operator and in [43] for arbitrary elements in a Banach algebra.

Theorem 6.10 *Let $x, y \in \mathscr{A}$ and*

$$x = \begin{bmatrix} a & c \\ 0 & b \end{bmatrix}_p, \quad y = \begin{bmatrix} b & 0 \\ c & a \end{bmatrix}_{(1-p)}$$

relative to the idempotent $p \in \mathscr{A}$.

(1) *If $a \in (p\mathscr{A}p)^d$ and $b \in ((1-p)\mathscr{A}(1-p))^d$, then x and y are Drazin invertible and*

$$x^d = \begin{bmatrix} a^d & u \\ 0 & b^d \end{bmatrix}_p, \quad y^d = \begin{bmatrix} b^d & 0 \\ u & a^d \end{bmatrix}_{(1-p)} \qquad (6.28)$$

where $u = \sum_{n=0}^{\infty} (a^d)^{n+2} cb^n b^{\pi} + \sum_{n=0}^{\infty} a^{\pi} a^n c(b^d)^{n+2} - a^d cb^d$.

(2) *If $x \in \mathscr{A}^d$ and $a \in (p\mathscr{A}p)^d$, then $b \in ((1-p)\mathscr{A}(1-p))^d$ and x^d, y^d are given by (6.28).*

We will need the following auxiliary result.

Lemma 6.2 *Let $a, b \in \mathscr{A}^{qnil}$. If $ab = ba$ or $ab = 0$, then $a + b \in \mathscr{A}^{qnil}$.*

Proof If $ab = ba$, we have that

$$\rho(a + b) \leq \rho(a) + \rho(b),$$

which gives $a + b \in \mathscr{A}^{\text{qnil}}$. The case when $ab = 0$ follows from the equation

$$(\lambda - a)(\lambda - b) = \lambda(\lambda - (a + b))$$

$\square$

In view of the previous lemma, the first approach to the problem addressed in this section was to replace the condition $ab = 0$ used in [9, 10] by $ab = ba$. As expected, this alone was not enough to derive a formula for $(a + b)^{\text{d}}$. We will thus impose the following three conditions on $a, b \in \mathscr{A}^{\text{d}}$:

$$a = ab^\pi, \quad b^\pi ba^\pi = b^\pi b, \quad b^\pi a^\pi ba = b^\pi a^\pi ab. \tag{6.29}$$

Instead of the condition $ab = ba$ we are thus assuming the weaker condition $b^\pi a^\pi ba = b^\pi a^\pi ab$. Notice that

$$a = ab^\pi \Leftrightarrow ab^{\text{d}} = 0 \Leftrightarrow \mathscr{A}a \subseteq \mathscr{A}b^\pi, \tag{6.30}$$

$$b^\pi ba^\pi = b^\pi b \Leftrightarrow b^\pi ba^{\text{d}} = 0 \Leftrightarrow \mathscr{A}b^\pi b \subseteq \mathscr{A}a^\pi, \tag{6.31}$$

$$b^\pi a^\pi ba = b^\pi a^\pi ab \Leftrightarrow (ba - ab)\mathscr{A} \subseteq (b^\pi a^\pi)^\circ, \tag{6.32}$$

where for $u \in \mathscr{A}, u^\circ = \{x \in \mathscr{A} : ux = 0\}$.

For matrices and bounded linear operators on a Banach space the conditions (6.30)–(6.32) are equivalent to

$$\mathscr{N}(b^\pi) \subseteq \mathscr{N}(a), \quad \mathscr{N}(a^\pi) \subseteq \mathscr{N}(b^\pi b), \quad \mathscr{R}(ba - ab) \subseteq \mathscr{N}(b^\pi a^\pi).$$

Remark that, unlike the conditions (3.1) from [43], the conditions (6.29) are not symmetric in a, b so our expression for $(a + b)^{\text{d}}$ will not be symmetric in a, b.

In the next theorem, under the assumption that (6.29) holds, we can give an expression for $(a + b)^{\text{d}}$ as follows.

Theorem 6.11 *Let* $a, b \in \mathscr{A}^d$ *be such that (6.29) is satisfied. Then* $a + b \in \mathscr{A}^d$ *and*

$$(a + b)^d = (b^d + \sum_{n=0}^{\infty} (b^d)^{n+2} a(a + b)^n) a^\pi \qquad (6.33)$$

$$- \sum_{n=0}^{\infty} \sum_{k=0}^{\infty} (b^d)^{n+2} a(a + b)^n (a^d)^{k+2} b(a + b)^{k+1}$$

$$+ \sum_{n=0}^{\infty} (b^d)^{n+2} a(a + b)^n a^d b - \sum_{n=0}^{\infty} b^d a(a^d)^{n+2} b(a + b)^n$$

Before proving Theorem 6.11, we first have to prove the special case of it given below.

Theorem 6.12 *Let* $a \in \mathscr{A}^{qnil}$, $b \in \mathscr{A}^d$ *satisfy* $b^\pi ab = b^\pi ba$ *and* $a = ab^\pi$. *Then (6.29) is satisfied,* $a + b \in \mathscr{A}^d$ *and*

$$(a + b)^d = b^d + \sum_{n=0}^{\infty} (b^d)^{n+2} a(a + b)^n. \qquad (6.34)$$

Proof First, suppose that $b \in \mathscr{A}^{qnil}$. Then $b^\pi = 1$ and from $b^\pi ab = b^\pi ba$ we obtain $ab = ba$. Using Lemma 6.2, $a + b \in \mathscr{A}^{qnil}$ and (6.28) holds. Now, we assume that b is not quasi-nilpotent and consider the matrix representations of a and b relative to $p = 1 - b^\pi$. We have

$$b = \begin{bmatrix} b_1 & 0 \\ 0 & b_2 \end{bmatrix}_p, \quad a = \begin{bmatrix} a_{11} & a_{12} \\ a_{21} & a_{22} \end{bmatrix}_p,$$

where $b_1 \in (p\mathscr{A}p)^{-1}$ and $b_2 \in ((1 - p)\mathscr{A}(1 - p))^{qnil} \subset \mathscr{A}^{qnil}$. From $a = ab^\pi$, it follows that $a_{11} = 0$ and $a_{21} = 0$. We denote $a_1 = a_{12}$ and $a_2 = a_{22}$. Hence

$$a + b = \begin{bmatrix} b_1 & a_1 \\ 0 & a_2 + b_2 \end{bmatrix}_p.$$

The condition $b^\pi ab = b^\pi ba$ implies that $a_2 b_2 = b_2 a_2$. Hence, using Lemma 6.2, we get $a_2 + b_2 \in ((1 - p)\mathscr{A}(1 - p))^{qnil}$. Now, by Theorem 6.10, we obtain $a + b \in \mathscr{A}^d$ and

$$(a + b)^d = \begin{bmatrix} b_1^{-1} \sum_{n=0}^{\infty} b_1^{-(n+2)} a_1 (a_2 + b_2)^n & \\ 0 & 0 \end{bmatrix}_p$$

$$= b^d + \sum_{n=0}^{\infty} (b^d)^{n+2} a(a + b)^n.$$

$\square$

Let us observe that the expression for $(a+b)^d$ in (6.28) and that in (3.6) of Theorem 3.3 in [43] are exactly the same. If we assume that $ab = ba$ instead of $b^\pi ab = b^\pi ba$, we get a much simpler expression for $(a+b)^d$.

Corollary 6.4 *Suppose $a \in \mathscr{A}^{qnil}$, $b \in \mathscr{A}^d$ satisfy $ab = ba$ and $a = ab^\pi$. Then $a + b \in \mathscr{A}^d$ and*

$$(a+b)^d = b^d.$$

Proof From $a = ab^\pi$, as we mentioned before, it follows that $ab^d = 0$. Because the Drazin inverse b^d is a double commutant of a, we have

$$(b^d)^{n+2}a(a+b)^n = a(b^d)^{n+2}(a+b)^n = 0.$$

$\square$

Proof of Theorem 6.11: If b is quasi-nilpotent we can apply Theorem 6.12. Hence, we assume that b is neither invertible nor quasi-nilpotent and consider the matrix representations of a and b relative to $p = 1 - b^\pi$:

$$b = \begin{bmatrix} b_1 & 0 \\ 0 & b_2 \end{bmatrix}_p, \quad a = \begin{bmatrix} a_{11} & a_{12} \\ a_{21} & a_{22} \end{bmatrix}_p,$$

where $b_1 \in (p\mathscr{A}p)^{-1}$ and $b_2 \in ((1-p)\mathscr{A}(1-p))^{qnil}$. As in the proof of Theorem 6.12, from $a = ab^\pi$ it follows that

$$a = \begin{bmatrix} 0 & a_1 \\ 0 & a_2 \end{bmatrix}_p, \quad a+b = \begin{bmatrix} b_1 & a_1 \\ 0 & a_2 + b_2 \end{bmatrix}_p.$$

From the conditions $b^\pi a^\pi ba = b^\pi a^\pi ab$ and $b^\pi ba^\pi = b^\pi b$, we obtain $a_2^\pi b_2 a_2 = a_2^\pi a_2 b_2$ and $b_2 = b_2 a_2^\pi$. Now, from Theorem 6.12 it follows that $(a_2 + b_2) \in ((1-p)\mathscr{A}(1-p))^d$ and

$$(a_2 + b_2)^d = a_2^d + \sum_{n=0}^{\infty} (a_2^d)^{n+2} b_2 (a_2 + b_2)^n. \tag{6.35}$$

By Theorem 6.10, we get

$$(a+b)^d = \begin{bmatrix} b_1^{-1} & u \\ 0 & (a_2 + b_2)^d \end{bmatrix}_p,$$

where $u = \sum_{n=0}^{\infty} b_1^{-(n+2)} a_1 (a_2 + b_2)^n (a_2 + b_2)^\pi - b_1^{-1} a_1 (a_2 + b_2)^d$ and b_1^{-1} is the inverse of b_1 in the algebra $p\mathscr{A}p$. Using (6.35), we have

$$u = \sum_{n=0}^{\infty} b_1^{-(n+2)} a_1 (a_2 + b_2)^n = a_2^{\pi} - \sum_{n=0}^{\infty} b_1^{-(n+2)} a_1 (a_2 + b_2)^n a_2^{d} b_2$$

$$\sum_{n=0}^{\infty} \sum_{k=0}^{\infty} (b_1)^{-(n+2)} a_1 (a_2 + b_2)^n (a_2^{d})^{k+2} b_2 (a_2 + b_2)^{k+1} - b_1^{-1} a_1 a_2^{d}$$

$$- \sum_{n=0}^{\infty} b_1^{-1} a_1 (a_2^{d})^{n+2} b_2 (a_2 + b_2)^n.$$

By a straightforward manipulation, (6.33) follows.   □

Corollary 6.5 *Suppose* $a, b \in \mathscr{A}^{d}$ *are such that* $ab = ba$, $a = ab^{\pi}$ *and* $b^{\pi} = ba^{\pi} = b^{\pi} b$. *Then* $a + b \in \mathscr{A}^{d}$ *and*

$$(a + b)^{d} = b^{d}.$$

If a is invertible and b is group invertible, then conditions (6.31) and (6.32) are satisfied, so we only have to assume $a = ab^{\pi}$. In the remaining case when b is invertible we get $a = 0$.

It is interesting to remark that conditions (6.26) and (6.29) are independent, i.e., neither of them implies the other, but in some cases the same expressions for $(a + b)^{d}$ are obtained.

If we consider the algebra $\mathscr{A}$ of all complex 3×3 matrices and $a, b \in \mathscr{A}$ which are given in the Example 3.1 [43], we can see that condition (6.26) is satisfied, whereas condition (6.29) fails. In the following example we have the opposite case. We construct a, b in the algebra $\mathscr{A}$ of all complex 3×3 matrices such that (6.29) is satisfied but (6.26) is not. If we assume that $ab = ba$ in Theorem 6.11 the expression for $(a + b)^{d}$ will be exactly the same as that in [43, Theorem 3.5] (which is Corollary 6.7 there).

Example 6.1 Let

$$a = \begin{pmatrix} 1 & 0 & 0 \\ 0 & 0 & 0 \\ 0 & 0 & 0 \end{pmatrix}, \quad b = \begin{pmatrix} 0 & 1 & 0 \\ 0 & 0 & 0 \\ 0 & 0 & 0 \end{pmatrix}.$$

Then

$$a^{\pi} = \begin{pmatrix} 0 & 0 & 0 \\ 0 & 1 & 0 \\ 0 & 0 & 1 \end{pmatrix}$$

and $b^{\pi} = 1$. We can see that $a = ab^{\pi}$, $a^{\pi} ab = a^{\pi} = ba$ and $ba^{\pi} = b$, i.e., (6.29) holds. Also, $a^{\pi} b = 0 \neq b$, so (6.26) is not satisfied.

In the rest of the section, we present a generalization of the results from [43]. We use some weaker conditions than those in [43]. For example in the next theorem, which generalizes [43, Theorem 3.3], we assume that $e = (1 - b^{\pi})(a +$

$b)(1 - b^\pi) \in \mathscr{A}^d$ instead of $ab^\pi = a$. If $ab^\pi = a$, then $e = (1 - b^\pi)b = \begin{bmatrix} b_1 & 0 \\ 0 & 0 \end{bmatrix}_p$,

for $p = 1 - b^\pi$ and $e^d = b^d$.

Theorem 6.13 *Let $b \in \mathscr{A}^d$, $a \in \mathscr{A}^{\text{qnil}}$ be such that*

$$e = (1 - b^\pi)(a + b)(1 - b^\pi) \in \mathscr{A}^d, \quad b^\pi ab = 0.$$

Then $a + b \in \mathscr{A}^d$ and

$$(a + b)^d = e^d + \sum_{n=0}^{\infty} (e^d)^{n+2} ab^\pi (a + b)^n.$$

Proof The case when $b \in \mathscr{A}^{\text{qnil}}$ follows from Lemma 6.2. Hence, we assume that b is not quasi-nilpotent. Then

$$b = \begin{bmatrix} b_1 & 0 \\ 0 & b_2 \end{bmatrix}_p, \quad a = \begin{bmatrix} a_{11} & a_{12} \\ a_{21} & a_{22} \end{bmatrix}_p,$$

where $p = 1 - b^\pi$. From $b^\pi ab = 0$ we have $b^\pi a(1 - b^\pi) = 0$, i.e., $a_{21} = 0$. Put $a_1 = a_{11}, a_{22} = a_2$ and $a_{12} = a_3$. Then,

$$a + b = \begin{bmatrix} a_1 + b_1 & a_3 \\ 0 & a_2 + b_2 \end{bmatrix}_p.$$

Also, $b^\pi ab = 0$ implies that $a_2 b_2 = 0$, so $a_2 + b_2 \in ((1 - p)\mathscr{A}(1 - p))^{\text{qnil}}$, according to Lemma 6.2. Applying Theorem 6.10, we obtain

$$(a + b)^d = \begin{bmatrix} (a_1 + b_1)^d & u \\ 0 & 0 \end{bmatrix}_p,$$

where $u = \sum_{n=0}^{\infty} ((a_1 + b_1)^d)^{n+2} a_3 (a_2 + b_2)^n$.

By direct computation, we verify that

$$(a + b)^d = e^d + \sum_{n=0}^{\infty} (e^d)^{n+2} ab^\pi (a + b)^n.$$

$\square$

Now, as a corollary we obtain Theorem 3.3 from [43].

Corollary 6.6 *Let $b \in \mathscr{A}^d$, $a \in \mathscr{A}^{\text{qnil}}$ and $ab^\pi = a$, $b^\pi ab = 0$. Then $a + b \in \mathscr{A}^d$ and*

$$(a+b)^{\mathsf{d}} = b^{\mathsf{d}} + \sum_{n=0}^{\infty} (b^{\mathsf{d}})^{n+2} a(a+b)^n.$$

The next result is a generalization of [43, Theorem 3.5]. For simplicity, we use the following notation:

$$
\begin{aligned}
e &= (1 - b^{\pi})(a+b)(1 - b^{\pi}) \in \mathscr{A}^{\mathsf{d}}, \\
f &= (1 - a^{\pi})(a+b)(1 - a^{\pi}), \\
\mathscr{A}_1 &= (1 - a^{\pi})\mathscr{A}(1 - a^{\pi}), \\
\mathscr{A}_2 &= (1 - b^{\pi})\mathscr{A}(1 - b^{\pi}),
\end{aligned}
$$

for given $a, b \in A^{\mathsf{d}}$.

Theorem 6.14 *Let $a, b \in \mathscr{A}^{\mathsf{d}}$ be such that $(1 - a^{\pi})b(1 - a^{\pi}) \in \mathscr{A}^{\mathsf{d}}$, $f \in \mathscr{A}_1^{-1}$ and $e \in \mathscr{A}_2^{\mathsf{d}}$. If*

$$(1 - a^{\pi})ba^{\pi} = 0, \quad b^{\pi}aba^{\pi} = 0, \quad a^{\pi} = a(1 - b^{\pi})a^{\pi} = 0,$$

then $a + b \in \mathscr{A}^{\mathsf{d}}$ and

$$
\begin{aligned}
(a+b)^{\mathsf{d}} = {}& \Big(b^{\mathsf{d}} + \sum_{n=0}^{\infty} (b^{\mathsf{d}})^{n+2} a(a+b)^n\Big)a^{\pi} + \sum_{n=0}^{\infty} b^{\pi}(a+b)^n a^{\pi} b(f)_{\mathscr{A}_1}^{-(n+2)} \\
& - \sum_{n=0}^{\infty}\sum_{k=0}^{\infty} (b^{\mathsf{d}})^{k+1} a(a+b)^{n+k} a^{\pi} b(f)_{\mathscr{A}_1}^{-(n+2)} - b^{\mathsf{d}} a^{\pi} b(f)_{\mathscr{A}_1}^{-1} \\
& - \sum_{n=0}^{\infty} (b^{\mathsf{d}})^{n+2} a(a+b)^n a^{\pi} b(f)_{\mathscr{A}_1}^{-1} + (f)_{\mathscr{A}_1}^{-1},
\end{aligned}
$$

where by $(f)_{\mathscr{A}_1}^{-1}$ we denote the inverse of f in $\mathscr{A}_1$.

Proof Obviously, if a is invertible, then the statement of the theorem holds. If a is quasi-nilpotent, then the result follows from Theorem 6.13. Hence, we assume that a is neither invertible nor quasi-nilpotent. As in the proof of Theorem 6.11, we have

$$
a = \begin{bmatrix} a_1 & 0 \\ 0 & a_2 \end{bmatrix}_p, \quad b = \begin{bmatrix} b_{11} & b_{12} \\ b_{21} & b_{22} \end{bmatrix}_p,
$$

where $p = 1 - a^{\pi}$, $a_1 \in (p\mathscr{A}p)^{-1}$ and $a_2 \in ((1-p)\mathscr{A}(1-p))^{\mathsf{qnil}}$. From $(1 - a^{\pi})ba^{\pi} = 0$, we have that $b_{12} = 0$. Let $b_1 = b_{11}$, $b_{22} = b_2$ and $b_{21} = b_3$. Then,

$$
a + b = \begin{bmatrix} a_1 + b_1 & 0 \\ b_3 & a_2 + b_2 \end{bmatrix}_p.
$$

The condition $a^\pi b^\pi aba^\pi = 0$ expressed in the matrix form yields

$$a^\pi b^\pi aba^\pi = \begin{bmatrix} 0 & 0 \\ 0 & b_2^\pi \end{bmatrix} \begin{bmatrix} a_1 & 0 \\ 0 & a_2 \end{bmatrix} \begin{bmatrix} 0 & 0 \\ 0 & b_2 \end{bmatrix}$$
$$= \begin{bmatrix} 0 & 0 \\ 0 & b_2^\pi a_2 b_2 \end{bmatrix} = \begin{bmatrix} 0 & 0 \\ 0 & 0 \end{bmatrix}.$$

Similarly, $a^\pi a(1 - b^\pi) = 0$ implies that $a_2 b_2^\pi = a_2$. From Corollary 6.6, we get $a_2 + b_2 \in \mathscr{A}^d$ and

$$(a_2 + b_2)^d = b_2^d + \sum_{n=0}^{\infty} (b_2^d)^{n+2} a_2 (a_2 + b_2)^n.$$

Using Theorem 6.10, we obtain $a + b \in \mathscr{A}^d$ and

$$(a + b)^d = \begin{bmatrix} (a_1 + b_1)^d & 0 \\ u & (a_2 + b_2)^d \end{bmatrix}_p,$$

where

$$u = \sum_{n=0}^{\infty} b_2^\pi (a_2 + b_2)^n b_3(f)_{\mathscr{A}_1}^{-(n+2)}$$
$$- \sum_{n=0}^{\infty} \sum_{k=0}^{\infty} (b_2^d)^{k+1} a_2(a_2 + b_2)^{n+k} b_3(f)_{\mathscr{A}_1}^{-(n+2)} - b_2^d b_3(f)_{\mathscr{A}_1}^{-1}$$
$$- \sum_{n=0}^{\infty} (b_2^d)^{n+2} a_2(a_2 + b_2)^n b_3(f)_{\mathscr{A}_1}^{-1}.$$

By straightforward computation, the desired result follows.   □

Corollary 6.7 *Suppose* $a, b \in \mathscr{A}^d$ *satisfy condition (6.26). Then* $a + b \in \mathscr{A}^d$ *and*

$$(a + b)^d = \left(b^d + \sum_{n=0}^{\infty} (b^d)^{n+2} a(a + b)^n\right) a^\pi + \sum_{n=0}^{\infty} b^\pi (a + b)^n b(a^d)^{(n+2)}$$
$$- \sum_{n=0}^{\infty} \sum_{k=0}^{\infty} (b^d)^{k+1} a(a + b)^{n+k} b(a^d)^{(n+2)} + b^\pi a^d$$
$$- \sum_{n=0}^{\infty} (b^d)^{n+2} a(a + b)^n b a^d$$

Proof We have that $f = (1 - a^\pi)a$, so $(f)_{\mathscr{A}_1}^{-1} = a^d$.   □

Next we generalize the results from [45] to the Banach algebra case.

Theorem 6.15 *Let $a, b \in \mathscr{A}^d$ and $ab = ba$. Then $a + b \in \mathscr{A}^d$ if and only if $1 + a^d b \in \mathscr{A}^d$. In this case, we have*

$$(a + b)^d = a^d (1 + a^d b)^d b b^d + (1 - b b^d) \left[\sum_{n=0}^{\infty} (-b)^n (a^d)^n \right] a^d$$

$$+ b^d \left[\sum_{n=0}^{\infty} (b^d)^n (-a)^n \right] (1 - a a^d),$$

and

$$(a + b)(a + b)^d = (a a^d + b a^d)(1 + a^d b)^d b b^d + (1 - b b^d) a a^d$$
$$+ b b^d (1 - a a^d).$$

Moreover, if $\|b\| \|a^d\| < 1$ and $\|a\| \|b^d\| < 1$, then we have

$$\|(a + b)^d - a^d\| \leq \|b b^d\| \|a^d\| \left[\|(1 + a^d b)^d\| + 1 \right]$$

$$+ \|1 - b b^d\| \left[\sum_{n=1}^{\infty} \|(-b)^n (a^d)^n\| \right] \|a^d\|$$

$$+ \|b^d\| \left[\sum_{n=0}^{\infty} \|(b^d)^n (-a)^n\| \right] \|1 - a a^d\|,$$

and

$$\|(a + b)(a + b)^d - a a^d\| \leq \left[\|a a^d + b a^d\| \|(1 + a^d b)^d\| + \|1 - 2 a a^d\| \right] \|b b^d\|.$$

Proof Since a is generalized Drazin invertible, and

$$a = \begin{bmatrix} a_{11} & 0 \\ 0 & a_{22} \end{bmatrix}_p,$$

relative to $p = 1 - a^\pi$, where a_{11} is invertible in the algebra $p \mathscr{A} p$ and a_{22} is a quasi-nilpotent element of the algebra $(1 - p)\mathscr{A}(1 - p)$. Let $b = \begin{bmatrix} b_{11} & b_{12} \\ b_{21} & b_{22} \end{bmatrix}_p$.

From $ab = ba$, we have $b_{12} = (a_{11})^{-1}_{p\mathscr{A}p} b_{12} a_{22}$ which implies that $b_{12} = (a_{11})^{-n}_{p\mathscr{A}p} b_{12} a_{22}^n$, for arbitrary $n \in \mathbb{N}$. Since a_{22} is a quasi-nilpotent, we obtain $b_{12} = 0$. Similarly, from $ab = ba$ it follows that $b_{21} = a_{22} b_{21} (a_{11})^{-1}_{p\mathscr{A}p}$, i.e., $b_{21} = 0$. Also, $a_{11} b_{11} = b_{11} a_{11}$ and $a_{22} b_{22} = b_{22} a_{22}$.

Since, $b \in \mathscr{A}^d$ and $\sigma(b) = \sigma(b_1)_{p\mathscr{A}p} \cup \sigma(b_2)_{(1-p)\mathscr{A}(1-p)}$, using Theorem 4.2 from [38], we deduce $b_1 \in p\mathscr{A}p$ and $b_2 \in (1-p)\mathscr{A}(1-p)$, so $b_{11}, b_{22} \in \mathscr{A}^d$ and we can represent b_{11} and b_{22} as

$$b_{11} = \begin{bmatrix} b'_{11} & 0 \\ 0 & b'_{22} \end{bmatrix}_{p_1}, \quad b_{22} = \begin{bmatrix} b''_{11} & 0 \\ 0 & b''_{22} \end{bmatrix}_{p_2},$$

where $p_1 = 1 - b_{11}^{\pi}$ and $p_2 = 1 - b_{22}^{\pi}$, b'_{11}, b''_{11} are invertible in the algebras $p_1\mathscr{A}p_1$ and $p_2\mathscr{A}p_2$ respectively, and b'_{22}, b''_{22} are quasi-nilpotent. Since b_{11} commutes with an invertible a_{11} and b_{22} with a quasi-nilpotent a_{22}, we prove as before that

$$a_{11} = \begin{bmatrix} a'_{11} & 0 \\ 0 & a'_{22} \end{bmatrix}_{p_1}, \quad a_{22} = \begin{bmatrix} a''_{11} & 0 \\ 0 & a''_{22} \end{bmatrix}_{p_2}.$$

Since $p_1 p = pp_1 = p_1$, from the fact that a_{11} is invertible in the sub-algebra $p\mathscr{A}p$, we prove that a'_{11} and a'_{22} are invertible in the algebras $p_1\mathscr{A}p_1$ and $(p - p_1)\mathscr{A}(p - p_1)$, respectively. Also, a''_{11} and a''_{22} are quasi-nilpotent, thus a''_{ii} commutes with b'_{ii} and a''_{ii} with b''_{ii}, for $i = 1, 2$.

Since a'_{22} is invertible and b'_{22} is quasi-nilpotent and they commute, we have that $(a'_{22})^{-1}_{(1-p_1)\mathscr{A}(1-p_1)}b'_{22}$ is quasi-nilpotent, so $(1 - p_1) + (a'_{22})^{-1}_{(1-p_1)\mathscr{A}(1-p_1)}b'_{22}$ is invertible in $(1 - p_1)\mathscr{A}(1 - p_1)$ and $a'_{22} + b'_{22} \in \mathscr{A}^d$.

Similarly, we conclude that $a''_{11} + b''_{11} \in \mathscr{A}^d$. Also, $a''_{22} + b''_{22}$ is generalized Drazin invertible.

Now, we obtain

$$a + b = a'_{11} + b'_{11} + a'_{22} + b'_{22} + a''_{11} + b''_{11} + a''_{22} + b''_{22}.$$

Since, $a'_{11} + b'_{11} \in p_1\mathscr{A}p_1$ and $b'_{22} + a''_{11} + b''_{11} + a''_{22} + b''_{22} \in (1 - p_1)\mathscr{A}(1 - p_1)$ we have

$$a + b \in \mathscr{A}^d \Leftrightarrow \left(a'_{11} + b'_{11} \in \mathscr{A}^d, \quad a'_{22} + b'_{22} + a''_{11} + b''_{11} + a''_{22} + b''_{22} \in \mathscr{A}^d \right).$$

Next, we inspect generalized Drazin invertibility of $y = a'_{22} + b'_{22} + a''_{11} + b''_{11} + a''_{22} + b''_{22}$. From $p_2 y p_2 = a''_{11} + b''_{11}$ and $(1 - p_2)y(1 - p_2)y = a'_{22} + b'_{22} + a''_{22} + b''_{22}$, we conclude

$$y \in \mathscr{A}^d \Leftrightarrow \left(a''_{11} + b''_{11} \in \mathscr{A}^d \text{ and } a'_{22} + b'_{22} + a''_{22} + b''_{22} \in \mathscr{A}^d \right).$$

Previously, we showed that $a''_{11} + b''_{11} \in \mathscr{A}^d$, so $y \in \mathscr{A}^d$ if and only if $z = a'_{22} + b'_{22} + a''_{22} + b''_{22} \in \mathscr{A}^d$. Notice that $z = pzp + (1 - p)z(1 - p)$, where $pzp = a'_{22} + b'_{22} \in \mathscr{A}^d$ and $(1 - p)z(1 - p) = a''_{22} + b''_{22} \in \mathscr{A}^d$, so $z \in \mathscr{A}^d$. Hence, $y \in \mathscr{A}^d$ and we obtain $a + b \in \mathscr{A}^d$ if and only if $a'_{11} + b'_{11} \in \mathscr{A}^d$.

Now,

$$(a'_{11} + b'_{11})^{\mathsf{d}} = a'_{11}(p_1 + (a'_{11})^{-1}_{p_1 \mathscr{A} p_1} b'_{11})^{\mathsf{d}} = p_1 p a^{\mathsf{d}}(1 + a^{\mathsf{d}}b)^{\mathsf{d}} bb^{\mathsf{d}} p p_1.$$

From the first equation, we obtain

$$(a + b)^{\mathsf{d}} - a^d = a^{\mathsf{d}}(1 + a^{\mathsf{d}}b)^{\mathsf{d}} bb^{\mathsf{d}} + (1 - bb^{\mathsf{d}}) \left[\sum_{n=0}^{\infty} (-b)^n (a^{\mathsf{d}})^n \right] a^{\mathsf{d}}$$

$$+ b^{\mathsf{d}} \left[\sum_{n=0}^{\infty} (b^{\mathsf{d}})^n (-a)^n \right] (1 - aa^{\mathsf{d}}) - a^d$$

$$= a^{\mathsf{d}}(1 + a^{\mathsf{d}}b)^{\mathsf{d}} bb^{\mathsf{d}} - bb^{\mathsf{d}} a^{\mathsf{d}} + (1 - bb^{\mathsf{d}}) \left[\sum_{n=1}^{\infty} (-b)^n (a^{\mathsf{d}})^n \right] a^{\mathsf{d}}$$

$$+ b^{\mathsf{d}} \left[\sum_{n=0}^{\infty} (b^{\mathsf{d}})^n (-a)^n \right] (1 - aa^{\mathsf{d}}).$$

Consequently, we have the estimates

$$\|(a + b)^{\mathsf{d}} - a^d\| \leq \|bb^{\mathsf{d}}\| \|a^{\mathsf{d}}\| \left[\|(1 + a^{\mathsf{d}}b)^{\mathsf{d}}\| + 1 \right]$$

$$+ \|(1 - bb^{\mathsf{d}})\| \left[\sum_{n=1}^{\infty} \|(-b)^n (a^{\mathsf{d}})^n\| \right] \|a^{\mathsf{d}}\|$$

$$+ \|b^{\mathsf{d}}\| \left[\sum_{n=0}^{\infty} \|(b^{\mathsf{d}})^n (-a)^n\| \right] \|(1 - aa^{\mathsf{d}})\|,$$

and

$$\|(a + b)(a + b)^{\mathsf{d}} - aa^d\| = \|(aa^{\mathsf{d}} + ba^{\mathsf{d}})(1 + a^{\mathsf{d}}b)^{\mathsf{d}} bb^{\mathsf{d}} - bb^{\mathsf{d}} aa^{\mathsf{d}} + bb^{\mathsf{d}}(1 - aa^{\mathsf{d}})\|$$

$$\leq \left[\|aa^{\mathsf{d}} + ba^{\mathsf{d}}\| \|(1 + a^{\mathsf{d}}b)^{\mathsf{d}}\| + \|1 - 2aa^{\mathsf{d}}\| \right] \|bb^{\mathsf{d}}\|.$$

$\square$

Corollary 6.8 Let $a, b \in \mathscr{A}^{\mathsf{d}}$ be such that $ab = ba$ and $1 + a^{\mathsf{d}}b \in \mathscr{A}^{\mathsf{d}}$.

(1) If b is quasi-nilpotent, then

$$(a + b)^{\mathsf{d}} = \sum_{n=0}^{\infty} (a^{\mathsf{d}})^{n+1} (-b)^n = (1 + a^{\mathsf{d}}b)^{-1} a^{\mathsf{d}}.$$

(2) If $b^k = 0$, then $(a + b)^{\mathsf{d}} = \sum_{n=0}^{k-1} (a^{\mathsf{d}})^{n+1} (-b)^n = (1 + a^{\mathsf{d}}b)^{-1} a^{\mathsf{d}}.$

(3) *If $b^k = b$ $(k \geq 3)$, then $b^{\mathrm{d}} = b^{k-2}$ and*

$$(a+b)^{\mathrm{d}} = a^{\mathrm{d}}(1 + a^{\mathrm{d}}b)^{\mathrm{d}}b^{k-1} + (1 - b^{k-1})a^{\mathrm{d}} + b^{k-2}\left[\sum_{n=0}^{\infty}(b^{\mathrm{d}})^n(-a)^n\right](1 - aa^{\mathrm{d}})$$
$$= a^{\mathrm{d}}(1 + a^{\mathrm{d}}b)^{\mathrm{d}}b^{k-1} + (1 - b^{k-1})a^{\mathrm{d}} + b^{k-2}(1 + ab^{k-2})^{\mathrm{d}}(1 - aa^{\mathrm{d}}).$$

(4) *If $b^2 = b$, then $b^{\mathrm{d}} = b$ and*

$$(a+b)^{\mathrm{d}} = a^{\mathrm{d}}(1 + a^{\mathrm{d}}b)^{\mathrm{d}}b + (1 - b)a^{\mathrm{d}} + b\left[\sum_{n=0}^{\infty}(-a)^n\right](1 - aa^{\mathrm{d}})$$
$$= a^{\mathrm{d}}(1 + a^{\mathrm{d}}b)^{\mathrm{d}}b + (1 - b)a^{\mathrm{d}} + b(1 + a)^{\mathrm{d}}(1 - aa^{\mathrm{d}}).$$

(5) *If $a^2 = a$ and $b^2 = b$, then $1 + ab$ is invertible and $a(1 + ab)^{-1}b = \frac{1}{2}ab$. In this case,*

$$(a+b)^{\mathrm{d}} = a(1 + ab)^{-1}b + b(1 - a) + (1 - b)a$$
$$= a + b - \frac{3}{2}ab.$$

Theorem 6.16 *Let $a, b \in \mathscr{A}^{\mathrm{d}}$ be such that $\|a^{\mathrm{d}}b\| < 1$, $a^{\pi}ba^{\pi} = a^{\pi}b$ and $a^{\pi}ab = a^{\pi}ba$. If $a^{\pi}b \in \mathscr{A}^{\mathrm{d}}$, then $a + b \in \mathscr{A}^{\mathrm{d}}$. In this case,*

$$(a+b)^{\mathrm{d}} = (1 + a^{\mathrm{d}}b)^{-1}a^{\mathrm{d}} + (1 + a^{\mathrm{d}}b)^{-1}(1 - aa^{\mathrm{d}})\sum_{n=0}^{\infty}(b^{\mathrm{d}})^{n+1}(-a)^n$$

$$+ \left[\sum_{n=0}^{\infty}\left((1 + a^{\mathrm{d}}b)^{-1}a^{\mathrm{d}}\right)^{n+2}b(1 - aa^{\mathrm{d}})(a+b)^n\right](1 - aa^{\mathrm{d}})$$

$$\times\left[1 - (a+b)(1 - aa^{\mathrm{d}})\sum_{n=0}^{\infty}(b^{\mathrm{d}})^{n+1}(-a)^n\right].$$

Moreover, if $\|a\|\|b^{\mathrm{d}}\| < 1$, $\|b\|\|a^{\mathrm{d}}\| < 1$ and $\frac{\|a^{\mathrm{d}}\|\|a^{\mathrm{d}}b\|}{1 - \|a^{\mathrm{d}}b\|}\|a + b\| < 1$, then

$$\|(a+b)^{\mathrm{d}} - a^{\mathrm{d}}\| \leq \frac{\|a^{\mathrm{d}}\|\|a^{\mathrm{d}}b\|}{1 - \|a^{\mathrm{d}}b\|} + \frac{\|1 - aa^{\mathrm{d}}\|}{1 - \|a^{\mathrm{d}}b\|}\sum_{n=0}^{\infty}\|b^{\mathrm{d}}\|^{n+1}\|a\|^n$$

$$+ \left[\sum_{n=0}^{\infty}\left(\frac{\|a^{\mathrm{d}}\|\|a^{\mathrm{d}}b\|}{1 - \|a^{\mathrm{d}}b\|}\right)^{n+2}\|b\|\|a + b\|^n\right]\|1 - aa^{\mathrm{d}}\|^2$$

$$+ \|1 - aa^{\mathrm{d}}\|^3\left[\sum_{n=0}^{\infty}\left(\frac{\|a^{\mathrm{d}}\|\|a^{\mathrm{d}}b\|}{1 - \|a^{\mathrm{d}}b\|}\right)^{n+2}\|b\|\|a + b\|^{n+1}\right]$$

$$\times\left[\sum_{n=0}^{\infty}\|b^{\mathrm{d}}\|^{n+1}\|a\|^n\right].$$

Proof Since $a \in \mathscr{A}^d$ and $a^\pi b(I - a^\pi) = 0$, we have that for $p = 1 - a^\pi$

$$a = \begin{pmatrix} a_1 & 0 \\ 0 & a_2 \end{pmatrix}_p, \quad b = \begin{pmatrix} b_1 & b_3 \\ 0 & b_2 \end{pmatrix}_p \tag{6.36}$$

where a_1 is invertible in the algebra $p\mathscr{A}p$ and a_2 is a quasi-nilpotent element of the algebra $(1 - p)\mathscr{A}(1 - p)$. Also from $a^\pi ab = a^\pi ba$ and the fact that $a^\pi b \in \mathscr{A}^d$, we conclude that $a_2 b_2 = b_2 a_2$ and $b_2 \in \mathscr{A}^d$. It follows from $\|a^d b\| < 1$ that $1 + a^d b$ is invertible. Now, from Theorem 6.15, we have

$$(a_2 + b_2)^d = \sum_{n=0}^{\infty} (b_2^d)^{n+1}(-a_2)^n.$$

Using Theorem 6.10, we get

$$(a + b)^d = \begin{pmatrix} (a_1 + b_1)^{-1} & S \\ 0 & \sum_{n=0}^{\infty} (b_2^d)^{n+1}(-a_2)^n \end{pmatrix}_p,$$

where

$$S = \left[\sum_{n=0}^{\infty} (a_1 + b_1)^{-n-2} b_3 (a_2 + b_2)^n\right] \left[1 - p - (a_2 + b_2) \sum_{n=0}^{\infty} (b_2^d)^{n+1}(-a_2)^n\right]$$

$$-(a_1 + b_1)^{-1} b_3 \sum_{n=0}^{\infty} (b_2^d)^{n+1}(-a_2)^n.$$

We know that

$$\begin{bmatrix} (a_1 + b_1)^{-1} & 0 \\ 0 & 0 \end{bmatrix}_p = (1 + a^d b)^{-1} a^d$$

and

$$\begin{bmatrix} 0 & 0 \\ 0 & \sum_{n=0}^{\infty} (b_2^d)^{n+1}(-a_2)^n \end{bmatrix}_p = a^\pi \sum_{n=0}^{\infty} (b^d)^{n+1}(-a)^n.$$

By computation we obtain

$$\begin{pmatrix} 0 & S \\ 0 & 0 \end{pmatrix}_p = \left[\sum_{n=0}^{\infty} \left((1 + a^d b)^{-1} a^d \right)^{n+2} ba^\pi (a+b)^n \right] a^\pi$$

$$\times \left[1 - (a+b)a^\pi \sum_{n=0}^{\infty} (b^d)^{n+1}(-a)^n \right]$$

$$-(1 + a^d b)^{-1} a^d ba^\pi \sum_{n=0}^{\infty} (b^d)^{n+1}(-a)^n.$$

Hence, we have

$$(a+b)^d = (1 + a^d b)^{-1} a^d + (1 + a^d b)^{-1} a^\pi \sum_{n=0}^{\infty} (b^d)^{n+1}(-a)^n$$

$$+ \left[\sum_{n=0}^{\infty} \left((1 + a^d b)^{-1} a^d \right)^{n+2} ba^\pi (a+b)^n \right] a^\pi$$

$$\times \left[1 - (a+b)a^\pi \sum_{n=0}^{\infty} (b^d)^{n+1}(-a)^n \right].$$

If $\|a\| \|b^d\| < 1$, $\|b\| \|a^d\| < 1$ and $\frac{\|a^d\| \|a^d b\|}{1 - \|a^d b\|} \|a+b\| < 1$, we obtain

$$\|(a+b)^d - a^d\| = \left\| \sum_{n=1}^{\infty} (a^d b)^n a^d + \sum_{n=0}^{\infty} (a^d b)^n (1 - aa^d) \sum_{n=0}^{\infty} (b^d)^{n+1}(-a)^n \right.$$

$$+ \left[\sum_{n=0}^{\infty} \left(\sum_{n=0}^{\infty} (a^d b)^n a^d \right)^{n+2} b(1 - aa^d)(a+b)^n \right] (1 - aa^d)$$

$$\times \left. \left[1 - (a+b)(1 - aa^d) \sum_{n=0}^{\infty} (b^d)^{n+1}(-a)^n \right] \right\|$$

$$\leq \frac{\|a^d\| \|a^d b\|}{1 - \|a^d b\|} + \frac{\|1 - aa^d\|}{1 - \|a^d b\|} \sum_{n=0}^{\infty} \|(b^d)\|^{n+1} \|(-a)\|^n$$

$$+ \left[\sum_{n=0}^{\infty} \left(\frac{\|a^d\| \|a^d b\|}{1 - \|a^d b\|} \right)^{n+2} \|b\| \|a+b\|^n \right] \|(1 - aa^d)\|^2$$

$$+ \|(1 - aa^d)\|^3 \left[\sum_{n=0}^{\infty} \left(\frac{\|a^d\| \|a^d b\|}{1 - \|a^d b\|} \right)^{n+2} \|b\| \|a+b\|^{n+1} \right]$$

$$\times \left[\sum_{n=0}^{\infty} \|(b^d)\|^{n+1} \|a\|^n \right].$$

□

Corollary 6.9 *Let* $a \in \mathscr{A}^{d}$ *and* $b \in \mathscr{A}$ *be such that* $\|ba^{d}\| < 1, a^{\pi}b(1 - a^{\pi}) = 0$ *and* $a^{\pi}ab = a^{\pi}ba$,

(1) *If* $baa^{d} = 0$ *and* b *is quasi-nilpotent, then* $a + b \in \mathscr{A}^{d}$ *and*

$$(a + b)^{d} = \sum_{n=0}^{\infty} (a^{d})^{n+2} b(a + b)^{n} + a^{d}.$$

(2) *If* $a^{\pi}b = ba^{\pi}, \sigma(a^{\pi}b) = 0$, *then* $a + b \in \mathscr{A}^{d}$ *and*

$$(a + b)^{d} = (1 + a^{d}b)^{-1}a^{d} = a^{d}(1 + ba^{d})^{-1}.$$

The following theorem is a generalization of Theorem 6.16 and Theorem 6 from [45].

Theorem 6.17 *Let* $a, b \in \mathscr{A}^{d}$ *and* q *be an idempotent such that* $aq = qa$, $(1 - q)bq = 0$, $(ab - ba)q = 0$ *and* $(1 - q)(ab - ba) = 0$. *If* $(a + b)q$ *and* $(1 - q)(a + b)$ *are generalized Drazin invertible, then* $a + b \in \mathscr{A}^{d}$ *and*

$$(a + b)^{d} = \sum_{n=0}^{\infty} S^{n+2}qb(1 - q)(a + b)^{n}(1 - q)\left[1 - (a + b)S\right]$$

$$+ \left[1 - (a + b)S\right]q \sum_{n=0}^{\infty}(a + b)^{n}qb(1 - q)S^{n+2}$$

$$+ (1 - Sqb)(1 - q)S + Sq,$$

where

$$S = a^{d}(1 + a^{d}b)^{d}bb^{d} + (1 - bb^{d})\left[\sum_{n=0}^{\infty}(-b)^{n}(a^{d})^{n+1}\right]$$

$$+ \left[\sum_{n=0}^{\infty}(b^{d})^{n+1}(-a)^{n}\right](1 - aa^{d}).$$

(6.37)

Proof The proof is a similar to that of Theorem 6.16. □

References

1. Drazin, M.P.: Pseudoinverse in associative rings and semigroups. Amer. Math. Monthly **65**, 506–514 (1958)
2. Deng, C.Y.: The Drazin inverses of products and differences of orthogonal projections. J. Math. Anal. Appl. **335**, 64–71 (2007)
3. Eisenstat, S.C.: A perturbation bound for the eigenvalues of a singular diagonalizable matrix. Linear Algebra Appl. **416**, 742–744 (2006)
4. Hartwig, R.E., Li, X., Wei, Y.: Representations for the Drazin inverse of 2 × 2 block matrix. SIAM J. Matrix Anal. Appl. **27**, 757–771 (2006)

5. Simeon, B., Fuhrer, C., Rentrop, P.: The Drazin inverse in multibody system dynamics. Numer. Math. **64**, 521–539 (1993)

6. Xu, Q., Song, C., Wei, Y.: The stable perturbation of the Drazin inverse of the square matrices. SIAM J. Matrix Anal. Appl. **31**, 1507–1520 (2010)

7. Castro-González, N., Koliha, J., Rakočević, V.: Continuity and general perturbation of the Drazin inverse for closed linear operators. Abstr. Appl. Anal. **7**, 335–347 (2002)

8. Cvetković-Ilić, D.S.: A note on the representation for the Drazin inverse of 2×2 block matrices. Linear Algebra Appl. **429**, 242–248 (2008)

9. Djordjević, D.S., Wei, Y.: Additive results for the generalized Drazin inverse. J. Austral Math. Soc. **73**, 115–125 (2002)

10. Hartwig, R.E., Wang, G., Wei, Y.: Some additive results on Drazin inverse. Linear Algebra Appl. **322**, 207–217 (2001)

11. Ben-Isreal, A., Greville, T.N.E.: Generalized Inverse: Theory and Applications, 2nd edn. Springer, New York (2003)

12. Cline, R.E.: An application of representation of a matrix, MRC Technical Report, # 592 (1965)

13. Meyer, C.D.: The condition number of a finite Markov chains and perturbation bounds for the limiting probabilities. SIAM J. Algebraic Discrete Methods **1**, 273–283 (1980)

14. Meyer, C.D., Shoaf, J.M.: Updating finite Markov chains by using techniques of group inversion. J. Statist. Comput. Simulation **11**, 163–181 (1980)

15. Wei, Y., Wang, G.: The perturbation theory for the Drazin inverse and its applications. Linear Algebra Appl. **258**, 179–186 (1997)

16. Wei, Y.: On the perturbation of the group inverse and the oblique projection. Appl. Math. Comput. **98**, 29–42 (1999)

17. Campbell, S.L., Meyer, C.D.: Continuality properties of the Drazin inverse. Linear Algebra Appl. **10**, 77–83 (1975)

18. Wei, Y., Deng, C.: A note on additive results for the Drazin inverse. Linear Multilinear Algebra **59**, 1319–1329 (2011)

19. Hartwig, R.E., Shoaf, J.M.: Group inverse and Drazin inverse of bidiagonal and triangular Toeplitz matrices. Austral J. Math. **24A**, 10–34 (1977)

20. Meyer, C.D., Rose, N.J.: The index and the Drazin inverse of block triangular matrices. SIAM J. Appl. Math. **33**, 1–7 (1977)

21. Wei, Y.: Expressions for the Drazin inverse of a 2×2 block matrix. Linear Multilinear Algebra **45**, 131–146 (1998)

22. Castro-González, N.: Additive perturbation results for the Drazin inverse. Linear Algebra Appl. **397**, 279–297 (2005)

23. Bauer, F., Fike, C.: Norms and exclusion theorem. Numer. Math. **2**, 137–141 (1960)

24. Eisensat, S.C., Ipsen, I.C.F.: Three absolute perturbation bounds for matrix eigenvalues imply relative bounds. SIAM J. Matrix Anal. Appl. **20**, 149–158 (1998)

25. Prasolov, V.V.: Problems and Theorems in Linear Algebra. American Mathematical Society, Providence, RI (1994)

26. Wei, Y., Wu, H.: The perturbation of the Drazin inverse and oblique projection. Appl. Math. Lett. **13**, 77–83 (2000)

27. Wei, Y., Wu, H.: Convergence properties of Krylov subspace methods for singular linear systems with arbitrary index. J. Comput. Appl. Math. **114**, 305–318 (2000)

28. Wei, Y.: The Drazin inverse of a modified matrix. Appl. Math. Comput. **125**, 295–301 (2002)

29. Wei, Y., Li, X.: An improvement on perturbation bounds for the Drazin inverse. Numer Linear Algebra Appl. **10**, 563–575 (2003)

30. Wei, Y., Li, X., Bu, F.: A perturbation bound of the Drazin inverse of a matrix by separation of simple invariant subspaces. SIAM J. Matrix Anal. Appl. **27**, 72–81 (2005)

31. Wei, Y., Diao, H.: On group inverse of singular Toeplitz matrices. Linear Algebra Appl. **399**, 109–123 (2005)

32. Wei, Y., Diao, H., Ng, M.K.: On Drazin inverse of singular Toeplitz matrix. Appl. Math. Comput. **172**, 809–817 (2006)

33. Wei, Y., Li, X., Bu, F., Zhang, F.: Relative perturbation bounds for the eigenvalues of diagonal-izable and singular matrices-application of perturbation theory for simple invariant subspaces. Linear Algebra Appl. **419**, 765–771 (2006)
34. Castro-González, N., Dopazo, E., Robles, J.: Formulas for the Drazin inverse of special block matrices. Appl. Math. Comput. **174**, 252–270 (2006)
35. Du, H., Deng, C.: The representation and characterization of Drazin inverses of operators on a Hilbert space. Linear Algebra Appl. **407**, 117–124 (2005)
36. Koliha, J.J.: The Drazin and Moore-Penrose inverse in C^*-algebras. Mathematical Proceedings of the Royal Irish Academy **99A**, 17–27 (1999)
37. Cvetković-Ilić, D.S., Djordjević, D.S., Wei, Y.: Additive results for the generalized Drazin inverse in a Banach algebra. Linear Algebra Appl. **418**, 53–61 (2006)
38. Koliha, J.J.: A generalized Drazin inverse. Glasgow Math. J. **38**, 367–381 (1996)
39. Harte, R.E.: Invertibility and Singularity for Bounded Linear Operators. Marcel Dekker, New York (1988)
40. Harte, R.E.: On quasinilpotents in rings. PanAm. Math. J. **1**, 10–16 (1991)
41. Harte, R.E.: Spectral projections. Irish Math. Soc. Newsletter **11**, 10–15 (1984)
42. Han, J.K., Lee, H.Y., Lee, W.Y.: Invertible completions of 2×2 upper triangular operator matrices. Proc. Amer. Math. Soc. **128**, 119–123 (1999)
43. Castro-González, N., Koliha, J.J.: New additive results for the g-Drazin inverse. Proc. Royal Soc. Edinburgh **134A**, 1085–1097 (2004)
44. Djordjević, D.S., Stanimirović, P.S.: On the generalized Drazin inverse and generalized resol-vent. Czechoslovak Math. J. **51**(126), 617–634 (2001)
45. Deng, C.Y., Wei, Y.: New additive results for the generalized Drazin inverse. J. Math. Anal. Appl. **370**, 313–321 (2009)

Index

© Springer Nature Singapore Pte Ltd. 2017
D. Cvetković Ilić and Y. Wei, *Algebraic Properties of Generalized Inverses*,
Developments in Mathematics 52, DOI 10.1007/978-981-10-6349-7

Printed in the United States
By Bookmasters